D1671008

Microcavity Semiconductor Lasers

Microcavity Semiconductor Lasers

Principles, Design, and Applications

Yong-zhen Huang
Yue-de Yang

Authors

Prof. Yong-zhen Huang
State Key Lab of Integrated
Optoelectronics
Institute of Semiconductors
Chinese Academy of Sciences and
College of Materials Sciences and
Optoelectronic Technology
University of Chinese Academy of
Sciences
No.A35, QingHua East Road
Haidian District
100083 Beijing
China

Prof. Yue-de Yang
State Key Lab of Integrated
Optoelectronics
Institute of Semiconductors
Chinese Academy of Sciences and
College of Materials Sciences and
Optoelectronic Technology
University of Chinese Academy of
Sciences
No.A35, QingHua East Road
Haidian District
100083 Beijing
China

Cover Image: © Supphachai Salaeman/Shutterstock

Library of Congress Card No.: applied for

British Library Cataloguing-in-Publication Data A catalogue record for this book is available from the British Library.

Bibliographic information published by the Deutsche Nationalbibliothek The Deutsche Nationalbibliothek lists this publication in the Deutsche Nationalbibliografie; detailed bibliographic data are available on the Internet at <http://dnb.d-nb.de>.

Print ISBN: 978-3-527-34546-5
ePDF ISBN: 978-3-527-82018-4
ePub ISBN: 978-3-527-82020-7
oBook ISBN: 978-3-527-82019-1

Typesetting SPi Global, Chennai, India
Printing and Binding CPI Group (UK) Ltd, Croydon, CR0 4YY

Printed on acid-free paper

C090432_050521

Contents

Preface

As typical whispering-gallery microcavities, microdisks, and microrings have been studied for applications in integrated optics over a half century. The study of semiconductor microdisk lasers has become a distinct subject of optoelectronics, and deformed microdisk lasers have attracted great attention for realizing directional emission microlasers. In addition to circular microcavities, polygonal microcavities can also support high *Q*-confined modes that rely on the total internal reflection, similar to whispering-gallery modes in microdisks. We have investigated microcavity lasers for the past two decades, mainly focusing on mode analysis, design, processing technique, photonic integration, and applications of microcavity lasers. This book summarizes the research on semiconductor microcavity lasers based on whispering-gallery modes. Although there are several books on optical microcavities, this book provides unique descriptions of directional emission microcavity lasers by directly connecting an output waveguide, mode behaviors based on group theory for polygonal microcavities, and hybrid-cavity lasers with integrated microcavity and waveguide.

The book is organized into 11 chapters: introduction emphasized on mode *Q* factor, multilayer optical waveguides, FDTD method and Padé approximation, deformed and chaos microdisk lasers, unidirectional emission microdisk lasers, equilateral triangle resonator microlasers, square microcavity lasers, hexagonal microcavity lasers and polygonal microcavities, vertical loss for 3D microcavities, nonlinear dynamics for microcavity lasers, and hybrid-cavity lasers.

We want to thank Huang's students, especially Dr. Wei-Hua Guo, at Institute of Semiconductors; their works have contributed to main parts of this book.

Yong-zhen Huang
Yue-de Yang
State Key Lab of Integrated Optoelectronics
Institute of Semiconductors
Chinese Academy of Sciences and College of
Materials Sciences and Optoelectronic Technology
University of Chinese Academy of Sciences

Beijing, China
November 2020

1

Introduction

1.1 Whispering-Gallery-Mode Microcavities

Optical resonant cavities, composed of two or more mirrors, are essential part of ordinary lasers and have been utilized in almost all branches of modern optics and photonics. Optical energy is recirculated inside the cavities due to the reflection on the mirrors, and one basic property of the optical cavities is the quality (Q) factor related to the mode lifetime for describing the light-confining ability. Mode volume (V) is another important parameter of an optical cavity and a small V is of great importance for realizing a compact-size integrated device. A suitable parameter, *finesse*, which is defined as the ratio of the free spectral range to the resonance linewidth, takes both the mode Q factor and the resonator size into account. For certain applications, high-finesse microcavity with a large value of Q/V, which is also related to the electromagnetic field enhancement factor of an optical cavity, is very important. Compared with conventional lasers, microcavity lasers with a large Q/V can promise lower lasing threshold. Moreover, light–matter interactions can be greatly enhanced by storing optical energy in a small mode volume [1, 2]. The ability to concentrate light is important to both fundamental science studies and practical device applications [3], such as strong-coupling cavity quantum electrodynamics, enhancement and suppression of spontaneous emission, high-sensitivity sensors, low-threshold light sources, and compact optical add-drop filters in optical communication.

To obtain high Q modes in optical cavities with a small V, a high reflectivity close to unity is necessary, which can be realized by utilizing a periodic structure to construct a photonic forbidden band, such as that in vertical-cavity surface-emitting lasers and photonic crystal microcavities, or simply by total internal reflection (TIR) at the dielectric boundary with a high-low refractive index contrast in whispering-gallery (WG)-mode optical microcavities [4]. The idea of WG mode was born out of the observation of acoustical phenomenon in [5] where sound waves were efficiently reflected with minimal diffraction and struck the wall again at the same angle and thereby traveled along the gallery surface. Similarly, classical electromagnetic waves can undergo reflection, refraction, and diffraction like the sound waves when the wavelengths of the waves are smaller than the bending radius of a reflection mirror. Among various kinds of optical microcavities, WG-mode microcavities with simple

Microcavity Semiconductor Lasers: Principles, Design, and Applications, First Edition.
Yong-zhen Huang and Yue-de Yang.

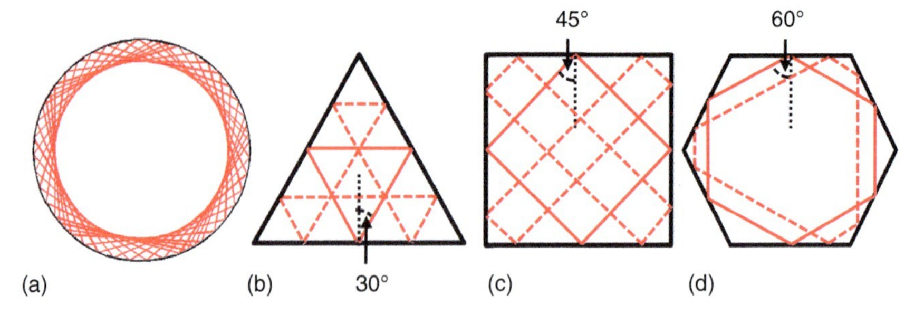

Figure 1.1 Schematic diagrams of (a) circular, (b) triangular, (c) square, and (d) hexagonal microcavities with the confined TIR light rays.

cavity geometries and suitability for planar integration play an important role in photonics integration nowadays [6–8]. WG-mode optical microcavities formed by various materials have been studied, such as liquid droplet, glass, crystal, polymer, and semiconductor [8].

The concept of WG mode was subsequently extended to the radiofrequency and optical domains for the electromagnetic waves. For WG-mode microcavities, the optical modes will experience ultra-low loss as the light rays are guided by continuous TIR at the boundaries. In fact, light guidance by continuous TIR is quite common in modern optics and photonics, such as the propagating optical modes in fibers and waveguides. The incident angles of the light rays in WG-mode microcavities are greater than the TIR criticality at all boundaries, and the light rays are mostly like to propagate along the WG surface. There are various kinds of WG-mode optical microcavities to maintain continuous TIR for the confined light rays. Figure 1.1 schematically shows circular, triangular, square, and hexagonal microcavities with the confined TIR light rays. As a natural choice, WG-mode optical microcavities with circular shapes, which maintain a perfect rotational symmetry, have attracted most research interest for the demonstration of low-power-consumption compact-size photonic devices, e.g. microdisk lasers [9]. In circular microcavities, the light ray confined inside the cavity has a conserved incident angle above the TIR critical angle resulting in ultrahigh Q factors and isotropic near- and far-field patterns. However, for a practical device, efficient input or output coupling is crucial but can be hardly achieved in the circular-shaped microcavities. By deforming the WG microcavity to a noncircular cavity shape, the far-field emission patterns can be modulated while high Q factor mode is maintained [10, 11].

1.2 Applications of Whispering-Gallery-Mode Microcavities

WG-mode optical microcavities with the light rays confined by continuous TIR at the cavity boundaries have unique mode properties, including high Q factors, small mode volume, and planar integration capability. Due to these mode properties, there has been a wide range of applications for WG-mode microcavities in both fundamental physics and practical devices [12]. In this section, part of the applications, including photonic filters, sensors, and microlasers, based on both passive and active WG-mode microcavities, is briefly summarized.

One basic application of microcavities is photonic filter based on the wavelength selectivity property of WG modes, including all-pass and add-drop filters. A common microcavity-based filter includes a WG-mode microcavity and optical coupler for coupling light into or out from the WG modes [13–15]. In an optical filter, the filtering response is relative to both the intrinsic mode Q factors and the coupling coefficients for the WG modes. The high-order transverse WG modes have lower Q factors but stronger coupling, which is undesirable for most filtering applications. Hence, microrings are utilized for suppressing high-order transverse modes and realizing high-performance optical filter. Microrings have similar mode properties with the microdisks, as the modes are confined by the continuous TIR at the outer boundaries [14]. A typical on-chip all-optical four-port add-drop filter includes a microring cavity and two evanescent-field coupled waveguides. The add-drop filter exhibits pass-band filtering characteristics with the on-resonance light dropping through the drop port. The microring-based filters and corresponding active devices have the advantages of compact size, narrow band, and a large free spectral range, and have been widely studied in the silicon photonics for on-chip optical interconnection application. The pass-band characteristics can be further improved by using higher-order filter structure with multiple coupled microrings. Based on the microring photonic filters, cascaded microring-based matrix switches have been demonstrated for on-chip optical networks [16]. The networks-on-chip can be passive networks with fixed-wavelength assignment and switching networks with the resonance wavelength tuning by thermal effect or carrier injection. With the structure of all-pass microring filter, silicon-based microring electro-optic modulators were demonstrated with the carrier injection or depletion to change the resonance wavelength of the microring cavity [17].

Another important application of WG optical microcavities is photonic sensing. WG microcavities have been extensively investigated for their applications in chemicals and biosensing. Strong light–matter interactions and high optical energy intensity in the optical microcavities with a large value of Q/V can help to achieve ultrasensitive and label-free detection. The sensing principle is to measure the spectral changes of a WG mode in response to changes in the environment, e.g. refractive index shift of surrounding media or nanoparticles onto the cavity surface [18, 19]. The key feature is the strong evanescent field of the WG mode that propagates along and extends from the surface of the microcavity leading to strong interaction between the internal field and the external environment. The measured transmission spectra will experience a wavelength shift and/or splitting for sensing. The resonance shift in a microcavity is a more direct detection scheme, but it can be easily perturbed by environmental noises resulting in a reduction of the sensing resolution. The environmental noises are minimized in the mode splitting–based detection scheme as the two split modes suffer the same noises. The variation of mode splitting carries the information of particles to be measured. The detection resolution for a passive microcavity-based sensing is limited by the linewidth of the WG modes. In a microcavity laser–based active sensing devices, the stimulated emission will narrow down the linewidth, and hence the sensing resolution is greatly enhanced [20]. For a cold cavity with a Q factor of 10^8, the laser linewidth can be as narrow as a few Hertz allowing ultrasensitive detection [8].

Light sources, such as microlasers and quantum sources, are an extremely important research direction in optics and photonics. Compared to other optical cavities for laser application, the WG-mode microcavities have extraordinarily high Q factor and small V, which lead to diverse applications in the study of laser physics and the realization of compact-size microlasers. In WG-mode microcavities, the optical density of states can be modulated by designing the cavity structure and matching the resonance wavelength to the emission wavelength of the active material. Thus, the Purcell factor can be enhanced greatly in optical microcavities [1]. Semiconductor quantum dot is a quasi-atom gain material; the coupling between the quantum dot and the optical mode can be enhanced in the high Q microcavity with a ultrasmall V. Strong coupling of a single GaAs quantum dot to a WG mode of a microdisk has been observed, facilitating the investigations of cavity quantum electrodynamics and single photon source [21]. High Q factor and small V also allow the demonstration of conventional low-threshold semiconductor microlasers [9]. The high Q factor of a WG mode guarantees a low-threshold current density and a small V leads to a compact size for the microlaser for achieving low-power consumption. Continuous wave lasing with a threshold of 40 μA was realized in an InGaAsP microdisk laser at room temperature [22]. However, the nearly perfect confinement of the mode light ray and the rotational symmetry of a circular microcavity led to low-output power and isotropic emission to free space despite a low-lasing threshold. This is a serious problem for most practical applications of WG microcavity lasers. Evanescent wave coupling of a waveguide is one traditional scheme to couple lasing light out from the circular microcavity lasers, but it has extremely high requirements for fabrication processing technology and parameter control. Experimental results show that a small variation in the coupling gap will reduce the output optical power by several orders of magnitude. In addition, the competition between the clockwise and counterclockwise modes in the circular microcavities will cause instability of the output optical power in the waveguide. To realize directional lasing emission, various deformed microcavities, such as adding local boundary defects or using smoothly deformed cavity shapes, have been proposed and demonstrated [3]. By carefully designing the cavity geometries, directional emission, or even unidirectional emission with low divergence angle in free space was achieved for deformed microcavity lasers while preserving high-Q WG modes for low-threshold lasing. However, the directional or unidirectional emission of asymmetric microcavities is still limited to free space, and the application to on-chip photonic integration requires waveguide-coupled output. Moreover, regular-polygonal-shaped microcavities have distinct mode properties, as the WG modes distribute nonuniformly along the cavity boundaries. A waveguide directly connecting to the position with weak mode field can be used for realizing a waveguide-coupled microcavity laser without strong perturbation to the corresponding high Q WG mode. Especially, a quasi-analytical solution can be obtained for the equilateral-triangular and square microcavities with integrable internal dynamics, making them a reliable solution to demonstrate waveguide-coupled unidirectional-emission semiconductor microlasers [23].

1.3 Ultra-High *Q* Whispering-Gallery-Mode Microcavities

Spherical optical microcavities of liquid droplets and highly transparent silica have been extensively investigated, which can have nearly perfect microspheres due to surface tension of liquid and fused silica [1]. Based on liquid droplet microcavities, cavity quantum electrodynamics with modified spontaneous and stimulated emission spectra were studied and ultralow threshold of nonlinear optical processes was observed with fluorescent dyes. The effects of droplet deformation on the resonance frequencies and Q factors were investigated experimentally and theoretically using first-order perturbation theory, and the dye-lasing spectra from liquid droplet optical microcavities were observed under perturbations. By using the CO_2 laser fusion process, high-Q silica microspheres were fabricated by fusing the end of a silica fiber. WG modes with Q factors up to $10^9 \sim 10^{10}$ were observed, and low-threshold microlasers based on silica microspheres with doped irons were realized. Droplet microlasers as easily replaced coherent light sources were investigated for potential applications in integrated lab-on-a-chip systems [24]. The droplet-based microlasers can be prepared in microfluidic chip with different active media, such as live bacteria. In addition, intracellular droplet microlasers were studied by injecting oil doped with a dye gain medium inside biological cells as luminescent probes [25].

Ultra-high-Q microcavities can also be fabricated on a silicon wafer using wafer-scale processing, in the form of a microcavity on-a-chip suitable for photonic integrated circuits [2]. The fabrication processing is simply summarized in the following. First, silica circular patterns were transformed from photoresist layer to the thermally oxidized surface layer of a silicon wafer using lithography and etching technique processes, and then silica disks were used as an etching mask for selectively removing the underneath silicon. Finally, silica microdisks on a silicon post were fabricated with a vertical optical confined by air for avoiding a leaking loss into high-index silicon substrate. The WG modes with Q-factors up to 3×10^6 were measured for such microdisks under optimal processing conditions. As the mode-field distributions located near the disk periphery, the Q-factors were mainly limited by scattering loss due to disk roughness caused by lithography and etching. To further increase mode Q factors, a reflow process for the silica microdisks was applied under the surface-normal irradiation of CO_2 laser by improving surface smooth of the microdisks without affecting the underneath silicon post. The reflow process under the laser irradiation can lead to melting and collapse of the silica at the disk periphery and form silica microtoroid on a silicon chip. The ultra-high Q factor based on linewidth measurement is a challenge as a loaded cavity Q factor is measured with coupled waveguide. In addition, WG mode splits into doublets caused by weak back scattering in the microcavity, and thermal effects due to input optical power induce distortion of the resonance peak. An intrinsic cavity Q factor of 4.3×10^8 was obtained by cavity ring-down measurement for a microtoroid cavity mode. The WG mode loss is negligible for microtoroids with principal tori-radii larger than 15 μm, and measured Q values are more than 10^8 in the wavelength of 1550-nm band. The ultra-high-Q microcavities are especially suitable to study optical nonlinear processes, such as Raman and Kerr nonlinearities. Under fiber evanescently coupling with a low-input power at resonant frequency, high-mode field intensity can be stored inside an ultra-high-Q microcavity for ultra-low-threshold fiber-compatible Raman lasers and parameter oscillators. Furthermore, optical frequency combs were realized through cascaded four-wave mixing process in a high Q microresonator [26].

In addition, chemically etched wedge resonators on-a-chip were fabricated using conventional semiconductor processing, with a Q factor of 875 million surpassing microtoroids [27]. The smoothness of wedge resonators was improved using post exposure bake method to cure the roughness of photoresist patterns and extend the chemically etched time to form wedge profiles for the resonator perimeter. Without the reflow process of laser irradiation, the wedge resonators are of easy-to-control size and can be integrated with other photonic devices.

1.4 Mode *Q* Factors for Semiconductor Microlasers

1.4.1 Output Efficiency and Mode *Q* Factor

For a microcavity laser, as shown in Figure 1.2, with a passive cavity mode Q factor Q_R related to planar and vertical radiation losses α_r and α_v and an output coupling loss α_o, we can have a modified mode lifetime varied with a mode gain Γg as:

$$\tau_p = \frac{1}{v_g\alpha_i + \omega/Q_R - \Gamma v_g g}, \tag{1.1}$$

where Γ is the optical confinement factor, α_i is an internal material absorption loss, $v_g = c/n_g$ is the light group speed with a group index n_g, and ω is the mode angular frequency. Mode Q_T factor, including the absorption loss, can be defined as:

$$\frac{1}{Q_T} = \frac{1}{Q_R} + \frac{\alpha_i v_g}{\omega} = \frac{1}{Q_R} + \frac{1}{Q_A}. \tag{1.2}$$

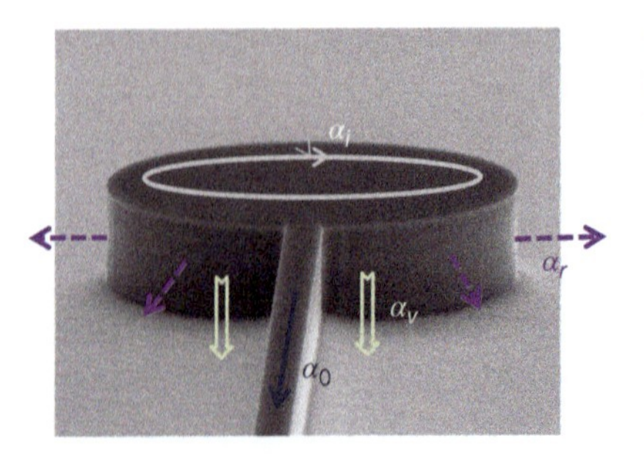

Figure 1.2 Scanned electron micrograph image of a microdisk connected with an output waveguide.

For a silica microdisk without gain and $\alpha_i \approx 0$, it is easy to measure mode Q factor from the transmission linewidth because $Q_T = Q_R$. However, semiconductor lasers usually have an absorption loss, which limited mode Q factors. Taking the absorption loss $\alpha_i = 1$ and $10\,\text{cm}^{-1}$, which corresponds to the magnitude of the absorption loss for GaAs and InP system semiconductor lasers, respectively, we have mode Q_A factor of 1.4×10^5 and 1.4×10^4 at $n_g = 3.5$ and mode wavelength of 1550 nm.

Accounting for the internal absorption loss α_i related to Q_A, output coupling loss through the output waveguide α_o, the vertical loss α_v into the substrate, and the other radiation loss, including scattering loss due to rough perimeter α_r, we can define an output efficiency as

$$\eta = \frac{\alpha_o}{\alpha_i + \alpha_o + \alpha_r + \alpha_v} = \frac{\alpha_o}{\alpha_o + \alpha_v + \alpha_r} \frac{Q_T}{Q_R} = \frac{\alpha_o}{\alpha_o + \alpha_v + \alpha_r} \frac{Q_A}{Q_A + Q_R}. \tag{1.3}$$

The laser output efficiency will be very low for an ultra-high Q microcavity with $Q_R \gg Q_A$. The material of low absorption loss with a high Q_A is important for realizing high-output efficiency for a microlaser.

1.4.2 Measurement of Mode *Q* Factor

The mode Q factors of a microlaser are usually measured as the ratio of mode wavelength to the full-width at half maximum (FWHM) of the resonator peak at the threshold. The mode linewidth is described by the Schawlow–Townes linewidth formula below the threshold [28]:

$$\Delta\lambda = \frac{\lambda^2}{2\pi c \tau_p} = \frac{\lambda^2 \Gamma \beta B n^2}{2\pi c s}. \tag{1.4}$$

Due to the gain-refractive index coupling effect with a linewidth enhancement factor α and carrier density clamping above threshold, lasing mode linewidth above the threshold is given by the modified Schawlow–Townes linewidth formula [29]:

$$\Delta\lambda = \frac{\lambda^2(1+\alpha^2)}{4\pi c \tau_p}, \tag{1.5}$$

which is equal to (1.4) multiplying by a factor of $(1+\alpha^2)/2$. The laser linewidth enhancement near threshold was observed experimentally for semiconductor microlasers [30].

The output characteristics of semiconductor microcavity lasers can be described by the following single-mode rate equations

$$\frac{dn}{dt} = \frac{\eta_i I}{q V_a} - An - Bn^2 - Cn^3 - v_g g(n) s, \tag{1.6}$$

$$\frac{ds}{dt} = v_g[\Gamma g(n) - \alpha_i]s - \frac{s}{\tau_{pc}} + \Gamma \beta B n^2, \tag{1.7}$$

where s is the photon density, n is the carrier density, I is the injection current, η_i is the injection efficiency, q is the electron charge, V_a is the volume of the active region, A, B, and C are the defect, bimolecular, and Auger recombination coefficients, respectively, β is the spontaneous emission factor, and $\tau_{pc} = Q_R/\omega$ is the passive cavity mode lifetime. The threshold gain of semiconductor lasers is usually expressed as

$$\Gamma g = \alpha_i + \frac{\omega}{Q_R v_g}, \tag{1.8}$$

which is only approached in the steady state. From Eq. (1.7), we can obtain the output term in steady state as

$$\frac{s}{\tau_{\text{pc}}} = v_g[\Gamma g(n) - \alpha_i]s + \Gamma\beta Bn^2. \tag{1.9}$$

The mode lifetime is determined by the passive cavity Q_R factor as $\Gamma g = \alpha_i$ from (1.1), so the passive mode Q_R factor should be measured at the following condition

$$\Gamma g = \alpha_i, \tag{1.10}$$

instead of threshold gain of (1.8). However, it is difficult to determine the condition of (1.10) from the curve of the output power vs. injection current.

In the following part of this section, we give numerical results of the rate Eqs. (1.6) and (1.7). The gain coefficient is assumed to be a logarithmic function as [31]

$$g(n) = g_0 \ln\left(\frac{n + N_s}{N_{\text{tr}} + N_s}\right), \tag{1.11}$$

where g_0 is the material gain parameter, N_{tr} is the transparency carrier density, and N_s is a gain parameter. Taking the parameters as $\lambda = 1550\,\text{nm}$, $N_{\text{tr}} = 1.2\times10^{18}\,\text{cm}^{-3}$, $N_s = 1.1\times10^{18}\,\text{cm}^{-3}$, mode group index $n_g = 3.5$, $\eta_i = 0.8$, $\Gamma = 0.1$, $A = 1\times10^{8}\,\text{s}^{-1}$, $B = 1\times10^{-10}\,\text{cm}^{-3}\,\text{s}^{-1}$, $C = 1\times10^{-28}\,\text{cm}^{6}\,\text{s}^{-1}$, $g_0 = 1500\,\text{cm}^{-1}$, we numerically calculate the steady solutions of Eqs. (1.6) and (1.7), and plot the output powers and linewidths of Eqs. (1.4) and (1.5) vs. the injection current in Figure 1.3 at $Q_R = 60\,000$, 20 000, and 6000, and $\alpha_i = 4\,\text{cm}^{-1}$, for microdisk lasers with a circular radius of

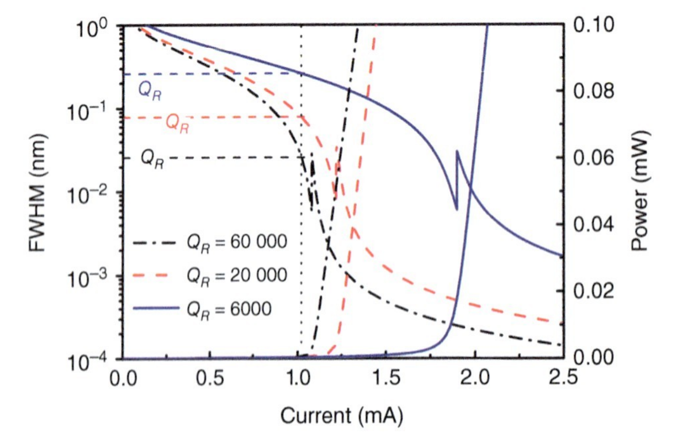

Figure 1.3 Laser-mode linewidth and output power vs. injection current for semiconductor microdisk lasers with $\beta = 10^{-3}$ at $Q_R = 60\,000$, 20 000, and 6000, and $\alpha_i = 4\,\text{cm}^{-1}$. The three horizontal dashed lines correspond to the FWHMs determined by the Q_R factors, and vertical dotted line is at the condition of $\Gamma g = \alpha_i$.

10 μm, $\beta = 10^{-3}$, and a linewidth enhancement factor $\alpha = 3$. The output power is related to the mode photon density as

$$P = \frac{h\nu V_a S}{\Gamma \tau_{pc}}, \tag{1.12}$$

where $h\nu$ is the photon energy. The FWHMs determined by the Q_R factors are marked by the horizontal dashed lines. The FWHMs vary rapidly around the threshold, and Q factor measured using the FWHM at the threshold will overestimate the passive mode Q_R factor greatly.

The FWHMs at the vertical dashed line in Figure 1.3 should be used to calculate Q_R factors, which is difficult to determine from measured curve of output power vs. injection current. If the lasing mode output power can be divided into the first term and the second term of the right side of Eq. (1.9) from lasing spectra, we can calculate the derivative of the first term with respect to the output power and determine the condition of $\Gamma g = \alpha_i$ by finding the zero of the derivative.

Finally, we present experimental results for a microdisk laser as shown in Figure 1.1, with a radius of 8 μm and a 2-μm-wide output waveguide. The output powers vs. injection currents at 288, 290, and 296 K are presented in Figure 1.4a with magnified curves around threshold current in Figure 1.4b, where extrapolated lines are plotted as dashed lines. The practical output powers are limited by heating effect; even a thermoelectric cooler was used to control the temperature. The lasing spectra at 288 K and 10 mA are plotted in Figure 1.4c with single-mode operation

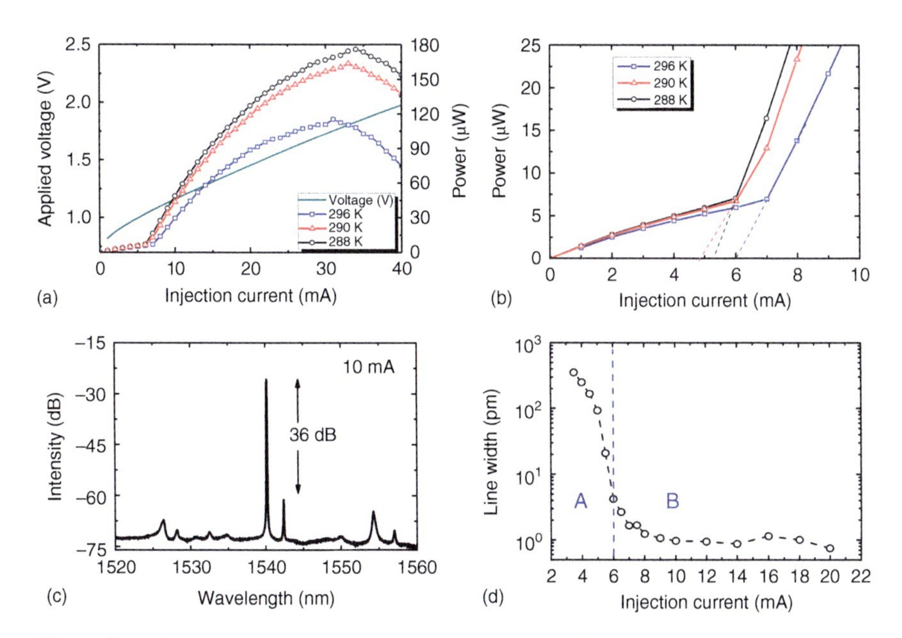

Figure 1.4 (a) Output power coupled into a multiple-mode fiber vs. the injection current, (b) the power vs. the injection current around threshold, (c) lasing spectra at 10 mA, and (d) lasing-mode linewidth vs. injection current, for a microdisk laser connected to a 2-μm-width output waveguide with a radius of 8 μm.

at 1540.3 nm. The FWHM of the lasing mode vs. the injection current at 288 K is shown in Figure 1.4d, where the FWHMs in A and B regions divided by a vertical solid line are directly measured using an optical spectrum analyzer at the finest resolution of 0.02 nm, and estimated from the FWHM of the beating microwave signal by mixing the outputs of the microdisk laser and a tunable laser [32]. The measured FWHM of the beating microwave can be considered as the FWHM of the microdisk laser, as its linewidth is much larger than that of the tunable laser. The FWHMs of the lasing mode are 250, 166, 92.5, 20.8, 4.2, and 1.0 pm at 4, 4.5, 5, 5.5, 6, and 10 mA, respectively, which varies greatly around the threshold. We can estimate mode Q factors of 1.7×10^4, 7.4×10^4, and 3.7×10^5 from the FWHMs at the currents of 5, 5.5, and 6 mA. The mode Q factor of 1.7×10^4 measured at 5 mA, about the intercept of the extrapolated dashed line with the current axis in Figure 1.4b, is in agreement with that obtained by numerical simulation for two-dimensional microcavity under effective index approximation. The agreement indicates that the choice of the intercept current is a better approximation for measuring passive mode Q factor than using the FWHM at the kink position of the power vs. current.

1.5 Book Overview

In this book, we present an overview of the principle, design, and application of semiconductor microlasers, especially for directional emission microlasers based on polygonal optical microcavities. In Chapter 2, we derive an eigenvalue equation for multilayer complex slab optical waveguides, give an optical confinement factor based on the relation between mode gain and material gain, and discuss the effective index method for reducing three-dimensional (3D) waveguide to a two-dimensional (2D) problem. In Chapter 3, we simply introduce finite-difference time-domain method and Padé approximation with Baker's algorithm for simulating optical microcavities, and give some numerical results for simulating mode frequencies and mode Q factors of microcavities and the transmission coefficient for optical microring add-drop filters. Chapter 4 presents an eigenvalue equation for 2D microdisk and deformed and chaotic microcavity lasers for directional emission. In Chapter 5, we summarize unidirectional emission microdisk lasers based on mode coupling due to connection with an output waveguide, and propose to realize unidirectional emission hybrid microlaser on silicon wafer by wafer bonding using locally deformed microring resonator. In Chapter 6, we derive analytical mode solution for equilateral triangle resonator, and compare with numerical simulated results and lasing spectra of fabricated devices. In Chapter 7, we present analytical-mode solution for square microcavities and discuss the formation of high-Q coupled modes. Furthermore, the enhancement of mode Q factors by circular-sided square microcavities is discussed, and dual-mode lasing microlasers are designed and realized in addition to single-mode microlasers. In Chapter 8, we discuss mode characteristics for polygonal microcavities based on group theory especially for hexagonal microcavity, and present lasing characteristics for hexagonal and octagonal and circular-sided hexagonal microlasers. In Chapter 9, we consider the vertical radiation loss for 3D semiconductor microcavities with vertical semiconductor waveguiding, and discuss lateral size limit for such microcavities due to the vertical radiation loss. In Chapter 10, we summarize nonlinear dynamics for microlasers subject to optical injection and integrated microlasers with mutually optical injection. Finally, in Chapter 11, we demonstrate a hybrid cavity composed of a Fabry–Pérot (FP) cavity and a square microcavity for mode selection. Stable single-mode operation with high coupling efficiency to a single mode fiber is realized, and controllable optical bistability is achieved for all-optical signal processing.

References

1 Chang, R.K. and Campillo, A.J. (eds.) (1996). *Optical Processes in Microcavities*. Singapore: World Scientific.
2 Vahala, K.J. (ed.) (2004). *Optical Microcavities*. Singapore: World Scientific.
3 Cao, H. and Wiersig, J. (2015). Dielectric microcavities: model systems for wave chaos and non-Hermitian physics. *Rev. Mod. Phys.* 87: 61–111.
4 Vahala, K.J. (2003). Optical Microcavities. *Nature* 424: 839–846.
5 Rayleigh, L. (1910). The problem of the whispering gallery. *Philos. Mag.* 20: 1001–1004.
6 Matsko, A.B. and Ilchenko, V.S. (2006). Optical resonators with whispering-gallery modes – part I: basics. *IEEE J. Sel. Top. Quantum Electron.* 12: 3–14.
7 Ward, J. and Benson, O. (2011). WG microresonators: sensing, lasing and fundamental optics with microspheres. *Laser Photonics Rev.* 5: 553–570.
8 He, L.N., Ozdemir, S.K., and Yang, L. (2013). Whispering gallery microcavity lasers. *Laser Photonics Rev.* 7: 60–82.
9 Mccall, S.L., Levi, A.F.J., Slusher, R.E. et al. (1992). Whispering-gallery mode microdisk lasers. *Appl. Phys. Lett.* 60: 289–291.
10 Nockel, J.U. and Stone, A.D. (1997). Ray and wave chaos in asymmetric resonant optical cavities. *Nature* 385: 45–47.
11 Yang, Y.D. and Huang, Y.Z. (2016). Mode characteristics and directional emission for square microcavity lasers. *J. Phys. D: Appl. Phys.* 49: 253001.
12 Ilchenko, V.S. and Matsko, A.B. (2006). Optical resonators with whispering-gallery modes – part II: applications. *IEEE J. Sel. Top. Quantum Electron.* 12: 15–32.
13 Little, B.E., Chu, S.T., Haus, H.A. et al. (1997). Microring resonator channel dropping filters. *J. Lightwave Technol.* 15: 998–1005.
14 Schwelb, O. (2004). Transmission, group delay, and dispersion in single-ring optical resonators and add/drop filters – a tutorial overview. *J. Lightwave Technol.* 22: 1380–1394.
15 Heebner, J., Grover, R., and Ibrahim, T. (2008). *Optical Microresonators: Theory, Fabrication, and Applications*. London: Springer-Verlag.

16 Poon, A.W., Luo, X.S., Xu, F., and Chen, H. (2009). Cascaded microresonator-based matrix switch for silicon on-chip optical interconnection. *Proc. IEEE* 97: 1216–1238.

17 Xu, Q.F., Schmidt, B., Pradhan, S., and Lipson, M. (2005). Micrometre-scale silicon electro-optic modulator. *Nature* 435: 325–327.

18 Vollmer, F. and Arnold, S. (2008). Whispering-gallery-mode biosensing: label-free detection down to single molecules. *Nat. Methods* 5: 591–596.

19 Zhu, J.G., Ozdemir, S.K., Xiao, Y.F. et al. (2010). On-chip single nanoparticle detection and sizing by mode splitting in an ultrahigh-*Q* microresonator. *Nat. Photonics* 4: 46–49.

20 He, L.N., Ozdemir, K., Zhu, J.G. et al. (2011). Detecting single viruses and nanoparticles using whispering gallery microlasers. *Nat. Nanotechnol.* 6: 428–432.

21 Peter, E., Senellart, P., Martrou, D. et al. (2005). Exciton-photon strong-coupling regime for a single quantum dot embedded in a microcavity. *Phys. Rev. Lett.* 95: 067401.

22 Fujita, M., Ushigome, R., and Baba, T. (2000). Continuous wave lasing in GaInAsP microdisk injection laser with threshold current of 40 mu A. *Electron. Lett.* 36: 790–791.

23 Yang, Y.D. and Huang, Y.Z. (2007). Symmetry analysis and numerical simulation of mode characteristics for equilateral-polygonal optical microresonators. *Phys. Rev. A* 76: 023822.

24 Jonaš, A., McGloin, D., and Kiraz, A. (2015). Droplet lasers. *Opt. Photonics News* 26 (5): 36–43.

25 Humar, M. and Yun, S.H. (2015). Intracellular microlasers. *Nat. Photonics* 9: 572–576.

26 Del'Haye, P., Schliesser, A., Arcizet, O. et al. (2007). Optical frequency comb generation from a monolithic microresonator. *Nature* 450: 1214–1217.

27 Lee, H., Chen, T., Li, J. et al. (2012). Chemically etched ultrahigh-*Q* wedge-resonator on a silicon chip. *Nat. Photonics* 6: 369–373.

28 Schawlow, A.L. and Townes, C.H. (1958). Infrared and optical masers. *Phys. Rev.* 112: 1940–1949.

29 Henry, C.H. (1982). Theory of the linewidth of semiconductor-lasers. *IEEE J. Quantum Electron.* 18: 259–264.

30 Bagheri, M., Shih, M.H., Choi, S.J. et al. (2009). Microcavity laser linewidth close to threshold. *IEEE J. Quantum Electron.* 45: 935–939.

31 Coldren, L.A. and Corzine, S.W. (1995). *Diode Lasers and Photonic Integrated Circuits*. New York: Wiley.

32 Zou, L.X., Huang, Y.Z., Liu, B.W. et al. (2015). Nonlinear dynamics for semiconductor microdisk laser subject to optical injection. *IEEE J. Sel. Top. Quantum Electron.* 21 (6): 1800408.

2

Multilayer Dielectric Slab Waveguides

2.1 Introduction

Semiconductor lasers are usually grown in a substrate with multiple-layer semiconductor materials, such as AlGaAs/GaAs and InGaAsP/InP material systems, which can form a multilayer slab waveguide. The multilayer slab waveguide is the simplest case and has easy-to-analyze mode characteristics for guided and radiation modes [1–3]. The modes are solutions of Maxwell's equations under boundary conditions imposed on the mode fields at the dielectric interfaces of the waveguide. For a multiple-layer slab waveguide with interfaces parallel to y–z plane, we can model z-direction propagating modes under uniform distribution condition in y-direction, which correspond to transverse mode distribution for edge-emitting semiconductor lasers. Under the y-direction uniform condition, the z-direction propagating modes can be classified as transverse-electric or TE modes and transverse-magnetic or TM modes, with the z-direction component of electric field and magnetic field considered to be zero.

In Section 2.2, TE and TM modes are defined and wave equations are given based on Maxwell's equations. In Section 2.3, we present mode solutions and eigenvalue equations for TE- and TM-guided modes in a symmetric three-layer slab waveguide. Furthermore, guided modes and radiation modes are discussed. In Section 2.4, we derive a general form of eigenvalue equations for guided modes in multilayer complex slab waveguides, and give the phase shift and Goos–Hanchen shift of total internal reflection based on the eigenvalue equations. In Section 2.5, the above eigenvalue equations are applied to one-dimensional multilayer waveguides, such as vertical-cavity surface-emitting lasers and Fabry–Perot cavity. In Section 2.6, the relations of mode gain and material gain are derived from wave equations for complex slab waveguides and compared with the optical confinement factor based on the power flow density. Finally, some numerical results are presented in Section 2.7 and the effective index method is introduced in Section 2.8.

Microcavity Semiconductor Lasers: Principles, Design, and Applications, First Edition.
Yong-zhen Huang and Yue-de Yang.

2.2 TE and TM Modes in Slab Waveguides

In the absence of electric charges, Maxwell's equations in uniform medium of the slab waveguide at light wave frequency can be written as

$$\nabla \times E = -\frac{\partial}{\partial t}\mu H, \tag{2.1}$$

$$\nabla \times H = \frac{\partial}{\partial t}\varepsilon E, \tag{2.2}$$

$$\nabla \cdot \varepsilon E = 0, \tag{2.3}$$

$$\nabla \cdot \mu H = 0. \tag{2.4}$$

In a dielectric slab waveguide, we have different dielectric permittivity ε in different layer and magnetic permeability μ of that in free space μ_0 at optical frequency. For a multilayer slab waveguide with the x-coordinate perpendicular to the interfaces in the Cartesian coordinate system, the electric and magnetic fields for mode with the radian frequency $\omega = 2\pi f$ propagating in the z-direction can be described as

$$\boldsymbol{E}(x, z, t) = (E_x(x), E_y(x), E_z(x)) \exp(i\beta z - i\omega t), \tag{2.5}$$

$$\boldsymbol{H}(x, z, t) = (H_x(x), H_y(x), H_z(x)) \exp(i\beta z - i\omega t), \tag{2.6}$$

where the mode-field patterns are assumed to be uniform along the y-direction, and the z-direction complex propagation constant is

$$\beta = \beta_r + i\beta_i. \tag{2.7}$$

Submitting (2.5) and (2.6) into (2.1), we obtain following equations:

$$\begin{cases} \omega\mu H_x(x) = -\beta E_y(x) \\ \omega\mu H_y(x) = \beta E_x(x) + i\frac{\partial E_z(x)}{\partial x} \\ \omega\mu H_z(x) = -i\frac{\partial E_y(x)}{\partial x} \end{cases}. \tag{2.8}$$

Based on (2.8), we can choose the solutions to be determined by $E_y(x)$ as $E_x(x)$ and $E_z(x)$ to be zero. The solutions are called TE modes with nonzero electromagnetic field components of E_y, H_x, and H_z, with the only nonzero electric field E_y being vertical to the field propagation z-direction. The nonzero magnetic fields are derived from $E_y(x)$ based on (2.8):

$$H_x(x) = -\frac{\beta}{\omega\mu} E_y(x), \tag{2.9}$$

$$H_z(x) = -\frac{i}{\omega\mu}\frac{\partial E_y(x)}{\partial x}. \tag{2.10}$$

Similarly, following equations are obtained by submitting (2.5) and (2.6) into (2.2):

$$\begin{cases} \omega\varepsilon E_x(x) = \beta H_y(x) \\ \omega\varepsilon E_y(x) = -\beta H_x(x) - i\frac{\partial H_z(x)}{\partial x} \\ \omega\varepsilon E_z(x) = i\frac{\partial H_y(x)}{\partial x} \end{cases}. \tag{2.11}$$

Taking H_x and H_z to be zero, we can get solutions for TM modes with nonzero electromagnetic fields of H_y, E_x, and E_z, with the only nonzero magnetic field H_y being vertical to the field propagation z-direction. The nonzero electric fields are determined by $H_y(x)$ from (2.11) as:

$$E_x(x) = \frac{\beta}{\omega\varepsilon} H_y(x), \tag{2.12}$$

$$E_z(x) = \frac{i}{\omega\varepsilon} \frac{\partial H_y(x)}{\partial x}. \tag{2.13}$$

Furthermore, we obtain the following equation by applying the curl operator to (2.1) with the help of (2.2)

$$\nabla \times (\nabla \times E) = -\mu \nabla \times \left(\frac{\partial}{\partial t} H \right) = -\mu\varepsilon \frac{\partial^2 E}{\partial t^2}. \tag{2.14}$$

Using the vector identity $\nabla \times (\nabla \times E) = \nabla(\nabla \cdot E) - \nabla^2 E$ to simplify (2.14), we can obtain the following wave equation

$$\nabla^2 E + k_0^2 n^2 E = 0, \tag{2.15}$$

which is called Helmholtz equation. The free space wavenumber k_0 is related to wavelength λ by

$$k_0 = \frac{\omega}{c} = \frac{2\pi}{\lambda}. \tag{2.16}$$

Dielectric permittivity is $\varepsilon = \varepsilon_0 n^2$ with ε_0 dielectric permittivity in free space, and light speed in free space c is $1/c^2 = \varepsilon_0 \mu_0$. The magnetic field also satisfies the same wave equation as (2.15).

2.3 Modes in Symmetric Three-Layer Slab Waveguides

In this section, eigenvalue equations are given for TE and TM modes in a three-layer slab waveguide as shown in Figure 2.1, and some numerical examples are presented.

2.3.1 TE Modes in Three-Layer Slab Waveguides

For TE-confined modes in the waveguide with a thickness of d, we can express $E_y(x)$ as:

$$\text{Symmetry} \quad E_y(x) = \begin{cases} \cos\frac{\kappa d}{2} \exp\left[-\gamma\left(x - \frac{d}{2}\right)\right] & x > d/2 \\ \cos \kappa x & -d/2 \le x \le d/2\,, \\ \cos\frac{\kappa d}{2} \exp\left[\gamma\left(x + \frac{d}{2}\right)\right] & x < -d/2 \end{cases} \tag{2.17}$$

$$\text{Antisymmetry} \quad E_y(x) = \begin{cases} \sin\frac{\kappa d}{2} \exp\left[-\gamma\left(x - \frac{d}{2}\right)\right] & x > d/2 \\ \sin \kappa x & -d/2 \le x \le d/2\,, \\ -\sin\frac{\kappa d}{2} \exp\left[\gamma\left(x + \frac{d}{2}\right)\right] & x < -d/2 \end{cases} \tag{2.18}$$

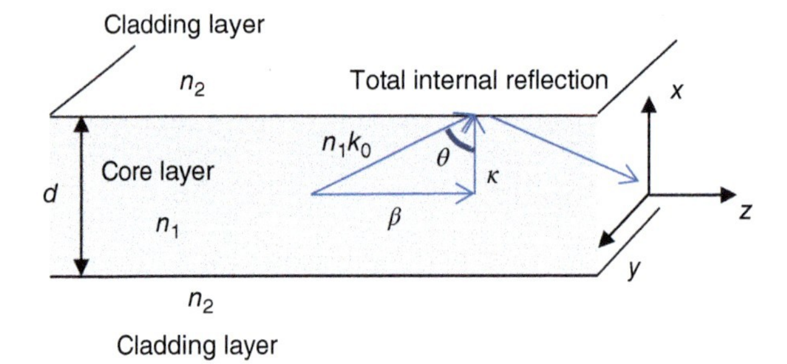

Figure 2.1 A three-layer slab waveguide with a thickness of d and refractive indices of n_1 and n_2, and the illustration of mode light ray with an incident angle determined by propagation constants.

where a term of $\exp(i\beta z - i\omega t)$ is omitted and the continuous conditions of $E_y(x)$ at the interfaces $x = \pm d/2$ are satisfied. The transverse propagation constant κ and decaying constant γ are given by:

$$\begin{cases} \kappa = \sqrt{k_0^2 n_1^2 - \beta^2} \\ \gamma = \sqrt{\beta^2 - {k_0}^2 {n_2}^2} \end{cases}. \tag{2.19}$$

In addition, we need H_z to be continuous functions across the interfaces, which result in the continuity of dE_y/dx for TE modes based on (2.10). The following eigenvalue equations can be obtained from the boundary conditions

$$\begin{cases} \kappa \tan \frac{\kappa d}{2} = \gamma & \text{symmetry} \\ \kappa \tan \left(\frac{\kappa d}{2} - \frac{\pi}{2} \right) = \gamma & \text{antisymmetry} \end{cases}. \tag{2.20}$$

The equations can also be written as

$$\begin{cases} \frac{\kappa d}{2} = \tan^{-1} \frac{\gamma}{\kappa} + m\pi & \text{symmetry} \\ \frac{\kappa d}{2} = \tan^{-1} \frac{\gamma}{\kappa} + \left(m + \frac{1}{2} \right) \pi & \text{antisymmetry} \end{cases}, \tag{2.21}$$

with $m = 0, 1, 2, \ldots$, for different order transverse modes. Multiplying (2.21) by a factor of 4, we can combine the eigenvalue equations of the symmetric and antisymmetric wavefunctions as

$$2\kappa d - 4\tan^{-1} \frac{\gamma}{\kappa} = 2m\pi, \quad m = 0,\ 1,\ 2,\ 3,\ \ldots., \tag{2.22}$$

where mode number $m = 0, 2, 4, \ldots$ corresponding to even modes, and $m = 1, 3, 5, \ldots$ for odd mode. By taking a mode wavelength, we can solve the z-direction propagation constant β from the eigenvalue with the relations (2.19).

2.3.2 TM Modes in Three-Layer Slab Waveguides

For TM-confined modes, the z-direction magnetic field can be written as $H_y(x)\exp(i\beta z - i\omega t)$ with:

$$\text{Symmetry} \quad H_y(x) = \begin{cases} \cos\frac{\kappa d}{2}\exp\left[-\gamma\left(x-\frac{d}{2}\right)\right] & x > d/2 \\ \cos\kappa x & -d/2 \le x \le d/2\,, \\ \cos\frac{\kappa d}{2}\exp\left[\gamma\left(x+\frac{d}{2}\right)\right] & x < -d/2 \end{cases} \tag{2.23}$$

$$\text{Antisymmetry} \quad H_y(x) = \begin{cases} \sin\frac{\kappa d}{2}\exp\left[-\gamma\left(x-\frac{d}{2}\right)\right] & x > d/2 \\ \sin\kappa x & -d/2 \le x \le d/2\,, \\ -\sin\frac{\kappa d}{2}\exp\left[\gamma\left(x+\frac{d}{2}\right)\right] & x < -d/2 \end{cases} \tag{2.24}$$

where the continuous conditions of $H_y(x)$ at the interfaces $x = \pm d/2$ are already applied, and the transverse propagation constant κ and decaying constant γ are given by (2.19).

The continuous condition of E_z across the interfaces requires continuity of $1/n^2 \cdot dH_y(x)/dx$ for TM modes based on (2.13), which yields following eigenvalue equations

$$\begin{cases} n_2^2\kappa\tan\frac{\kappa d}{2} = n_1^2\gamma & \text{symmetry} \\ n_2^2\kappa\tan\left(\frac{\kappa d}{2}-\frac{\pi}{2}\right) = n_1^2\gamma & \text{antisymmetry} \end{cases}. \tag{2.25}$$

Similar to (2.22), we obtain the following eigenvalue equations for TM modes

$$2\kappa d - 4\tan^{-1}\frac{n_1^2\gamma}{n_2^2\kappa} = 2m\pi, \quad m = 0,\ 1,\ 2,\ 3,\ \ldots, \tag{2.26}$$

where mode number $m = 0, 2, 4, \ldots$ for even modes, and $m = 1, 3, 5, \ldots$for odd mode. The first term of Eqs. (2.22) and (2.26) is traveling phase shift as the wave transmits one period over the core layer, and the second term is the phase delay of the total reflection as discussed in Section 2.4.3. The sum of the two terms should be an integral multiple of 2π for resonant modes. TE and TM modes have different phase delays as they have boundary continuity of dE_y/dx and $1/n^2 \cdot dH_y(x)/dx$, respectively. By taking a mode wavelength, we can solve the z-direction propagation constant β from the eigenvalue with the transverse propagation constants as function of β in (2.19).

2.3.3 Guided and Radiation Modes

The above wave functions are guided modes with an oscillation function in the core layer and exponentially decay wave in the external layers if κ and γ are real quantities, which require

$$\frac{c}{n_1} < \frac{\omega}{\beta} < \frac{c}{n_2}, \tag{2.27}$$

where c is the light speed in vacuum. The eigenvalue β should be solved from (2.22) and (2.26) for TE and TM modes, respectively. In fact, the eigenvalue equations are transverse resonance conditions for mode transmission over one period inside the waveguide with the total phase shift of multiple of 2π, where the first term $2\kappa d$ is the propagation phase shift and the second term of the left side of (2.22) and (2.26) is phase delay of total internal reflection. As shown in Figure 2.1, the guided-mode light ray is total internal reflection on the interface with the incident angle $\theta = \sin^{-1}\frac{\beta}{n_1 k_0}$, which is larger than the critical angle of the total internal reflection $\sin^{-1}\frac{n_2}{n_1}$ for $\beta > k_0 n_2$ obtained from (2.27). The mode cutoff condition can be written as

$$\beta = k_0 n_2, \tag{2.28}$$

because the corresponding decaying constant $\gamma = 0$ and mode-field distribution is uniform in the external region. From the mode cutoff condition, we can determine the number of guided modes supported by the waveguide. For the fundamental transverse mode with $m = 0$ and $\gamma = 0$, we can obtain $d = 0$ from the eigenvalue Eqs. (2.22) and (2.26) for TE and TM modes, respectively. The results indicate the cut-off condition for the fundamental transverse mode is $d = 0$; in other words, the fundamental order transverse mode always exists in symmetry in three-layer waveguide. Furthermore, single-mode condition can be obtained from the mode cutoff condition of the first-order transverse mode by submitting $\gamma = 0$ and $m = 1$ into (2.22, 2.26), which yields $k_0 d\sqrt{n_1^2 - n_2^2} = \pi$. The corresponding single mode waveguide width is

$$d = \frac{\lambda}{2\sqrt{n_1^2 - n_2^2}}. \tag{2.29}$$

The single-mode condition requires a refractive index difference $\Delta n = n_1 - n_2$, which is inversely proportional to squared waveguide width. Taking $n_1 = 3.5$ and wavelength $\lambda = 1.5\,\mu\text{m}$, we obtain $\Delta n = 0.08$, 0.020, and 0.009 as the waveguide thickness $d = 1$, 2, and $3\,\mu\text{m}$ from the single-mode condition (2.29).

An imaginary propagation constant γ is obtained from (2.19) as $\beta < k_0 n_2$, and the corresponding wavefunctions are radiation modes, which are not decay to zero at an infinite distance. For the radiation modes, initial wavefunction should be a more general wavefunction than the guided modes. Different from the guided modes with terms in the external regions, four terms are accounted for in the external regions for radiation modes with incoming and outgoing traveling waves. The boundary conditions of the electromagnetic fields can yield an eigenvalue equation to solve the propagation constants. In fact, the radiation modes have continuous propagation constants. In addition, imaginary values of β can also yield radiation modes as evanescent waves, which influence the surface fine field shape without radiation power from the waveguide.

2.4 Eigenvalue Equations for Multilayer Slab Complex Waveguides

In this section, eigenvalue equations are derived for TE and TM modes propagating parallel to the interface in m-layer slab complex waveguides. Then, the eigenvalue equation is extrapolated to one-dimensional waveguide for mode propagating vertical to the interface of the slab waveguide, which is a simplified waveguide of a vertical-cavity surface-emitting laser.

2.4.1 Eigenvalue Equation for TE Modes

For TE modes in an m-layer slab complex waveguide, we express the z-direction propagating electric field $E_y(x,z,t)$ in the layer j ($j = 1, 2, \ldots, m$) as

$$E_{yj}(x, z, t) = [a_j \exp(ik_j x) + b_j \exp(-ik_j x)] \exp(i\beta z - i\omega t). \tag{2.30}$$

Submitting the wavefunction (2.30) into wave Eq. (2.15), we have the x-direction complex propagation constant k_j as

$$k_j = \begin{cases} (k_0^2 n_j^2 - \beta^2)^{1/2} & k_0 n_{jr} \geq \beta_r \\ i(\beta^2 - k_0^2 n_j^2)^{1/2} & k_0 n_{jr} < \beta_r \end{cases}. \tag{2.31}$$

The complex refractive index of the layer j is

$$n_j = n_{jr} + i\kappa_j, \tag{2.32}$$

with the extinction coefficient κ_j related to a material gain g_j by

$$\kappa_j = -g_j / 2k_0. \tag{2.33}$$

The boundary continuous conditions of E_y and H_z at $x = x_{j-1}$ and x_j require the electric field $E_y(x)$ and its derivative $dE_y(x)/dx$ to be continuous across the boundaries. The boundary conditions in the interfaces x_{j-1} and x_j are

$$a_{j-1} \exp(ik_{j-1} x_{j-1}) + b_{j-1} \exp(-ik_{j-1} x_{j-1}) \\ = a_j \exp(ik_j x_{j-1}) + b_j \exp(-ik_j x_{j-1}), \tag{2.34}$$

$$k_{j-1} [a_{j-1} \exp(ik_{j-1} x_{j-1}) - b_{j-1} \exp(-ik_{j-1} x_{j-1})] \\ = k_j [a_j \exp(ik_j x_{j-1}) - b_j \exp(-ik_j x_{j-1})], \tag{2.35}$$

and

$$a_j \exp(ik_j x_j) + b_j \exp(-ik_j x_j) \\ = a_{j+1} \exp(ik_{j+1} x_j) + b_{j+1} \exp(-ik_{j+1} x_j), \tag{2.36}$$

$$k_j [a_j \exp(ik_j x_j) - b_j \exp(-ik_j x_j)] \\ = k_{j+1} [a_{j+1} \exp(ik_{j+1} x_j) - b_{j+1} \exp(-ik_{j+1} x_j)]. \tag{2.37}$$

Dividing (2.35) by (2.34) yields the following relation

$$\frac{a_j \exp(ik_j x_{j-1})}{b_j \exp(-ik_j x_{j-1})} = \frac{(k_j - k_{j-1}) + (k_j + k_{j-1}) \dfrac{a_{j-1} \exp(ik_{j-1} x_{j-1})}{b_{j-1} \exp(-ik_{j-1} x_{j-1})}}{(k_j + k_{j-1}) + (k_j - k_{j-1}) \dfrac{a_{j-1} \exp(ik_{j-1} x_{j-1})}{b_{j-1} \exp(-ik_{j-1} x_{j-1})}}, \tag{2.38}$$

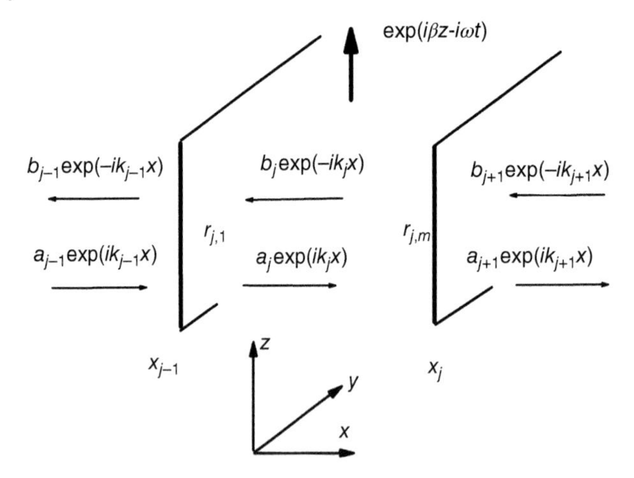

Figure 2.2 Wave functions propagating in layers $j-1, j$, and $j+1$, with only outgoing waves in external layers 1 and m, i.e. $a_1 = 0$ and $b_m = 0$.

The left term in (2.38) can be defined as

$$r_{j,1} \equiv \frac{a_j}{b_j} \exp(2ik_j x_{j-1}), \tag{2.39}$$

which is the electric amplitude reflectivity as it impinges the interface x_{j-1} from layer j as shown in Figure 2.2, including the influence of layer $j-1$, layer $j-2$, …, and layer 1. Based on (2.39), we have $a_{j-1} \exp(2ik_{j-1}x_{j-1})/b_{j-1} = r_{j-1,1} \exp[2ik_{j-1}(x_{j-1} - x_{j-2})] = r_{j-1,1} \exp(2ik_{j-1}d_{j-1})$, with the thickness of layer $j-1$ as

$$d_{j-1} = x_{j-1} - x_{j-2}, \tag{2.40}$$

So we can express (2.38) as following recursion amplitude reflectivity relation for multilayer structure

$$r_{j,1} = \frac{r_{j,j-1} + r_{j-1,1} \exp(2ik_{j-1}d_{j-1})}{1 + r_{j,j-1}r_{j-1,1} \exp(2ik_{j-1}d_{j-1})}, \tag{2.41}$$

with the amplitude reflectivity of only the interface effect as the electric field impinges the interface x_{j-1} from layer j

$$r_{j,j-1} = \frac{k_j - k_{j-1}}{k_j + k_{j-1}}. \tag{2.42}$$

Similarly, we obtain the following recursion relation by dividing (2.37) by (2.36):

$$r_{j,m} = \frac{r_{j,j+1} + r_{j+1,m} \exp(2ik_{j+1}d_{j+1})}{1 + r_{j,j+1}r_{j+1,m} \exp(2ik_{j+1}d_{j+1})}, \tag{2.43}$$

which is the electric amplitude reflectivity as it impinges the interface x_j from layer j as shown in Figure 2.1, including the influence of layer $j+1$, layer $j+2$, …, and layer m:

$$r_{j,m} \equiv \frac{b_j}{a_j} \exp(-2ik_j x_j). \tag{2.44}$$

The amplitude reflectivity $r_{j,j+1}$ satisfies the similar relation as (2.42)

$$r_{j,j+1} = \frac{k_j - k_{j+1}}{k_j + k_{j+1}}. \tag{2.45}$$

Multiplying $r_{j,1}$ and $r_{j,m}$ in (2.39) and (2.44), we get an eigenvalue equation as

$$r_{j,1} r_{j,m} \exp(2ik_j d_j) = 1, \tag{2.46}$$

with the thickness of layer j as in (2.40)$d_j = x_j - x_{j-1}$.

The x-direction wavefunctions are described as plane waves propagating in the positive and negative x-directions with complex amplitudes $a_j \exp(ik_j x)$ and $b_j \exp(-ik_j x)$ at layer j, but with only outgoing waves in the external regions $j = 1$ and m for the guided modes, i.e. $a_1 = 0$ and $b_m = 0$. The eigenvalue Eq. (2.46) is similar as that of Fabry–Perot cavity for mode propagating inside a cavity with a cavity length of d_j and two side reflectivities of $r_{j,1}$ and $r_{j,m}$.

2.4.2 Eigenvalue Equation for TM Modes

For TM modes in the m-layer slab complex waveguide, we express the z-direction propagating electric field $H_y(x,z,t)$ in the layer j ($j = 1, 2, \ldots, m$) as

$$H_{yj}(x, z, t) = [a_j \exp(ik_j x) + b_j \exp(-ik_j x)] \exp(i\beta z - i\omega t). \tag{2.47}$$

The boundary continuous conditions of H_y and E_z at $x = x_{j-1}$ require the magnetic field $H_y(x)$ and its derivative $dH_y(x)/dx$ derived by ε to be continuous across the boundaries. So the following equations are obtained from the boundary conditions:

$$\begin{aligned} &a_{j-1} \exp(ik_{j-1}x_{j-1}) + b_{j-1} \exp(-ik_{j-1}x_{j-1}) \\ &\qquad = a_j \exp(ik_j x_{j-1}) + b_j \exp(-ik_j x_{j-1}), \end{aligned} \tag{2.48}$$

$$\begin{aligned} &k_{j-1}[a_{j-1} \exp(ik_{j-1}x_{j-1})z - b_{j-1} \exp(-ik_{j-1}x_{j-1})]/\varepsilon_{j-1} \\ &\qquad = k_j[a_j \exp(ik_j x_{j-1}) - b_j \exp(-ik_j x_{j-1})]/\varepsilon_j, \end{aligned} \tag{2.49}$$

Comparing Eqs. (2.48) and (2.49) for TM modes with those (2.34) and (2.35) for TE modes, we can conclude that recursion reflection formulae (2.41) and (2.43) and eigenvalue Eq. (2.46) can be applied to TM modes. The only difference is that the amplitude reflectivities (2.42) and (2.45) should be replaced by the following relation for TM modes:

$$r_{j,j-1} = \frac{k_j/\varepsilon_j - k_{j-1}/\varepsilon_{j-1}}{k_j/\varepsilon_j + k_{j-1}/\varepsilon_{j-1}}. \tag{2.50}$$

2.4.3 Phase Shift of Total Internal Reflection

Applying the above eigenvalue equations of TE modes to three-layer symmetric waveguide of Figure 2.1, we have $k_2 = \kappa$ and $k_1 = k_3 = i\gamma$ from (2.19, 2.31),

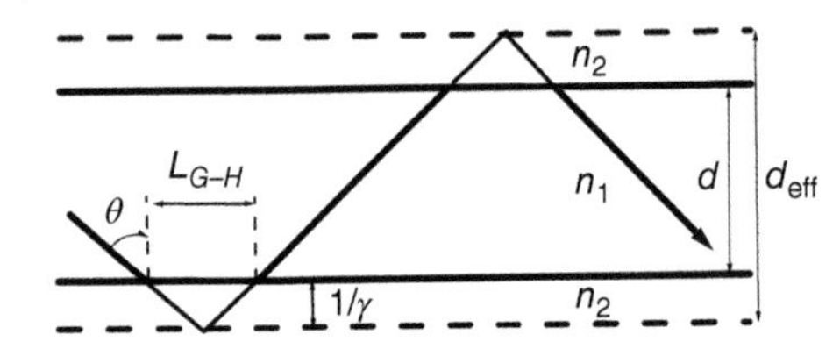

Figure 2.3 Schematic for zigzag propagating mode light ray and the effective width d_{eff} of mode intensity distribution in a three-layer symmetric waveguide.

and eigenvalue Eq. (2.46) as $\left(\frac{\kappa-i\gamma}{\kappa+i\gamma}\right)^2\exp(2i\kappa d)=1$, which can be reduced as $\exp[2i(\kappa d-\varphi)]=1$ with the solution of (2.22) and the reflection phase delay as

$$\varphi = 2\tan^{-1}\frac{\gamma}{\kappa}. \tag{2.51}$$

Using the reflection phase shift at interface $x = d/2$, we can express core layer wavefunction for three-layer symmetric waveguide based on (2.30)

$$E_y(x,z,t) = a\exp(i\beta z - i\omega t)[\exp(i\kappa x) + \exp(i\kappa d - i\kappa x - i\varphi)], \tag{2.52}$$

where subscript of layer number is omitted, and $\exp[-i(\omega t+\kappa x-\kappa d)+i\beta(z-\varphi/\beta)]$ is the reflected wave of the incident wave $\exp[-i(\omega t-\kappa x)+i\beta z)]$ at interface $x=d/2$. The reflected wave is phase lag behind the incident wave with a phase shift of ϕ, which corresponds to incident wave traveling into the cladding layer and then returning the core layer with the phase shift as shown in Figure 2.3. Similar to phase speed and group speed, we can expect mode intensity has a transmission delay along the z-direction as $d\phi/d\beta$, which can be expressed for TE modes as:

$$L_{G-H} = \frac{d\varphi}{d\beta} = \frac{2\beta}{\gamma\kappa} = \frac{2\tan\theta}{\gamma}, \tag{2.53}$$

where θ is the incident angle as shown in Figure 2.3. L_{G-H} is called Goos–Hanchen shift and is usually obtained by accounting total reflection of a phase packet. For TM modes, the κ_j and κ_{j-1} in the reflection phase formula are replaced by κ_j/n_j^2 and κ_{j-1}/n_{j-1}^2 as shown in (2.50), and corresponding reflection phase shift and L_{G-H} are more complicated.

2.5 Eigenvalue Equations for One-Dimensional Multilayer Waveguides

2.5.1 Eigenvalue Equation for Vertical-Cavity Surface-Emitting Lasers

Simplifying a vertical-cavity surface-emitting laser (VCSEL) as a one-dimensional m-layer slab waveguide as shown in Figure 2.4, we can treat resonant mode of the VCSEL as transverse electromagnetic (TEM) wave propagating in the positive and negative z-directions with the electromagnetic fields $(E_x(z),0,0)$ and $(0,H_y(z),0)$ with a periodic time dependence given by $\exp(-i\omega t)$ under the approximation of uniform distribution in the x–y plane.

Submitting the electromagnetic fields into Maxwell Eqs. (2.1, 2.2), we have

$$i\omega\mu H_y(z) = \frac{\partial E_x(z)}{\partial z}, \tag{2.54}$$

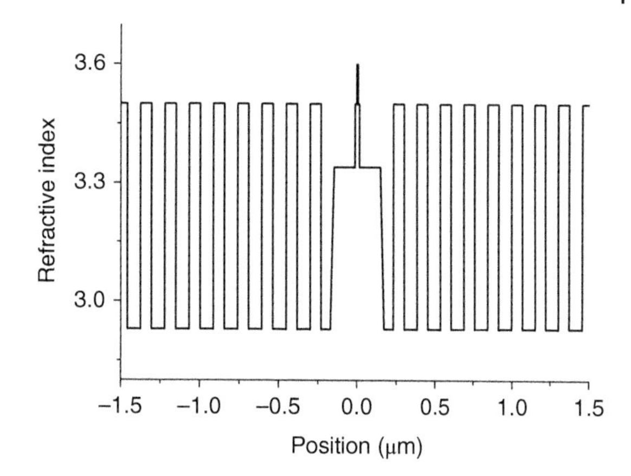

Figure 2.4 One-dimensional refractive index distribution for a vertical-cavity surface-emitting laser.

$$i\omega\varepsilon E_x(z) = \frac{\partial H_y(z)}{\partial z}. \tag{2.55}$$

As in Section 2.4, the wavefunction of the electric field in layer j can be written as

$$E_{xj}(z,t) = [a_j \exp(i\beta_j z) + b_j \exp(-i\beta_j z)] \exp(-i\omega t), \quad z_{j-1} < z < z_j, \tag{2.56}$$

with the complex propagation constants of

$$\beta_j = \frac{\omega}{c} n_j = k_0(n_{jr} + i\kappa_j). \tag{2.57}$$

The boundary continuous conditions of E_x and H_y at the interfaces require the electric field $E_x(z)$ and its derivative $dE_x(z)/dz$ to be continuous across the boundaries. Similar to the treatment for TE modes in Section 2.4.1, we can obtain recursion reflection formulae and eigenvalue equation as (2.41–2.46) with the complex propagation constant (2.31) replaced by (2.57). As (2.41, 2.43), recursion relations for amplitude reflectivity of multilayer structure satisfy:

$$r_{j,1} \equiv \frac{a_j}{b_j} \exp(2i\beta_j z_{j-1}) = \frac{r_{j,j-1} + r_{j-1,1} \exp(2i\beta_{j-1} d_{j-1})}{1 + r_{j,j-1} r_{j-1,1} \exp(2i\beta_{j-1} d_{j-1})}, \tag{2.58}$$

$$r_{j,m} \equiv \frac{b_j}{a_j} \exp(-2i\beta_j z_j) = \frac{r_{j,j+1} + r_{j+1,m} \exp(2i\beta_{j+1} d_{j+1})}{1 + r_{j,j+1} r_{j+1,m} \exp(2i\beta_{j+1} d_{j+1})}, \tag{2.59}$$

with the amplitude reflectivity of only the interface effect as the electric field impinges the interfaces between layer j and layer $j+1$ or $j-1$ as

$$r_{j,j\pm1} = \frac{\beta_j - \beta_{j\pm1}}{\beta_j + \beta_{j\pm1}} = \frac{n_j - n_{j\pm1}}{n_j + n_{j\pm1}}. \tag{2.60}$$

Multiplying (2.58, 2.59), we can obtain the eigenvalue equation for one-dimensional VCSEL waveguide structure as

$$r_{j,1} r_{j,m} \exp(2i\beta_j d_j) = 1, \tag{2.61}$$

where $d_j = z_j - z_{j-1}$ is the thickness of layer j.

A VCSEL grown on GaAs substrate with 24.5-period quarter-wavelength $Al_{0.9}Ga_{0.1}As$/GaAs n-DBR, a symmetric cavity consisting of 131.9 nm $Al_{0.3}Ga_{0.7}As$/

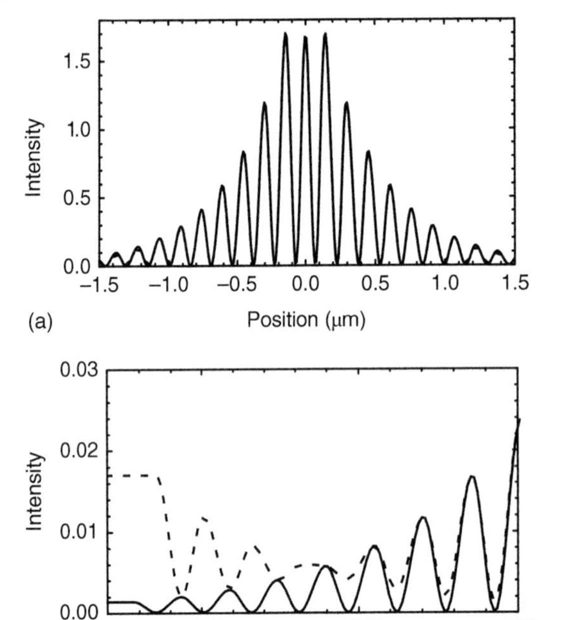

Figure 2.5 Normalized squared electric field at (a) center region and (b) external and p-DBR region for the VCSEL structure. Solid and dashed lines correspond to 20-period and 19.5-period p-DBR, dashed line: lack the up 70-nm GaAs layer of the p-DBR.

10 nm GaAs/8 nm InGaAs/10 nm GaAs/131.9 nm $Al_{0.3}Ga_{0.7}As$, and 20-period quarter-wavelength $Al_{0.9}Ga_{0.1}As$/GaAs p-DBR are considered with refractive indices of InGaAs, GaAs, $Al_{0.3}Ga_{0.7}As$, and $Al_{0.9}Ga_{0.1}As$ to be 3.6, 3.5, 3.34, and 2.93, and a material absorption loss of 5 cm^{-1}. Squared electric field distributions are plotted as solid and dashed lines in Figure 2.5 for the VCSEL with 20-period and 19.5-period p-DBR, respectively. In the central region, the intensity distributions are almost the same, but the intensity increases greatly near the output region as the upper GaAs layer thickness is zero instead of normal 70 nm, due to the reflection from air–AlAs interface is antiphase with that of the other interfaces of the p-DBR. The corresponding threshold gain for the VCSEL with a 19.5-period p-DBR is about four times that at 20-period p-DBR. The modification of upper-layer structure can be used for transverse-mode control for VCSELs.

2.5.2 Resonance Condition for the Fabry–Perot Cavity

The Fabry–Perot cavity typically consisted of two parallel reflecting mirrors, and is simply treated as a one-dimensional three-layer VCSEL. To model the Fabry–Perot cavity, we can rewrite (2.61) as

$$r_1 r_2 \exp(2i\beta L) = 1, \tag{2.62}$$

where r_1 and r_2 are amplitude reflectivities of two mirrors, which should satisfy (2.60) for simple interface mirrors between two dielectrics, and L is the cavity length between two mirrors. Using the relation of (2.33), we can express the complex propagation constant of (2.57) as

$$\beta = k_0\left(n_r - i\frac{g}{2k_0}\right) = n_r k_0 - i\frac{g}{2}. \tag{2.63}$$

Submitting (2.63) into (2.62), we have

$$\begin{cases} \text{Re}[r_1 r_2 \exp(2in_r k_0 L)] \exp(gL) = 1 \\ \text{Im}[r_1 r_2 \exp(2in_r k_0 L)] = 0 \end{cases}. \tag{2.64}$$

If $r_1 r_2$ is a real positive quantity, (2.64) can be reduced to

$$\begin{cases} r_1 r_2 \exp(gL) = 1 \\ 2n_r k_0 L = 2l\pi, \quad l = 1,\ 2,\ 3,\ \ldots \end{cases}, \tag{2.65}$$

which are threshold gain and phase condition with longitudinal mode number l for self-sustainable oscillation in the Fabry–Perot cavity. Submitting $k_0 = 2\pi/\lambda$ in the phase condition of (2.65) and taking the derivative of the equation with respect to the mode number l as l is much larger than unity, we can obtain following equation at $\delta l = -1$:

$$\frac{2n_r L}{\lambda^2}\delta\lambda - \frac{2L}{\lambda}\frac{dn_r}{d\lambda}\delta\lambda = 1. \tag{2.66}$$

So the longitudinal mode interval $\delta\lambda$ is

$$\delta\lambda = \frac{\lambda^2}{2n_g L}, \tag{2.67}$$

with a group index as

$$n_g = n_r - \lambda\frac{dn_r}{d\lambda}. \tag{2.68}$$

The group index can be measured from the longitudinal mode intervals as the cavity length L is also measured. Figure 2.6 shows amplified spontaneous emission spectrum near threshold for a 1550-nm Fabry–Perot InGaAsP/InP

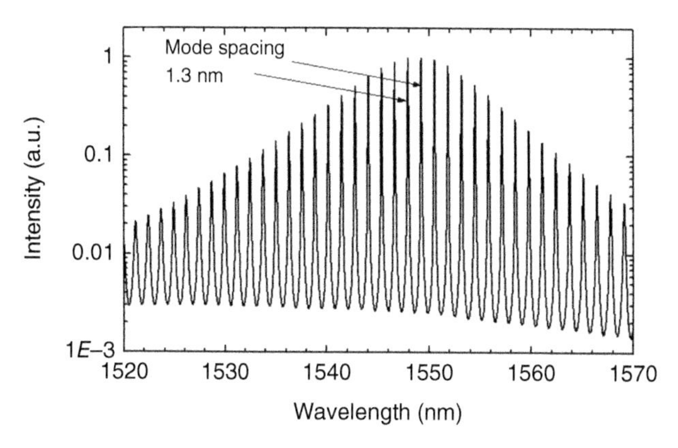

Figure 2.6 Lasing spectrum of a 1550-nm Fabry–Perot semiconductor laser with clear multiple longitudinal modes.

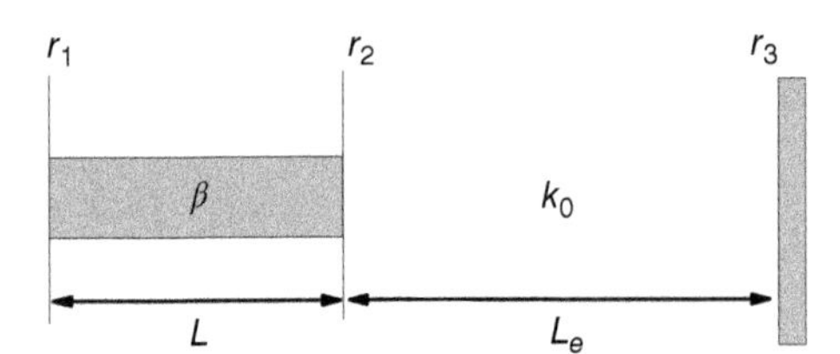

Figure 2.7 One-dimensional model for external cavity semiconductor lasers.

multiple-quantum-well semiconductor laser with a cavity length $L = 250\,\mu\text{m}$. A group index of $n_g = 3.7$ is obtained from (2.67) using the longitudinal-mode interval $\delta\lambda = 1.3\,\text{nm}$. Clear multiple longitudinal modes are observed without minor peaks of higher-order transverse modes, which are suitable for gain measurement. Furthermore, mode gain spectrum can be measured by Hakki–Paoli method based on the peak-valley ratio of the amplified spontaneous emission spectrum below the threshold [4], which corresponds to peak-valley ratio of the transmission spectrum of the Fabry–Perot cavity. However, the measured amplified spontaneous emission spectra are influenced by the resolution of the measurement system, which causes the decrease of the peaks and the increase of the valleys of the measured spectrum. Fourier series expansion method is proposed to measure mode gain based on the integrations of the product of amplified spontaneous emission spectrum and a phase function over one mode interval [5]. The whole resonance pattern is applied in the Fourier series expansion method for evaluating mode gain, so the method is less sensitive to noise compared to the Hakki–Paoli method. Furthermore, accuracy gain spectrum can be obtained by correcting the influence of the response function of low-resolution measurement system.

By adding an external mirror with a reflectivity r_3 and a distance of L_e to feed a part of laser output back into a Fabry–Perot cavity laser as shown in Figure 2.7, we can set up an external cavity semiconductor laser and write the corresponding eigenvalue equation according (2.46, 2.61) as

$$\frac{r_2 + r_3 \exp(2ik_0L_e)}{1 + r_2r_3 \exp(2ik_0L_e)} r_1 \exp(2i\beta L) = 1, \tag{2.69}$$

which can be further reduced to

$$r_1r_2 \exp(2i\beta L) + r_3 \exp(2ik_0L_e)[r_1 \exp(2i\beta L) - r_2] = 1, \tag{2.70}$$

where r_1, r_2, and β are the facet reflectivity and propagation constant of the Fabry–Perot cavity. The resonance condition (2.70) is a mixed effect of the Fabry–Perot cavity and external cavity. The first term corresponds to the resonance condition of the Fabry–Perot cavity, and the second term is a fast oscillation term relative to the first term in (2.70), as the external cavity length L_e is usually much larger than the Fabry–Perot cavity length L.

2.5.3 Mode Selection for Distributed Feedback Lasers

The extremely short cavity for VCSELs has a very large longitudinal mode interval, which guarantees a large gain difference between different longitudinal modes for realizing single-longitudinal-mode operation. But edge-emitting Fabry–Perot cavity semiconductor lasers usually have a cavity length in the order of 100 μm and are difficult to realize dynamic single-longitudinal-mode operation based on mode gain difference. Extra mode selection mechanisms are required for reaching single-longitudinal-mode operation edge-emitting lasers, such as using distributed Bragg reflectors (DBRs) as cavity mirrors. The DBRs, formed by grating with a small refractive index difference, can have a narrow band high reflectivity spectrum for the longitudinal-mode selection. In fact, VCSELs can be considered as a DBR laser with a very short cavity and a large refractive index difference, which results in a wide band high reflectivity spectrum.

Different from the DBR laser with passive DBR regions, distributed feedback (DFB) laser is a common single-mode semiconductor laser with whole cavity with gain and grating. Taking the grating as step variation of refractive index, we can model a DFB laser and obtain the eigenvalue equation of the DFB lasers as (2.61). As the optical thickness of each layer is a quarter of wavelength, we have $\exp(2i\beta_j d_j) \approx -1$ around the resonance wavelength and write the eigenvalue equation as

$$r_{j,1} r_{j,m} = -1. \tag{2.71}$$

For the DFB lasers with the grating of real refractive index difference, we have a real quantity for reflectivity $r_{j,1}$ and $r_{j,m}$ with the same sign from (2.58, 2.59). So (2.71) cannot be satisfied at the exact resonance wavelength of the grating. The solution corresponds to the wavelength for the reflectivity with a phase shift of $\pi/2$, which is at the two sides of high reflectivity band. Perfect DFB lasers with whole period grating at cleaved mirrors have two modes with the same threshold gain. The two degenerate modes can be distinguished for realizing single-mode operation in asymmetric DFB lasers under certain relative position between the cleaved facet and the grating interface. But the single-mode yield is only about 30% for cleaved mirrors with random reflectivity phase determined by the relative position. The lasing spectrum of a DFB laser in Figure 2.8 shows single-mode operation at 1306.1 nm with a

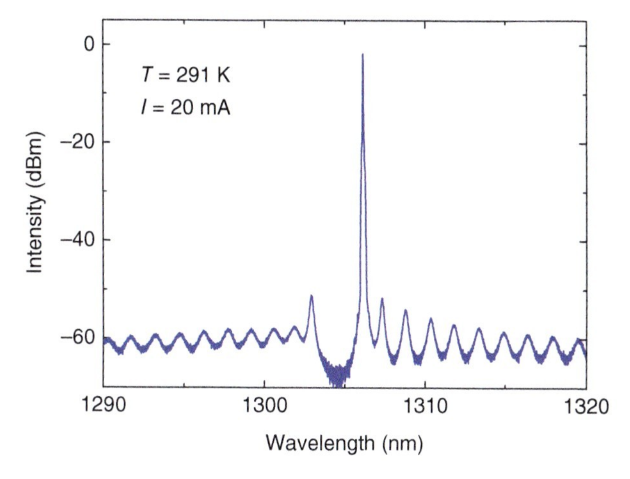

Figure 2.8 Lasing spectrum of a DFB laser with a stop band.

stop band from 1302.8 to 1306.1 nm. The magnitude of the stop band is determined by the refractive index step of the grating.

Introducing a quarter-wave-shift into layer j, we have $\exp(2i\beta_j d_j) \approx 1$ around the resonance wavelength and obtain the eigenvalue equation for quarter-wave-shifted DFB lasers as

$$r_{j,1} r_{j,m} = 1, \tag{2.72}$$

which can be satisfied at the center resonance wavelength of the grating. Quarter-wave-shifted DFB lasers are easy to realize single-mode operation irrespective of the cleaved facet phase shift. Furthermore, the DFB lasers with the grating of imagine refractive index difference caused by gain or loss can have an imaginary quantity for reflectivity $r_{j,1}$ and $r_{j,m}$ and satisfy (2.71) at the center resonance wavelength.

2.6 Mode Gain and Optical Confinement Factor

Most common semiconductor lasers are edge-emitting lasers with longitudinal modes satisfy phase condition of a Fabry–Perot cavity. In the material growth direction, the resonant modes are confined by multiple-layer waveguide for edge-emitting semiconductor lasers, consisting of active and passive regions. Resonant modes propagating in the z-direction inside the waveguide experience the material gain g of the active region and the absorption loss α_i of the passive regions, and corresponding mode intensity $I(z)$ increases with z by

$$I(z) = I_0 \exp(Gz). \tag{2.73}$$

The mode gain is usually expressed as

$$G = \Gamma g - (1-\Gamma)\alpha_i, \tag{2.74}$$

with an optical confinement factor Γ marked the contribution of material gain to the mode gain. The optical confinement factor is a key parameter for calculating mode gain and designing laser structure. A standard optical confinement factor Γ is defined as the ratio of z-component power flow in the active region to the total power flow. In this section, the formulae of standard optical confinement factors are compared with the ratio of mode gain to material gain derived from wave equation for slab waveguides. The results indicate that the standard optical confinement factors based on power flow will underestimate the mode gain of TE modes and overestimates that of TM modes. The correction is important to design polarization insensitive semiconductor optical amplifiers.

2.6.1 Optical Confinement Factor Based on Power Flow

The power flow density vector is $\boldsymbol{E} \times \boldsymbol{H}$, which is also called the Poynting vector. Using the real part of the electromagnetic fields, we can obtain time-averaged power flow as

$$\boldsymbol{S} = \frac{1}{2}\mathrm{Re}(\boldsymbol{E} \times \boldsymbol{H}^*). \tag{2.75}$$

For TE and TM modes, the power flow are $\boldsymbol{S}_{\mathrm{TE}} = (E_y H_z^*, 0, -E_y H_x^*)/2$ and $\boldsymbol{S}_{\mathrm{TM}} = (-E_z H_y^*, 0, E_x H_y^*)/2$, respectively. The standard optical confinement factors are defined as the ratio of z-direction power flow confined in active layer to the total power flow. Based on the definition, we have optical confinement factors for TE and TM modes as

$$\Gamma_{\mathrm{TE}} = \frac{\int_{\text{active}} |E_y(x)|^2 dx}{\int_{-\infty}^{\infty} |E_y(x)|^2 dx}, \tag{2.76}$$

and

$$\Gamma_{\mathrm{TM}} = \frac{\int_{\text{active}} |H_y(x)/n|^2 dx}{\int_{-\infty}^{\infty} |H_y(x)/n|^2 dx}. \tag{2.77}$$

The integral of denominator is over whole cross-section of semiconductor lasers. The optical confinement factor, defined as the proportion of the square of electric field in active region, was also used in the analysis of vertical-cavity surface-emitting semiconductor lasers. In Sections 2.6.2 and 2.6.3, we define optical confinement factors for TE and TM modes in dielectric slab waveguides based on mode gain directly as in [6].

2.6.2 Mode Gain for TE Modes

For the modes propagating in z-direction, the mode intensity variation with z can also be described by the complex propagation constant $\beta = \beta_r + i\beta_i$ in the z-direction. Submitting $E_y(x)\exp(i\beta z - i\omega t)$ into wave Eq. (2.15), we obtain:

$$-\beta^2 E_y(x) + \frac{d^2 E_y(x)}{dx^2} + k_0^2 n^2 E_y(x) = 0. \tag{2.78}$$

Multiplying (2.78) by $E_y^*(x)$, we get imaginary part of the equation as:

$$\beta_r \beta_i |E_y Ex)|^2 - \frac{1}{2}\mathrm{Im}\left[E_y^*(x)\frac{d^2 E_y(x)}{dx^2}\right] - k_0^2 n_r \kappa |E_y(x)|^2 = 0. \tag{2.79}$$

Integrating (2.79) over the x-axis and using the formulation of integration by parts, we can find the integral of the second term of the left side of (2.79) is zero, because $E_y^*(x)$ and $dE_y(x)/dx$ are continuous functions and equal to zero at infinite $x = \pm\infty$. So the integral yields:

$$\beta_i = \frac{k_0 \int_{-\infty}^{\infty} n_r \kappa |E_y(x)|^2 dx}{N_m \int_{-\infty}^{\infty} |E_y(x)|^2 dx}, \tag{2.80}$$

where the mode refractive index

$$N_m = \beta_r / k_0. \tag{2.81}$$

Accounting mode intensity is proportional to squared electric field and varied as $\exp(-2\beta_i z)$, so we can define a mode gain as

$$G = -2\beta_i. \tag{2.82}$$

If the material gain of (2.33) is only nonzero in the active region, we obtain the ratio of mode gain to the material gain from (2.80)

$$\frac{G}{g} = \frac{\int_{\text{active}} n_r |E_y(x)|^2 dx}{N_m \int_{-\infty}^{\infty} |E_y(x)|^2 dx}. \tag{2.83}$$

The ratio of (2.83) reduces to the standard optical confinement factor (2.76) for semiconductor lasers with a weak waveguiding structure at the approximation of $N_m \approx n_r$. In practical semiconductor lasers, the refractive index of active region is always larger than the mode index. So we can expect that (2.76) will underestimate the mode gain a little.

2.6.3 Mode Gain for TM Modes

For TM mode, we cannot derive the ratio of G/g by substituting $H_y(x, z, t)$ into wave equation and taking the same procedure as that for TE modes, because $dH_y(x)/dx$ is not a continuous function. Alternately, we substitute $E_x(x, z, t)$ obtained from (2.1, 2.12) into wave Eq. (2.15) and get:

$$-\frac{\beta^2}{n^2} H_y(x) + \frac{d}{dx}\left[\frac{1}{n^2}\frac{dH_y(x)}{dx} + H_y(x)\frac{dn^{-2}}{dx}\right] + k_0^2 H_y(x) = 0. \tag{2.84}$$

Multiplying (2.84) by $H_y^*(x)$ and then separating the real and imaginary parts, we get the imaginary part as:

$$\frac{2\beta_r(\beta_r\kappa - n_r\beta_i)}{n_r^3}|H_y(x)|^2 + \text{Im}\left[H_y^*(x)\frac{d}{dx}\left(\frac{1}{n^2}\frac{dH_y(x)}{dx}\right)\right]$$
$$+ \text{Im}\left[H_y^*(x)\frac{d}{dx}\left(H_y(x)\frac{dn^{-2}}{dx}\right)\right] = 0. \tag{2.85}$$

To obtain the relation between the mode gain and the material gain, we integrate (2.85) over the x-axis by ignoring the integral of the third term, because dn^{-2}/dx equals zero for slab waveguides except for some discrete points. Because $H_y^*(x)$ and $1/n^2 dH_y(x)/dx$ are continuous functions, we obtain the integral of the second term of (2.85) as $2\varepsilon_0^2\omega^2 \int_{-\infty}^{\infty} n_r\kappa|E_z(x)|^2 dx$ by the formulation of integration by parts. Finally, we get the following result from the integral:

$$\beta_i = \frac{N_m k_0 \int_{-\infty}^{\infty} n_r\kappa[|E_x(x)|^2 + |E_z(x)|^2]dx}{\int_{-\infty}^{\infty} n_r^2|E_x(x)|^2 dx}. \tag{2.86}$$

Similar to TE modes, we can obtain the ratio of mode gain to the material gain as the material gain is only nonzero in the active regions

$$\frac{G}{g} = \frac{N_m \int_{\text{active}} n_r[|E_x(x)|^2 + |E_z(x)|^2]dx}{\int_{-\infty}^{\infty} n_r^2|E_x(x)|^2 dx}. \tag{2.87}$$

By ignoring the term of $E_z(x)$ in (2.87), we approximately have the following relation based on (2.12) and (2.87)

$$\frac{G}{g} = \frac{N_m \int_{\text{active}} |H_y(x)/n^2|^2 n_r dx}{\int_{-\infty}^{\infty} |H_y(x)/n^2|^2 n_r^2 dx}. \tag{2.88}$$

Similarly, the ratio of (2.88) reduces to the standard optical confinement factor (2.77) for a weak waveguiding structure at the approximation of $N_m \approx n_r$.

2.7 Numerical Results of Optical Confinement Factors

2.7.1 Edge-Emitting Semiconductor Lasers

In this section, we compare the ratios (2.83, 2.87) with standard optical confinement factors of (2.76, 2.77) for different waveguide structures. The Powell method is applied for searching the minimum point of $|R_{j,1}R_{j,m}\exp(2ik_jd_j) - 1|^2$ under a mode wavelength and a given initial value of β. The minimum value of zero is the solution of β for the eigenvalue Eq. (2.46). After obtaining the eigenvalue β, we can calculate the mode-field amplitudes a_j and b_j based on the recursion relations of reflectivity and the optical confinement factors. Firstly, the mode gain and optical confinement factor are calculated for a mode at wavelength $\lambda = 860$ nm confined in a symmetric five-layer waveguide with refractive indices of 3.2/3.4/3.6/3.4/3.2. The layers with refractive indices of 3.2, 3.4, and 3.6 corresponding to $Al_{0.6}Ga_{0.4}As$, $Al_{0.3}Ga_{0.7}As$, and GaAs, respectively. Taking the thickness of 0.1 μm for the $Al_{0.3}Ga_{0.7}As$ layer, we calculate and compare the mode gains and optical confinement factors for TE and TM modes as functions of the thickness of GaAs active layer. As shown in Figure 2.9a, G_{TE}/g, Γ_{TE}, G_{TM}/g, and Γ_{TM} are 0.0192, 0.0175, 0.0135, and 0.0148 at the active layer thickness of 5 nm, with ratios $G_{TE}/G_{TM} = 1.42$ and $\Gamma_{TE}/\Gamma_{TM} = 1.18$ as shown in Figure 2.9b, and 0.721, 0.693, 0.660, and 0.676 with the ratios of 1.09 and 1.03 at the layer thickness of 200 nm.

Then, we consider a multiple-quantum-well laser, with the active region consisted of six 6-nm-thick 1.55Q InGaAsP quantum wells with the refractive index of 3.55- and 12-nm-thick 1.2Q InGaAsP barrier layers with a refractive index of 3.34. The cladding layers are two 80-nm 1.1Q InGaAsP with a refractive index of 3.28, and the external layers are InP with a refractive index of 3.17, where 1.55Q, 1.2Q, and 1.1Q indicate InGaAsP with the energy bandgap corresponding to a wavelength of 1.55,

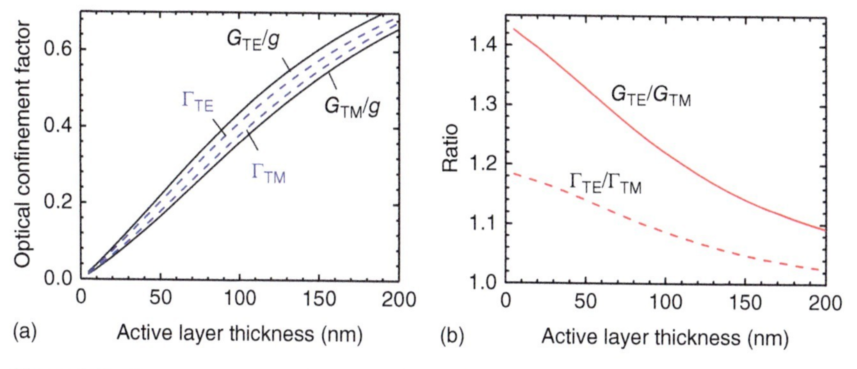

Figure 2.9 Comparison of mode gain and standard optical confinement factors for TE and TM modes at wavelength 860 nm in the five-layer waveguide.

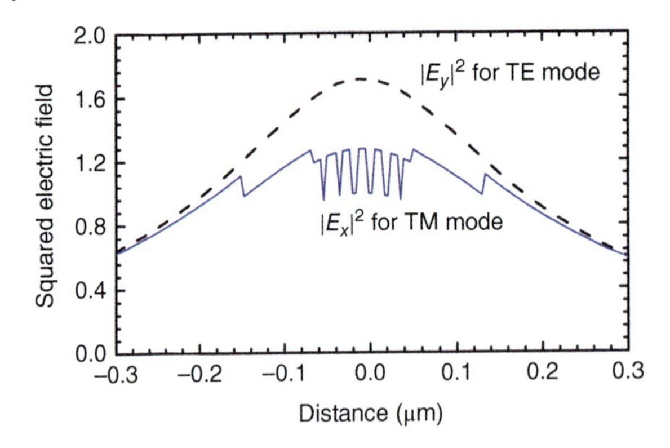

Figure 2.10 The squared electric fields for the multiple quantum well laser. The solid and the dashed lines are results for the TM and TE modes, respectively.

1.2, and 1.1 μm, respectively. The material gain of the quantum wells is taken to be $g = 10\,\text{cm}^{-1}$ and that of the other layers is taken to be zero in the calculation. Squared electric fields $|E_y(x)|^2$ and $|E_x(x)|^2$ are plotted as the solid and the dashed lines in Figure 2.10 for TE and TM modes at 1550 nm, respectively. For the TE mode, the ratio G_{TE}/g of mode gain to quantum well material gain is 0.0666, and Γ_{TE} of (2.76) is 0.0603. For the TM mode, we have $G_{\text{TM}}/g = 0.0427$ and $\Gamma_{\text{TM}} = 0.0472$ from (2.77). The corresponding ratios of $G_{\text{TE}}/G_{\text{TM}}$ and $\Gamma_{\text{TE}}/\Gamma_{\text{TM}}$ are 1.56 and 1.28, respectively. For realizing polarization-insensitive semiconductor optical amplifiers, we usually design tensile-strained quantum wells to enhance the material gain of TM modes. The difference of material gains between TM and TE modes necessary for polarization insensitive semiconductor optical amplifiers should be larger than that obtained from the standard optical confinement factor.

2.7.2 Si-on-SiO_2 Slab Waveguide

For a strong slab waveguide of Si-on-SiO_2 in air considered as three-layer waveguide of air/Si/SiO_2, we calculate the ratio of G/g and the optical confinement factor for TE modes. The real refractive index, gain, and thickness of the Si layer are taken to be 3.5, −1, and 0.5 μm, and the refractive index of 1.45 for SiO_2. The numerical results of the mode index, the ratio of G/g, and the standard optical confinement factor in the Si layer are presented in Table 2.1 for the fundamental, first- and second-order transverse modes TE_0, TE_1, and TE_2 at 1.3 μm.

Table 2.1 The mode index, G/g, and optical confinement factor for TE_0, TE_1, and TE_2.

Mode	N	G/g	Γ	θ (°)
TE_0	3.343 893	1.024 566	0.978 868	17.2
TE_1	2.843 007	1.112 983	0.904 062	35.7
TE_2	1.868 458	1.284 072	0.685 605	57.7

We find that the ratio of G/g is larger than unity in the strong waveguide and the higher-order transverse modes even have a larger ratio than the fundamental mode. The phenomena can be explained by the propagation of the mode light ray inside the waveguide. Based on the light ray concept as shown in Figure 2.3, mode light ray of (2.30) is zigzag propagation with an incident angle θ at the interface:

$$\theta = \cos^{-1}\frac{\beta_r}{n_1 k_0}, \tag{2.89}$$

where n_1 is the refractive index of the center layer. The mode light ray is zigzag propagation inside the center layer with the intersection angle θ as shown in Table 2.1. The zigzag path of the mode light ray is much longer than the transmitted length in the z-direction, so the ratio of G/g can be larger unity. Silicon-on insulator is widely used in the fabrication of passive photonic devices and light modulators using a single-mode waveguide. However, a strong waveguide with multiple guided modes can be applied to increase the absorption efficiency in a waveguide photodetector.

2.7.3 Vertical-Cavity Surface-Emitting Lasers

The eigenvalue Eq. (2.61) with propagation constants (2.57) is a function of complex variables, so two parameters can be determined by solving the equation. In addition to resonant mode wavelength, we can choose another parameter such as active layer gain or mode lifetime for solving the equation:

(1) Taking $k_j = k_0 n_j$ at constant complex refractive indices n_j and thickness d_j except active layer gain, we can obtain mode wavelength and threshold gain g_{th} by solving the eigenvalue equation with equation roots of wave number k_0 and active layer gain.
(2) Taking $k_j = \omega n_j/c$ at constant complex refractive indices n_j and thickness d_j, we can obtain mode wavelength and mode lifetime by solving the eigenvalue equation with equation roots of complex frequency. The real and imaginary parts of the complex frequency ω correspond to mode wavelength and lifetime, and the mode lifetime τ satisfies $1/\tau = -2\text{Im}(\omega)$ under the time term of $\exp.(-i\omega t)$.

The threshold condition of vertical-cavity surface-emitting lasers can be written as

$$\Gamma g_{\text{th}} - (1-\Gamma)\alpha_i = \alpha_m, \tag{2.90}$$

where Γ and g_{th} are optical confinement factor and threshold gain of the active region, α_i is the absorption loss of all the other layers, and α_m is the mirror loss. Assuming α_m is unaffected by the small variation of the gain and the absorption loss, we can obtain optical confinement factor according the variation of threshold gain Δg_{th} with that of absorption $\Delta\alpha_i$ as:

$$\Gamma = \frac{\Delta\alpha_i}{\Delta\alpha_i + \Delta g_{\text{th}}}, \tag{2.91}$$

The optical confinement factor as a function of active layer thickness is calculated for a VCSEL, which consisted of an $Al_{0.45}Ga_{0.55}As/GaAs/Al_{0.45}Ga_{0.55}As$ cavity sandwiched between two 20-period quarter-wavelength GaAs/AlAs DBRs. The refractive indices of GaAs, AlAs, and $Al_{0.45}Ga_{0.55}As$ are taken to be 3.6, 2.95, and 3.3, respectively, and the sum of the optical length of the cavity is 860 nm. The optical confinement factors calculated by (2.91) are plotted as function of the GaAs active layer thickness in Figure 2.11a. At the active layer thickness of 70, 100, and 120 nm, the optical confinement factor is 0.1145, 0.1217, and 0.1213, respectively, which even decreases a little as the active layer thickness increases from 10 to 12 nm due to the center peak drop for squared electric field distribution. Furthermore, mode wavelength vs. the active layer thickness is plotted in Figure 2.11b, which indicates the effect of reflection phase shift of GaAs/AlAs interface of the center cavity. In addition, the same numerical results of optical confinement factor are obtained from the following definition:

$$\Gamma = \frac{\int_{\text{activeregion}} n_r |E(x)|^2 dx}{\int_{\text{wholecavity}} n_r |E(x)|^2 dx}, \tag{2.92}$$

which is the ratio of the multiplication of photon density and optical velocity inside the active region.

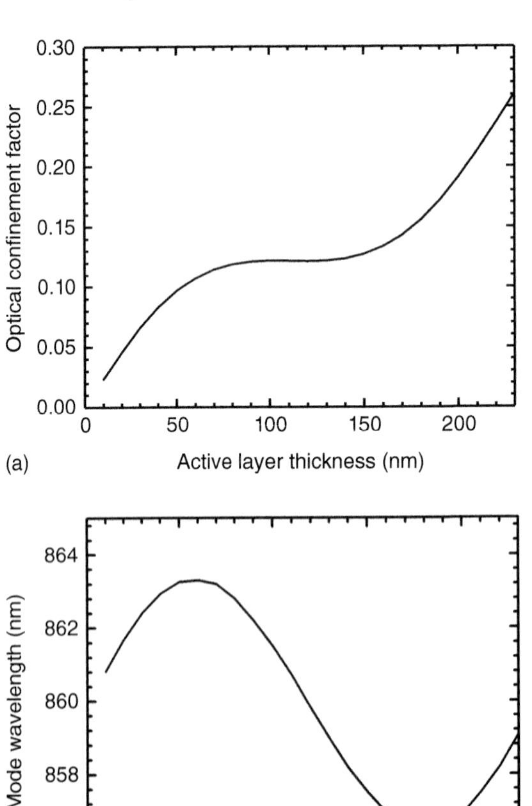

Figure 2.11 (a) Optical confinement factor and (b) mode wavelength vs. the active layer thickness for the VCSEL structure.

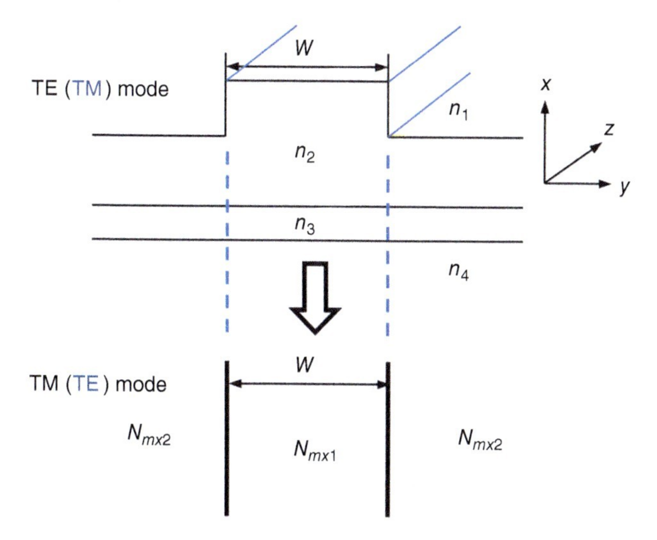

Figure 2.12 A simplified three-dimensional waveguide guide of a ridge waveguide and schematic for effective index method.

2.8 Effective Index Method

To simply model practical waveguide for semiconductor lasers, effective index approximation method is usually applied by reducing three-dimensional waveguide to two-dimensional slab waveguide. For a ridge waveguide semiconductor laser as shown in Figure 2.12, we can express z-direction propagating electric field by:

$$E(x, y, z, t) = F(x, y)\exp(i\beta z - i\omega t). \tag{2.93}$$

Submitting (2.93) into wave Eq. (2.15), we can obtain

$$\frac{\partial^2 F(x,y)}{\partial x^2} + \frac{\partial^2 F(x,y)}{\partial y^2} + {k_0}^2[n^2(x,y) - N_m^2]F(x,y) = 0, \tag{2.94}$$

where mode index $N_m = \beta/k_0$ as in (2.81), and $n(x, y)$ is refractive index distribution. Assuming the mode field can be approximately expressed as

$$F(x, y) = f(x)g(y), \tag{2.95}$$

we can obtain following equation form (2.94):

$$\frac{1}{f(x)}\frac{\partial^2 f(x)}{\partial x^2} + \frac{1}{g(y)}\frac{\partial^2 g(y)}{\partial y^2} + {k_0}^2[n^2(x,y) - N_m^2] = 0. \tag{2.96}$$

Assuming the mode-field distributions $f(x)$ and $g(y)$ can be determined by one-dimensional refractive index distribution as a slab waveguide, we can divide (2.96) into following equations:

$$\frac{d^2 f(x)}{dx^2} + {k_0}^2[n^2(x,y) - {N_{mx}}^2(y)]f(x) = 0, \tag{2.97}$$

$$\frac{d^2 g(y)}{dy^2} + {k_0}^2[N_{mx}^2(y) - N_m^2]g(y) = 0. \tag{2.98}$$

Eqs. (2.97, 2.98) are wave equations for slab waveguides in the x and y directions, respectively.

In practical three-dimensional waveguides, we cannot have perfect TE or TM modes but still can define quasi-TE or TM modes based on the main electromagnetic fields. According to the boundary conditions for electromagnetic fields in the x and y direction slab waveguides, we shall treat TE modes (TM modes) in the x-direction waveguide by solving (2.97) with the refractive index distribution $n(x,y)$, and then assign the modes as TM modes (TE modes) by solving (2.98) to obtain mode index N_m from the y-direction slab waveguide with obtained $N_{mx}(y)$ as shown in Figure 2.12. For a buried-heterostructure laser with the waveguide width and thickness of 1 and 0.2 μm and a complex refractive index of $3.52 + i0.0123$ surrounded by confined layers with a refractive index of 3.17, we obtain mode indices of 3.2302 and 3.2141 and optical confinement factors of 39% and 27% by effective index method. Using semivector beam propagation method, mode indices for the TE and TM modes are 3.2303 and 3.2141, respectively, and the optical confinement factors of 38% and 27%. The numerical results of the beam propagation method are in good agreement with the effective index method.

In addition, effective index method is also used to analyze vertical-cavity surface-emitting lasers with refractive index step Δn_{eff} determined by resonant wavelength difference $\Delta\lambda$ of one-dimensional waveguide of different lateral position of the vertical-cavity surface-emitting laser [7]:

$$\frac{\Delta n_{\text{eff}}}{n_{\text{eff}}} = \frac{\Delta\lambda}{\lambda}, \tag{2.99}$$

where n_{eff} is mode index and λ is resonant wavelength. Selectively oxidized AlAs layer is usually applied to form a current injection window and a lateral refractive index step for mode confinement. For an oxidized AlAs layer with a thickness of 20 nm, the corresponding refractive index step can be varied over two orders of magnitude by adjusting the oxide layer position.

References

1 Adams, M.J. (1981). *An Introduction to Optical Waveguides*. New York, NY: Wiley.

2 Marcuse, D. (1982). *Light Transmission Optics*, 2e. New Yok, NY: Van Nostrand Reinhold.

3 Coldren, L.A., Corzine, S.W., and Msanovic, M.L. (2012). *Diode Lasers and Photonic Integrated Circuits*, 2e. Hoboken, NJ: Wiley.

4 Hakki, B.W. and Paoli, T.L. (1975). Gain spectra in GaAs double-heterostructure injection lasers. *J. Appl. Phys.* 46 (3): 1299–1305.

5 Guo, W.H., Lu, Q.Y., Huang, Y.Z., and Yu, L.J. (2004). Fourier series expansion method for gain measurement from amplified spontaneous emission spectra of Fabry–Pérot semiconductor lasers. *IEEE J. Quantum Electron.* 40 (2): 123–129.

6 Huang, Y.Z., Pan, Z., and Wu, R.H. (1996). Analysis of the optical confinement factor in semiconductor lasers. *J. Appl. Phys.* 79: 3827–3830.

7 Hadley, G.R., Lear, K.L., Warren, M.E. et al. (1996). Comprehensive numerical modeling of vertical-cavity surface-emitting lasers. *IEEE J. Quantum Electron.* 32 (4): 607–616.

3

FDTD Method and Padé Approximation

3.1 Introduction

Finite-difference time-domain (FDTD) is a numerical method to deal with electromagnetic field in time domain, which is based on Maxwell's equation by using finite difference instead of differential. It was put forward by Kane Yee [1] and developed by Taflove and Hagness in the 1980s and 1990s [2]. With the rapid development of computing technology, the FDTD method has been developed into a powerful and widely used numerical method for solving the electromagnetic field problems. Compared with other numerical methods for simulating electromagnetic fields, such as beam propagation method (BPM), finite-element method (FEM), boundary element method (BEM), etc., the FDTD method has no approximation because it is directly derived from Maxwell's equation, and it may truly reflect the physical nature of the electromagnetic problem. In addition, the FDTD method has the flexibility that other methods do not have and can simulate electromagnetic field problems in various complex structures very conveniently. Therefore, the FDTD method is a highly desirable method in the study of optical modes in microcavities. However, the FDTD method also presents its own difficulties. For the simulation of a large-scale structure, a lot of computing memory and a long computing time are required, which will be gradually solved but with the development of computing technology.

The FDTD method is performed in time domain, and the spectrum in the frequency domain can be obtained by transforming the time-domain signal into the frequency domain. Fast Fourier transform (FFT) is one of the most familiar methods, but long time series data are typically required. Then, an efficient method is needed for converting the time-domain signals into frequency domain. We proposed Padé approximation method based on Baker algorithm [3], which can obtain high-resolution spectrum from short time series data and shorten the time required for the FDTD simulation. Therefore, the FDTD method associated with the Padé approximation can be an efficient tool for solving the electromagnetic problems in the optical microcavities and the corresponding system.

Microcavity Semiconductor Lasers: Principles, Design, and Applications, First Edition.
Yong-zhen Huang and Yue-de Yang.

In this chapter, we will introduce the basic principle of FDTD method and its application in microcavity research in Section 3.2. In Section 3.3, we present the Padé approximation with Baker's algorithm for processing FDTD simulation data. In Section 3.4, examples of FDTD and Padé approximation are presented. Finally, the summary will be given in Section 3.5.

3.2 Basic Principle of FDTD Method

3.2.1 Maxwell's Equation

Maxwell's equations include two divergence equations and two curl equations, and among them the two curl equations are more representative for the evolution of the electromagnetic fields with time. To study the electromagnetic fields in an optical system, only the two curl equations need to be considered. Maxwell's curl equations can be expressed as

$$\begin{cases} \nabla \times \vec{E}(\vec{r},t) = -\frac{\partial \vec{B}(\vec{r},t)}{\partial t} - \vec{J}_m(\vec{r},t) \\ \nabla \times \vec{H}(\vec{r},t) = \frac{\partial \vec{D}(\vec{r},t)}{\partial t} + \vec{J}(\vec{r},t) \end{cases} \tag{3.1}$$

where $\vec{E}$, $\vec{H}$, $\vec{D}$, and $\vec{B}$ are the electric field vector, the magnetic field vector, the electric displacement vector, and the magnetic inductive vector, $\vec{J}$ and $\vec{J}_m$ are the electric and magnetic current vectors, respectively, $\vec{r}$ is the position vector, $\nabla\times$ is the curl vector, which have different forms in different coordinate systems. In the isotropic media, the vectors satisfy the following relationship

$$\begin{cases} \vec{D} = \varepsilon_0 \varepsilon_r \vec{E} \\ \vec{B} = \mu_r \mu_0 \vec{H} \\ \vec{J} = \sigma \vec{E} \\ \vec{J}_m = \sigma_m \vec{H} \end{cases} \tag{3.2}$$

where $\varepsilon_0 = 8.8542 \times 10^{-12}$ F/m and $\mu_0 = 4\pi \times 10^{-7}$ H/m represent the free space permittivity and permeability, ε_r and μ_r represent the relative permittivity and permeability, respectively, σ and σ_m are the electric and magnetic conductivity of medium, which is equal to 0 in lossless medium. Eq. (3.1) can be simplified to the following form in a commonly considered passive lossless nonmagnetic medium

$$\begin{cases} \nabla \times \vec{E}(\vec{r},t) = -\frac{\partial \vec{B}(\vec{r},t)}{\partial t} \\ \nabla \times \vec{H}(\vec{r},t) = \frac{\partial \vec{D}(\vec{r},t)}{\partial t} \end{cases} \tag{3.3}$$

For simplicity, we take the lossless medium as an example when describing Yee's mesh for difference forms of Maxwell's equation.

3.2.2 2D FDTD Method in Cartesian Coordinate System

At first, we consider a two-dimensional (2D) structure, where the electromagnetic fields are uniformly distributed in z direction. In this case, Maxwell's equations in (3.3) can be divided into two independent groups, one is the so-called transverse-electric (TE) equation, the other is transverse-magnetic (TM) equation.

In the Cartesian coordinate systems with the coordinate axes in the x and y planes, Maxwell's equations for the TE and TM modes are

TE mode:

$$\begin{cases} \dfrac{\partial H_z}{\partial t} = \dfrac{1}{\mu_0}\left(\dfrac{\partial E_x}{\partial y} - \dfrac{\partial E_y}{\partial x}\right) \\ \dfrac{\partial E_x}{\partial t} = \dfrac{1}{\varepsilon_0 \varepsilon_r}\dfrac{\partial H_z}{\partial y} \\ \dfrac{\partial E_y}{\partial t} = -\dfrac{1}{\varepsilon_0 \varepsilon_r}\dfrac{\partial H_z}{\partial x} \end{cases} \tag{3.4}$$

TM mode:

$$\begin{cases} \dfrac{\partial E_z}{\partial t} = \dfrac{1}{\varepsilon_0 \varepsilon_r}\left(\dfrac{\partial H_y}{\partial x} - \dfrac{\partial H_x}{\partial y}\right) \\ \dfrac{\partial H_x}{\partial t} = -\dfrac{1}{\mu_0}\dfrac{\partial H_z}{\partial y} \\ \dfrac{\partial H_y}{\partial t} = \dfrac{1}{\mu_0}\dfrac{\partial E_z}{\partial x} \end{cases} \tag{3.5}$$

for the involved electromagnetic field components.

Kane Yee proposed to use finite-difference schemes to represent differential equations in Eqs. (3.4) and (3.5). The basic idea is to interleave electromagnetic components in space, and the electric and magnetic components also have half a time step on the time axis [1]. Under this configuration, the boundary conditions on the interface of the medium can be naturally satisfied. Figures 3.1 and 3.2 show Yee's grid of the electromagnetic field for the TE and TM modes, respectively.

In the division of grid, we adopt the following rules. The electric field component $E_l\ (l = x, y, z)$ is at half-integer grid point on the l coordinate axis, and is at integer grid point on the other coordinate axis. All the electric field components are at integer grid point on the time axis. The magnetic field component $H_l\ (l = x, y, z)$ is at integer step on the l coordinate axis and is at half-integer grid point on the other coordinate axis. All the magnetic field components are at half-integer time step on the time axis. The reason of the above division rules for the spatial grids is that we always use the tangential electric field as the boundary and set the boundary to the integer grid point.

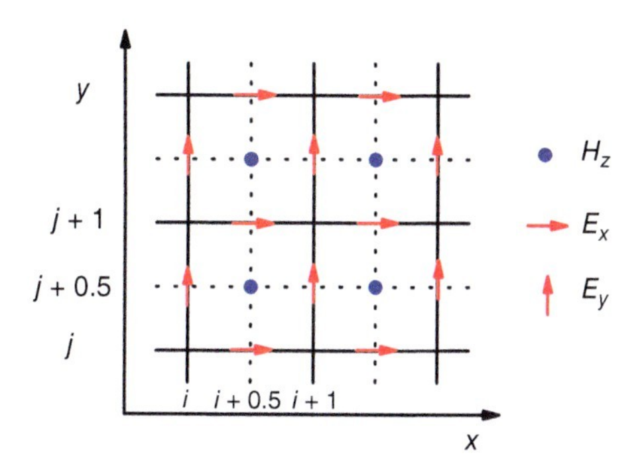

Figure 3.1 2D Yee's grid for TE mode.

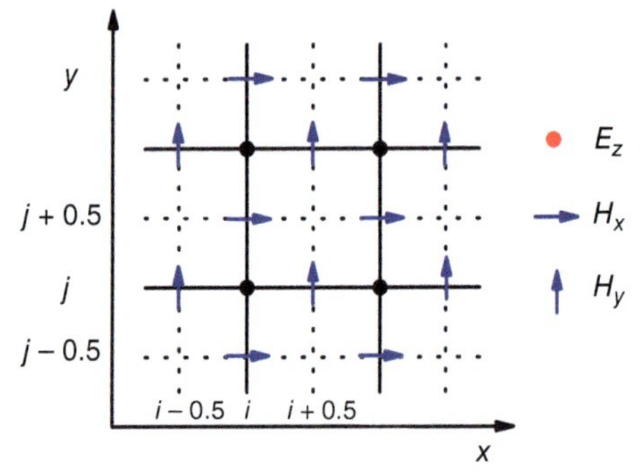

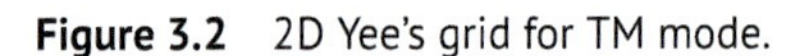
Figure 3.2 2D Yee's grid for TM mode.

For the TE mode, the differential equations in Eq. (3.4) can be expressed as the following finite-difference equations

$$H_z\Big|_{i+0.5,j+0.5}^{n+0.5} = H_z\Big|_{i+0.5,j+0.5}^{n-0.5} + \frac{\Delta t}{\mu_0}\left[\frac{E_x\Big|_{i+0.5,j+1}^{n} - E_x\Big|_{i+0.5,j}^{n}}{\Delta y} - \frac{E_y\Big|_{i+1,j+0.5}^{n} - E_y\Big|_{i,j+0.5}^{n}}{\Delta x}\right] \tag{3.6}$$

$$E_x\Big|_{i+0.5,j}^{n+1} = E_x\Big|_{i+0.5,j}^{n} + \frac{\Delta t}{\varepsilon_0\varepsilon_r\Big|_{i+0.5,j}}\frac{H_z\Big|_{i+0.5,j+0.5}^{n+0.5} - H_z\Big|_{i+0.5,j-0.5}^{n+0.5}}{\Delta y} \tag{3.7}$$

$$E_y\Big|_{i,j+0.5}^{n+1} = E_y\Big|_{i,j+0.5}^{n} - \frac{\Delta t}{\varepsilon_0\varepsilon_r\Big|_{i,j+0.5}}\frac{H_z\Big|_{i+0.5,j+0.5}^{n+0.5} - H_z\Big|_{i-0.5,j+0.5}^{n+0.5}}{\Delta x} \tag{3.8}$$

where Δt represents the time step, Δx and Δy represent the grid size in the x and y directions, respectively, n is the time step number on the time axis, and i and j are the grid point numbers on the spatial axis.

For the TM modes, the differential equations in (3.5) can be expressed as the following finite-difference equations similar to the TE modes

$$H_x\Big|_{i,j+0.5}^{n+0.5} = H_x\Big|_{i,j+0.5}^{n-0.5} - \frac{\Delta t}{\mu_0}\frac{E_z\Big|_{i,j+1}^{n} - E_z\Big|_{i,j}^{n}}{\Delta y} \tag{3.9}$$

$$H_y\Big|_{i+0.5,j}^{n+0.5} = H_y\Big|_{i+0.5,j}^{n-0.5} + \frac{\Delta t}{\mu_0}\frac{E_z\Big|_{i+1,j}^{n} - E_z\Big|_{i,j}^{n}}{\Delta x} \tag{3.10}$$

$$E_z\Big|_{i,j}^{n+1} = E_z\Big|_{i,j}^{n} + \frac{\Delta t}{\varepsilon_0\varepsilon_r\Big|_{i,j}}\left[\frac{H_y\Big|_{i+0.5,j}^{n+0.5} - H_y\Big|_{i-0.5,j}^{n+0.5}}{\Delta x} - \frac{H_x\Big|_{i,j+0.5}^{n+0.5} - H_x\Big|_{i,j-0.5}^{n+0.5}}{\Delta y}\right] \tag{3.11}$$

The specific iterative process is that if we know the electric field distributions at the nth time step and the magnetic field distributions at $n-0.5$ time step, the magnetic fields of the TE mode or the TM mode at the $n+0.5$th time step can be obtained from Eq. (3.6) or Eqs. (3.9) and (3.10), and then the electric field distributions at the $n+1$th time step of the TE mode or the TM mode can be obtained by

Eqs. (3.7) and (3.8) or Eq. (3.11), and the loop is iterated to obtain the electromagnetic field distribution on the whole time axis.

3.2.3 3D FDTD Method in Cartesian Coordinate System

In a full three-dimensional (3D) Cartesian coordinate system, the curl Eq. (3.3) can be written as the following scalar equations

$$\begin{cases} \dfrac{\partial H_x}{\partial t} = \dfrac{1}{\mu_0}\left(\dfrac{\partial E_y}{\partial z} - \dfrac{\partial E_z}{\partial y}\right) \\ \dfrac{\partial H_y}{\partial t} = \dfrac{1}{\mu_0}\left(\dfrac{\partial E_z}{\partial x} - \dfrac{\partial E_x}{\partial z}\right) \\ \dfrac{\partial H_z}{\partial t} = \dfrac{1}{\mu_0}\left(\dfrac{\partial E_x}{\partial y} - \dfrac{\partial E_y}{\partial x}\right) \\ \dfrac{\partial E_x}{\partial t} = \dfrac{1}{\varepsilon_0\varepsilon_r}\left(\dfrac{\partial H_z}{\partial y} - \dfrac{\partial H_y}{\partial z}\right) \\ \dfrac{\partial E_y}{\partial t} = \dfrac{1}{\varepsilon_0\varepsilon_r}\left(\dfrac{\partial H_x}{\partial z} - \dfrac{\partial H_z}{\partial x}\right) \\ \dfrac{\partial E_z}{\partial t} = \dfrac{1}{\varepsilon_0\varepsilon_r}\left(\dfrac{\partial H_y}{\partial x} - \dfrac{\partial H_x}{\partial y}\right) \end{cases} \tag{3.12}$$

In numerical simulations, the continuous equation can be discretized. Yee's grid and electromagnetic field components in the 3D Cartesian coordinate system are shown in Figure 3.3.

The time step and the spatial grid are divided by the same rules as those for the 2D Cartesian coordinate system. The electric field component E_l ($l = x, y, z$) is at half-integer grid point on the l coordinate axis, and at integer grid point on the other two coordinate axes. All the electrical field components are at integer time step on the time axis. The magnetic field component H_l ($l = x, y, z$) is at integer step on the

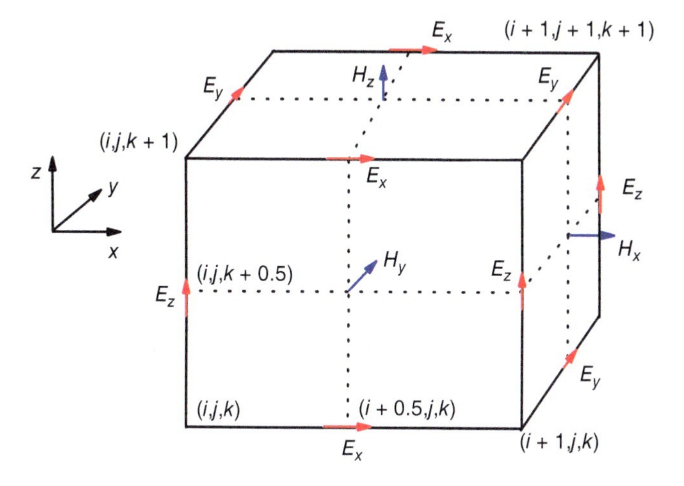

Figure 3.3 3D Yee's grids in Cartesian coordinate.

l coordinate axis, is at half-integer grid point on the other two coordinate axes. All the magnetic field components are at half-integer time step on the time axis. The differential equations in (3.12) can be written as difference equations as follows

$$\begin{aligned} H_x\Big|_{i,j+0.5,k+0.5}^{n+0.5} &= H_x\Big|_{i,j+0.5,k+0.5}^{n+0.5} + \frac{\Delta t}{\mu_0}\left[\frac{E_y\Big|_{i,j+0.5,k+1}^{n} - E_y\Big|_{i,j+0.5,k+1}^{n}}{\Delta z}\right. \\ &\quad \left. - \frac{E_z\Big|_{i,j+1,k+0.5}^{n} - E_z\Big|_{i,j,k+0.5}^{n}}{\Delta y}\right] \end{aligned} \tag{3.13}$$

$$\begin{aligned} H_y\Big|_{i+0.5,j,k+0.5}^{n+0.5} &= H_y\Big|_{i+0.5,j,k+0.5}^{n-0.5} + \frac{\Delta t}{\mu_0}\left[\frac{E_z\Big|_{i+1,j,k+0.5}^{n} - E_z\Big|_{i,j,k+0.5}^{n}}{\Delta x}\right. \\ &\quad \left. - \frac{E_x\Big|_{i+0.5,j,k+1}^{n} - E_x\Big|_{i+0.5,j,k}^{n}}{\Delta z}\right] \end{aligned} \tag{3.14}$$

$$\begin{aligned} H_z\Big|_{i+0.5,j+0.5,k}^{n+0.5} &= H_x\Big|_{i+0.5,j+0.5,k}^{n-0.5} + \frac{\Delta t}{\mu_0}\left[\frac{E_x\Big|_{i+0.5,j+1,k}^{n} - E_x\Big|_{i+0.5,j+1,k}^{n}}{\Delta y}\right. \\ &\quad \left. - \frac{E_y\Big|_{i+1,j+0.5,k}^{n} - E_y\Big|_{i,j+0.5,k}^{n}}{\Delta x}\right] \end{aligned} \tag{3.15}$$

$$\begin{aligned} E_x\Big|_{i+0.5,j,k}^{n+1} &= E_x\Big|_{i+0.5,j,k}^{n} + \frac{\Delta t}{\varepsilon_0\varepsilon_r\Big|_{i+0.5,j,k}}\left[\frac{H_z\Big|_{i+0.5,j+0.5,k}^{n+0.5} - H_z\Big|_{i+0.5,j-0.5,k}^{n+0.5}}{\Delta y}\right. \\ &\quad \left. - \frac{H_y\Big|_{i+0.5,j,k+0.5}^{n+0.5} - H_y\Big|_{i+0.5,j,k-0.5}^{n+0.5}}{\Delta z}\right] \end{aligned} \tag{3.16}$$

$$\begin{aligned} E_y\Big|_{i,j+0.5,k}^{n+1} &= E_y\Big|_{i,j+0.5,k}^{n} + \frac{\Delta t}{\varepsilon_0\varepsilon_r\Big|_{i,j+0.5,k}}\left[\frac{H_x\Big|_{i,j+0.5,k+0.5}^{n+0.5} - H_x\Big|_{i,j+0.5,k-0.5}^{n+0.5}}{\Delta z}\right. \\ &\quad \left. - \frac{H_z\Big|_{i+0.5,j+0.5,k}^{n+0.5} - H_z\Big|_{i-0.5,j+0.5,k}^{n+0.5}}{\Delta x}\right] \end{aligned} \tag{3.17}$$

$$\begin{aligned} E_z\Big|_{i,j,k+0.5}^{n+1} &= E_z\Big|_{i,j,k+0.5}^{n} + \frac{\Delta t}{\varepsilon_0\varepsilon_r\Big|_{i,j,k+0.5}^{n+1}}\left[\frac{H_y\Big|_{i+0.5,j,k+0.5}^{n+0.5} - H_y\Big|_{i-0.5,j,k+0.5}^{n+0.5}}{\Delta x}\right. \\ &\quad \left. - \frac{H_x\Big|_{i,j+0.5,k+0.5}^{n+0.5} - H_x\Big|_{i,j-0.5,k+0.5}^{n+0.5}}{\Delta y}\right] \end{aligned} \tag{3.18}$$

where Δt represents the time step, Δx, Δy, Δz represent the spatial step size in the x, y, and z axes, respectively, n is the time step number on time axis, and i, j, and k are the grid point numbers along x, y, and z directions on the spatial axes.

The specific iterative process is that if we know the electric field distributions at the nth time step and the magnetic field distributions at the $n - 0.5$th time step, the magnetic field distribution at the $n + 0.5$th time step can be obtained from Eqs. (3.13–3.15), and then the electric field distribution at the $n + 1$th time step can be obtained from Eqs. (3.16–3.18), and the electromagnetic field distributions on the whole time axis can be obtained by loop iteration.

3.2.4 3D FDTD Method in Cylindrical Coordinate System

In a 3D cylindrical coordinate system, Maxwell's curl Eq. (3.3) can be written as the following scalar equations

$$\begin{cases} \dfrac{\partial H_r}{\partial t} = \dfrac{1}{\mu_0}\left(\dfrac{\partial E_\phi}{\partial z} - \dfrac{1}{r}\dfrac{\partial E_z}{\partial \phi}\right) \\ \dfrac{\partial H_\phi}{\partial t} = \dfrac{1}{\mu_0}\left(\dfrac{\partial E_z}{\partial r} - \dfrac{\partial E_r}{\partial z}\right) \\ \dfrac{\partial H_z}{\partial t} = \dfrac{1}{\mu_0}\left(\dfrac{1}{r}\dfrac{\partial E_r}{\partial \phi} - \dfrac{1}{r}\dfrac{\partial (rE_\phi)}{\partial r}\right) \\ \dfrac{\partial E_r}{\partial t} = \dfrac{1}{\varepsilon_0\varepsilon_r}\left(\dfrac{1}{r}\dfrac{\partial H_z}{\partial \phi} - \dfrac{\partial H_\phi}{\partial z}\right) \\ \dfrac{\partial E_\phi}{\partial t} = \dfrac{1}{\varepsilon_0\varepsilon_r}\left(\dfrac{\partial H_r}{\partial z} - \dfrac{\partial H_z}{\partial r}\right) \\ \dfrac{\partial E_z}{\partial t} = \dfrac{1}{\varepsilon_0\varepsilon_r}\left(\dfrac{1}{r}\dfrac{\partial (rH_\phi)}{\partial r} - \dfrac{1}{r}\dfrac{\partial H_r}{\partial \phi}\right) \end{cases} \tag{3.19}$$

Due to rotational symmetry, the dependence of the above electromagnetic field components on the angular direction can be expressed as the function of the angular quantum number v. To ensure that the transformed Maxwell's equation is solved in the real number domain, the imaginary part caused by the derivation of the angular direction should be eliminated. Then, the electromagnetic field components can be defined as follows [4]

$$\begin{cases} E_r(r,\phi,z,t) = ie_r(r,z,t)\exp(iv\phi) \\ E_\phi(r,\phi,z,t) = e_\phi(r,z,t)\exp(iv\phi) \\ E_z(r,\phi,z,t) = ie_z(r,z,t)\exp(iv\phi) \\ H_r(r,\phi,z,t) = h_r(r,z,t)\exp(iv\phi) \\ H_\phi(r,\phi,z,t) = ih_\phi(r,z,t)\exp(iv\phi) \\ H_z(r,\phi,z,t) = e_z(r,z,t)\exp(iv\phi) \end{cases} \tag{3.20}$$

Substituting the components in (3.20) into (3.19), we can get

$$\begin{cases} \dfrac{\partial h_r}{\partial t} = \dfrac{1}{\mu_0}\left(\dfrac{\partial e_\phi}{\partial z} + \dfrac{v}{r}e_z\right) \\ \dfrac{\partial h_\phi}{\partial t} = \dfrac{1}{\mu_0}\left(\dfrac{\partial e_z}{\partial r} - \dfrac{\partial e_r}{\partial z}\right) \\ \dfrac{\partial h_z}{\partial t} = \dfrac{1}{\mu_0}\left(-\dfrac{v}{r}e_r - \dfrac{1}{r}\dfrac{\partial (re_\phi)}{\partial r}\right) \\ \dfrac{\partial e_r}{\partial t} = \dfrac{1}{\varepsilon_0\varepsilon_r}\left(\dfrac{v}{r}h_z - \dfrac{\partial h_\phi}{\partial z}\right) \\ \dfrac{\partial e_\phi}{\partial t} = \dfrac{1}{\varepsilon_0\varepsilon_r}\left(\dfrac{\partial h_r}{\partial z} - \dfrac{\partial h_z}{\partial r}\right) \\ \dfrac{\partial e_z}{\partial t} = \dfrac{1}{\varepsilon_0\varepsilon_r}\left(\dfrac{1}{r}\dfrac{\partial (rh_\phi)}{\partial r} - \dfrac{v}{r}h_r\right) \end{cases} \tag{3.21}$$

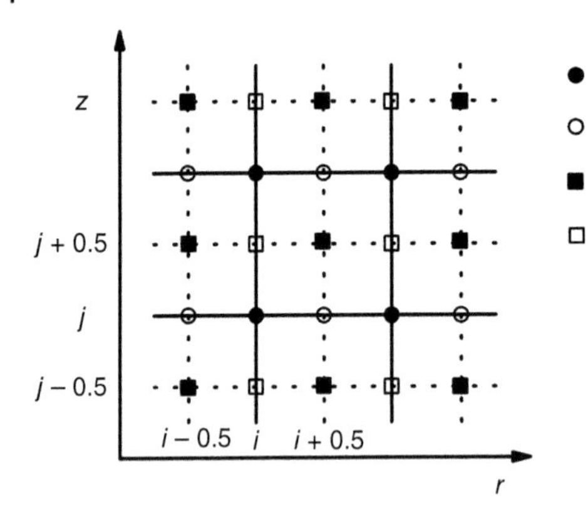

Figure 3.4 3D Yee's grids in cylindrical coordinate. The 3D system is simplified to 2D by assuming an angular dependence of exp($iv\phi$) for all the electromagnetic field components.

Yee's grid and the corresponding electromagnetic field components in the 3D cylindrical coordinate system with rotational symmetry are shown in Figure 3.4. The time step and the spatial grid are divided by a similar rule to that of two-dimensional rectangular coordinates, i.e. the electric field component E_l ($l = r, \phi, z$) is at half-integer grid point on the l coordinate axis, and at integer grid point on the other two coordinate axes, although there is no ϕ coordinate axis. The magnetic field component H_l ($l = r, \phi, z$) is at integer grid point on the l coordinate axis, and at half-integer grid point on the other two coordinate axes. Then, the differential Eq. (3.21) can be written as the difference form as follows.

$$h_r\Big|_{i,j+0.5}^{n+0.5} = h_r\Big|_{i,j+0.5}^{n-0.5} + \frac{\Delta t}{\mu_0}\left(\frac{ve_z\Big|_{i,j+0.5}^{n}}{r_i} + \frac{e_\phi\Big|_{i,j+1}^{n} - e_\phi\Big|_{i,j}^{n}}{\Delta z}\right) \tag{3.22}$$

$$h_\phi\Big|_{i+0.5,j+0.5}^{n+0.5} = h_\phi\Big|_{i+0.5,j+0.5}^{n-0.5} - \frac{\Delta t}{\mu_0}\left(\frac{e_r\Big|_{i+0.5,j+1}^{n} - e_r\Big|_{i+0.5,j}^{n}}{\Delta z} + \frac{e_z\Big|_{i+1,j+0.5}^{n} - e_z\Big|_{i,j+0.5}^{n}}{\Delta r}\right) \tag{3.23}$$

$$h_z\Big|_{i+0.5,j}^{n+0.5} = h_z\Big|_{i+0.5,j}^{n-0.5} - \frac{\Delta t}{\mu_0}\left(\frac{ve_r\Big|_{i+0.5,j}^{n}}{r_{i+0.5}} + \frac{r_{i+1}e_\phi\Big|_{i+1,j}^{n} - r_i e_\phi\Big|_{i,j}^{n}}{r_{i+0.5}\Delta r}\right) \tag{3.24}$$

$$e_r\Big|_{i+0.5,j}^{n+1} = e_r\Big|_{i+0.5,j}^{n} + \frac{\Delta t}{\varepsilon_0\varepsilon_r\Big|_{i+0.5,j}}\left(\frac{vh_z\Big|_{i+0.5,j}^{n+0.5}}{r_{i+0.5}} - \frac{h_\phi\Big|_{i+0.5,j+0.5}^{n+0.5} - h_\phi\Big|_{i+0.5,j-0.5}^{n+0.5}}{\Delta z}\right) \tag{3.25}$$

$$e_\phi\Big|_{i,j}^{n+1} = e_\phi\Big|_{i,j}^{n} + \frac{\Delta t}{\varepsilon_0\varepsilon_r\Big|_{i,j}}\left(\frac{h_r\Big|_{i,j+0.5}^{n+0.5} - h_r\Big|_{i,j-0.5}^{n+0.5}}{\Delta z} + \frac{h_z\Big|_{i+0.5,j}^{n+0.5} - h_z\Big|_{i-0.5,j}^{n+0.5}}{\Delta r}\right) \tag{3.26}$$

$$e_z\Big|_{i,j+0.5}^{n+1} = e_z\Big|_{i,j+0.5}^{n} - \frac{\Delta t}{\varepsilon_0 \varepsilon_r\Big|_{i,j+0.5}} \left(\frac{r_{i+1} h_\phi\Big|_{i+1,j+0.5}^{n} - r_i h_\phi\Big|_{i,j+0.5}^{n}}{\Delta r} - \frac{v h_r\Big|_{i,j+0.5}^{n}}{r_i} \right) \tag{3.27}$$

where Δt represents the time step, Δr, Δz represents the grid size in the r and the z directions, n is the time step number, and i and j are the grid point numbers in the r and the z directions on the coordinate axes. The specific iterative process is that if we know the spatial distribution of the electric field in the nth step and the spatial distribution of the magnetic field at the $n-0.5$th step, the magnetic field distribution at the $n+0.5$th steps can be obtained from Eqs. (3.22–3.24), and then the electric field distribution at the $n+1$th step can be obtained from Eqs. (3.25–3.27). By loop iteration, the relationship of electromagnetic field with time can be obtained.

3.2.5 Numerical Stability Condition

When performing FDTD simulations, it is important to maintain the convergence as time steps increase [2]. The stability of the numerical solution mainly depends on the relationship between the time step Δt and the grid size. Starting from Maxwell's equation, the homogeneous wave equation for the electromagnetic field can be obtained

$$\nabla^2 F(\vec{r}, t) - \frac{n^2}{c^2} \frac{\partial^2 F(\vec{r}, t)}{\partial t^2} = 0 \tag{3.28}$$

Taking the 3D Cartesian coordinate system as an example, Eq. (3.28) can be rewritten as

$$\frac{\partial^2 F(\vec{r}, t)}{\partial x^2} + \frac{\partial^2 F(\vec{r}, t)}{\partial y^2} + \frac{\partial^2 F(\vec{r}, t)}{\partial z^2} + \frac{n^2 \omega^2 F(\vec{r}, t)}{c^2} = 0 \tag{3.29}$$

where n is the refractive index of the medium and c is the speed of light in the vacuum. The wave function for the plane wave can be written as

$$F(\vec{r}, t) = F_0 \exp[i(k_x x + k_y y + k_z z - \omega t)] \tag{3.30}$$

Substituting Eq. (3.30) into Eq. (3.29), we can obtain the discretized equation as

$$\left(\frac{c\Delta t}{2n}\right)^2 \left[\frac{\sin^2\left(\frac{k_x \Delta x}{2}\right)}{\left(\frac{\Delta x}{2}\right)^2} + \frac{\sin^2\left(\frac{k_x \Delta y}{2}\right)}{\left(\frac{\Delta y}{2}\right)^2} + \frac{\sin^2\left(\frac{k_x \Delta z}{2}\right)}{\left(\frac{\Delta z}{2}\right)^2} \right] = \sin^2\left(\frac{\omega \Delta t}{2}\right) \tag{3.31}$$

where Δx, Δy, Δz represent grid size in the x, y, and z directions, respectively. $\sin \frac{\omega \Delta t}{2} \leq 1$ is required for all possible k, so the stability condition can be changed as

$$\left(\frac{c}{n}\Delta t\right)^2 \left[\frac{1}{(\Delta x)^2} + \frac{1}{(\Delta y)^2} + \frac{1}{(\Delta z)^2} \right] \leq 1 \tag{3.32}$$

This is the Courant stability condition [2], which can be written as:

$$\Delta t \leq \left(c\sqrt{\frac{1}{\Delta x^2} + \frac{1}{\Delta y^2} + \frac{1}{\Delta z^2}} \right)^{-1} \tag{3.33}$$

In the 2D simulation, the Courant stability condition can be written as:

$$\Delta t \leq \left(c\sqrt{\frac{1}{\Delta x^2} + \frac{1}{\Delta y^2}} \right)^{-1} \tag{3.34}$$

In the 3D cylindrical coordinates, the Courant stability condition can be written as

$$\Delta t \leq \left(c\sqrt{\frac{1}{\Delta r^2} + \frac{1}{\Delta z^2} + \frac{v^2}{4r^2}} \right)^{-1} \tag{3.35}$$

where c is the speed of light in vacuum, Δr, Δz represents the spatial step size in the r and z directions, and r represents the minimum radius of the calculated region.

3.2.6 Absorption Boundary Condition

When using the FDTD method to study the electromagnetic field problem in open systems, it is impossible to directly calculate the infinite structure due to the limitation of computer storage and calculation time. Therefore, an appropriate absorption boundary condition should be used to terminate the simulation window. The function of the absorption boundary is that the outpropagating wave of the computing space is absorbed as much as possible without generating pseudo-reflection, to achieve an approximation from a finite space to an infinite space. Therefore, how to design suitable absorbing boundary conditions has always been an important part of the development of FDTD method. At present, the most successful boundary absorption conditions are Mur's absorbing boundary conditions [2] and perfectly matched layer (PML) absorption boundaries [5].

Mur's absorption boundary condition is based on the traveling-wave equation to construct the boundary condition under the plane-wave approximation, so it will cause pseudo-reflection. The PML condition proposed by Berenger introduces an absorbing material on the boundary, and the electromagnetic wave is attenuated after entering the absorbing material without reflection. In practical calculations, PML is a very effective absorption boundary and is widely used in the FDTD simulation. We mainly discuss the PML absorption boundary below.

Taking a 2D structure as an example, as shown in Figure 3.5, it is assumed that an incident wave is incident on the PML layer from a lossless medium having a refractive index of n_1, and the incident angle is θ_i. The refractive index of the PML layer is n_2, and the electric conductivities in the x-direction and the y-direction are σ_x and σ_y, and the corresponding magnetic conductivities are σ_{mx} and σ_{my}. For the TM mode, Maxwell's Eq. (3.3) in the PML layer is changed to:

$$\begin{cases} \dfrac{\partial E_{zx}}{\partial t} = \dfrac{1}{\varepsilon_0 \varepsilon_r}\left(\dfrac{\partial H_y}{\partial x} - \sigma_x E_{zx}\right) \\ \dfrac{\partial E_{zy}}{\partial t} = \dfrac{1}{\varepsilon_0 \varepsilon_r}\left(-\dfrac{\partial H_x}{\partial y} - \sigma_y E_{zy}\right) \\ \dfrac{\partial H_x}{\partial t} = \dfrac{1}{\mu_0}\left(-\dfrac{\partial E_z}{\partial y} - \sigma_{my} H_x\right) \\ \dfrac{\partial H_y}{\partial t} = \dfrac{1}{\mu_0}\left(\dfrac{\partial E_z}{\partial x} - \sigma_{mx} H_y\right) \\ E_z = E_{zx} + E_{zy} \end{cases} \tag{3.36}$$

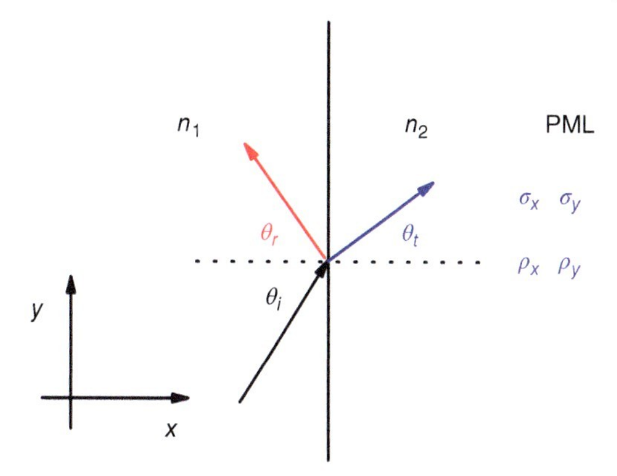

Figure 3.5 Schematic diagram of reflection and transmission of light at the boundary of the PML layer.

Assuming the incident wave has the following form:

$$E_{zi} = \exp\{i[n_1 k \cos(\theta_i)x + n_1 k \sin(\theta_i)y - \omega t]\} \tag{3.37}$$

where ω is the frequency, $k = \omega\sqrt{\mu_0\varepsilon_0}$ is the vacuum wave number, θ_i is the incident angle, then the reflected wave and the transmitted wave can be expressed as

$$E_{zr} = r\exp\{i[n_1 k \sin(\theta_r)y - n_1 k \cos(\theta_r)x - \omega t]\} \tag{3.38}$$

$$E_{zt} = t\exp[i(k_x x + k_y y - \omega t)] \tag{3.39}$$

where r and t are the reflection coefficient and the transmission coefficient, respectively. By substituting Eq. (3.39) into Maxwell's Eqs. (3.1) and (3.2), k_x and k_y in the PML layer satisfy

$$\begin{cases} k_x = kn_2\cos(\theta_t)\sqrt{\left(1 - \frac{\sigma_{mx}}{i\omega\mu_0}\right)\left(1 - \frac{\sigma_x}{i\omega\varepsilon_0 n_2^2}\right)} \\ k_y = kn_2\sin(\theta_t)\sqrt{\left(1 - \frac{\sigma_{my}}{i\omega\mu_0}\right)\left(1 - \frac{\sigma_y}{i\omega\varepsilon_0 n_2^2}\right)} \end{cases} \tag{3.40}$$

According to continuous condition of electric field on the boundary, we can get

$$\begin{cases} \sigma_y = \sigma_{my} = 0 \\ n_2\sin(\theta_t) = n_1\sin(\theta_i) = n_1\sin(\theta_r) \\ 1 + r = t \end{cases} \tag{3.41}$$

Combining the continuous condition of H_y on the boundary, it can be further deduced

$$r = \frac{n_1\cos(\theta_i)\sqrt{1 + \frac{\sigma_x}{i\omega\varepsilon_0 n_2^2}} - n_2\cos(\theta_t)\sqrt{1 + \frac{\sigma_{mx}}{i\omega\mu_0}}}{n_1\cos(\theta_i)\sqrt{1 + \frac{\sigma_x}{i\omega\varepsilon_0 n_2^2}} + n_2\cos(\theta_t)\sqrt{1 + \frac{\sigma_{mx}}{i\omega\mu_0}}} \tag{3.42}$$

According to Eq. (3.41), if

$$\begin{aligned} & n_1 = n_2 \\ & \frac{\sigma_x}{\varepsilon_0 n_2^2} = \frac{\sigma_{mx}}{\mu_0} \end{aligned} \tag{3.43}$$

the reflection coefficient r will be equal to zero. When the electric and magnetic conductivities satisfy Eqs. (3.36) and (3.38), the transmitted wave in the PML layer deduced from Eqs. (3.34)–(3.35) can be expressed as

$$E_z = \exp\{i[kn_2 \cos(\theta_t)x + kn_2 \sin(\theta_t)y - \omega t]\} \exp\left[-\frac{\cos(\theta_t)\sigma_x \eta_0 x}{n_2}\right] \tag{3.44}$$

where $\eta_0 = \sqrt{\mu_0/\varepsilon_0}$. The transmitted wave is exponentially decayed in the x direction in the PML layer. For TE mode, the reflectivity can also be obtained from Eqs. (3.40) and (3.41) as

$$r = \frac{\frac{\cos(\theta_i)}{n_1}\sqrt{1 - \frac{\sigma_x}{i\omega\varepsilon_0 n_2^2}} - \frac{\cos(\theta_t)}{n_2}\sqrt{1 - \frac{\sigma_{mx}}{i\omega\mu_0}}}{\frac{\cos(\theta_i)}{n_1}\sqrt{1 - \frac{\sigma_x}{i\omega\varepsilon_0 n_2^2}} + \frac{\cos(\theta_t)}{n_2}\sqrt{1 - \frac{\sigma_{mx}}{i\omega\mu_0}}} \tag{3.45}$$

It can be found that the zero-reflection condition of the TE mode is actually the same as the TM mode.

Since differential is replaced by the difference in the FDTD numerical simulation, if there is a large step on σ_x or σ_y, it will cause a large pseudo-reflection. Therefore, σ_x and σ_y are equal to zero at the inner boundary of the PML layer in the simulation, and gradually increase from the inside to the outside. Here, σ_x change follows $\sigma_x(x) = (x/L)^m \sigma_{\max}$. With this variation, the reflection coefficient is

$$\begin{aligned} r_p(\theta) &= \exp\left[-2\cos(\theta)\eta_0 \int_0^L \sigma_x(x)dx/\sqrt{\varepsilon_r}\right] \\ &= \exp[-2\cos(\theta)\eta_0\sqrt{\varepsilon_r}L\sigma_{\max}/(m+1)] \end{aligned} \tag{3.46}$$

The decrease in m and increase in $\sigma_{\max}$ in Eq. (3.46) reduce the theoretical reflection coefficient in the equation. However, the difference after discretization is increased, so that the pseudo-reflection introduced by discretization is increased, and a tradeoff is needed in actual simulation. The values of m are preferably 3 and 4, and the following values are better for $\sigma_{\max}$ [2].

$$\sigma_{\max} = \frac{m+1}{1.25\eta_0 \Delta x}\sqrt{\varepsilon_r} \tag{3.47}$$

In the above, we illustrate the PML absorption edge condition by taking 2D case as an example. In 3D case, we can still use similar method to set the PML layer, which will not be described in detail here.

3.2.7 FDTD Simulation of Microcavities

FDTD method can handle various complex resonator structures flexibly, for example, to simulate the microcavities with a series of resonant modes. The modes oscillate at specific frequencies, have a certain spatial distribution, and decay exponentially, which can be expressed as$F_m(\vec{r})\exp(-i2\pi f_m t)\exp(-\alpha_m t/2)$.

To excite the confined modes, excitation sources are required in the FDTD simulation. To ensure that all modes in the cavity can be excited, the excitation sources should be added to the positions where the mode field is not zero. For the microcavities with symmetry axis, the modes can also be divided into symmetric and antisymmetric modes relative to the axis. An excitation source with symmetry can be used to excite modes with the same symmetry, which can distinguish some modes with close frequencies.

To ensure that the frequency of the excitation source covers the frequency range of interest, Gaussian-modulated cosine pulses are usually applied to the electromagnetic field components at certain points

$$p(t) = \cos(2\pi f_0 t) \exp\left[-\left(\frac{t - t_0}{t_w}\right)^2\right] \tag{3.48}$$

where f_0 represents the central frequency of the pulse, t_0 represents the delay of the pulse, and $2t_w$ represents the $1/e$ width of the pulse. The pulse has a spectrum of the following form

$$P(f) \propto \exp[-\pi^2 t_w^2 (f - f_0)^2] \tag{3.49}$$

That is to say, its spectrum has the form of Gaussian function, with f_0 as the center, and $1/e$ width as $2/\pi t_w$. In the FDTD simulation, we record the change of an electromagnetic component with time $S(t, \vec{r})$, which should satisfy

$$S(t, \vec{r}) = \sum_m a_m F_m(\vec{r}) \exp(-i2\pi f_m t) \exp(-\alpha_m t/2) \tag{3.50}$$

The frequency spectrum can be obtained from Fourier transform of (3.50):

$$U(f, \vec{r}) \propto \sum_m a_m F_m(\vec{r}) \frac{1}{f - f_m + i\frac{\alpha_m}{4\pi}} \tag{3.51}$$

From Eq. (3.51), it can be seen that $U(f, \vec{r})$ is composed of a series of Lorentzian peaks. The intensity of one Lorentzian peak can be obtained as

$$|U_m(f, \vec{r})|^2 \propto \frac{1}{(f - f_m)^2 + \left(\frac{\alpha_m}{4\pi}\right)^2}. \tag{3.52}$$

It can be found that the central frequency and full width at half maximum (FWHM) are f_m and $\alpha_m/2\pi$, respectively. The quality factor can be calculated as

$$Q = 2\pi f_m / \alpha_m \tag{3.53}$$

To excite the field distribution of single mode, we can set the central frequency of the pulse near the resonant frequency of the simulated mode, and compress the spectrum of the pulse very narrowly to ensure that only one mode is excited. The FDTD method is iterated in the real number domain, so that Eq. (3.50) becomes

$$S(t, \vec{r}) = a_m \mid F_m(\vec{r}) \mid \cos(2\pi f_m t + \phi_m(\vec{r})) \exp(-\alpha_m t/2) \tag{3.54}$$

The phase $\phi_m(\vec{r})$ of the representative position $\vec{r}$ may be different for different points, so the field distribution obtained by single-mode excitation is still different at different computing times. For this reason, we calculate the sum of two sets of field distributions $S(0, \vec{r})$ and $S(\delta t, \vec{r})$ with a time distance of δt. When the decay of the break time δt is small enough to be negligible, the mode field distribution can be obtained from the two sets of field distributions.

$$F_m(\vec{r}) = |F_m(\vec{r})| \exp(-i\phi_m(\vec{r})) \\ = \frac{i}{\sin(2\pi f_m \delta t)} [\exp(-i2\pi f_m \delta t) S(0, \vec{r}) - S(\delta t, \vec{r})] \tag{3.55}$$

The field distribution expressed by Eq. (3.55) is independent of the choice of time. When the difference between different positions $\phi_m(\vec{r})$ is very small, the field distribution can be obtained by direct simulation to approximate the single-mode field distribution.

3.3 Padé Approximation for Time-Domain Signal Processing

3.3.1 Padé Approximation with Baker's Algorithm

Based on the discrete finite-time-domain data simulated by the FDTD method, the spectrum in frequency domain can be obtained by discrete Fourier transform (DFT) and implemented by FFT. For a signal with sequence length N as S $(n\Delta t)$, the DFT transform can be expressed as

$$U(m\Delta f) = \sum_{n=0}^{N-1} S(n\Delta t) \exp\left(-i2\pi \frac{mn}{N}\right) \tag{3.56}$$

where m and $\delta f = 1/N\Delta t$ represent the frequency spectrum point number and resolution, n and Δt represent the FDTD time sequence number and step, respectively. The frequency resolution is proportional to the length of the time series. To obtain high-frequency resolution, there must be a relatively long time series. Therefore, a fast and accurate method is needed to convert the time-domain signal to the frequency domain.

The Padé approximation with Baker algorithm is used to approach a shorter time sequence to an infinite time sequence. According to the definition of the Padé approximants [6], a given power series $f(z)$

$$f(z) = \sum_{n=0}^{\infty} c_n z^n \tag{3.57}$$

can be approximately expressed as

$$\sum_{n=0}^{\infty} c_n z^n - \frac{P(z)}{R(z)} = O(z^{M+N+1}), \tag{3.58}$$

where $P(z) = \sum_{n=0}^{M} a_n z^n$, $R(z) = 1 + \sum_{n=1}^{N} b_n z^n$. $P(z)/R(z)$ is defined as Padé approximant $[M, N]_{f(z)}$ of the given power series $f(z)$. Assuming $S(n\Delta t)$ is the time series of one of electromagnetic field components, where $n = 0, 1, \ldots, \infty$ is the sampling number and Δt is the sampling time interval in the FDTD simulation, we can obtain field spectrum from the FDTD output by FFT method:

$$U(\infty, f) = \sum_{n=0}^{\infty} S(n\Delta t) \exp(-i2\pi f n\Delta t). \tag{3.59}$$

Similar to (3.57), we can define a power series

$$F(z, f) = \sum_{n=0}^{\infty} C_n z^n, \tag{3.60}$$

with $C_n = S(n\Delta t)\exp(-i2\pi fn\Delta t)$ and $F(1,f) = U(\infty, f)$ to approach the field spectrum (3.59). In FDTD simulation, we only get a finite-time sequence of the field $S(n\Delta t)$ with $n = 0, 1, \ldots, N$. Assuming N is an even number and applying the Padé approximant $[N/2, N/2]_{F(z,f)}$ to the given power series (3.57) at $z = 1$, we can obtain the approximation of $U(\infty, f)$. According to Baker's algorithm, the Padé approximant $[N/2, N/2]_{F(z,f)}$ can be calculated by

$$\frac{\eta_{2j}(z)}{\theta_{2j}(z)} = [N - j, j]_{F(z,f)}, \tag{3.61}$$

$$\frac{\eta_{2j+1}(z)}{\theta_{2j+1}(z)} = [N - j - 1, j]_{F(z,f)}, \tag{3.62}$$

which satisfy the following recursion relations

$$\frac{\eta_{2j+2}(z)}{\theta_{2j+2}(z)} = \frac{\overline{\eta}_{2j+1}\eta_{2j}(z) - z\overline{\eta}_{2j}\eta_{2j+1}(z)}{\overline{\eta}_{2j+1}\theta_{2j}(z) - z\overline{\eta}_{2j}\theta_{2j+1}(z)}, \tag{3.63}$$

$$\frac{\eta_{2j+3}(z)}{\theta_{2j+3}(z)} = \frac{\overline{\eta}_{2j+2}\eta_{2j+1}(z) - \overline{\eta}_{2j+1}\eta_{2j+2}(z)}{\overline{\eta}_{2j+2}\theta_{2j+1}(z) - \overline{\eta}_{2j+1}\theta_{2j+2}(z)}, \tag{3.64}$$

where $\overline{\eta}_j$ is the coefficient of the highest power of z in $\eta_j(z)$. The initial values of the recursion are

$$\eta_0 = \sum_{n=0}^{N} C_n z^n, \theta_0 = 1.0,$$
$$\eta_1 = \sum_{n=0}^{N-1} C_n z^n, \theta_1 = 1.0. \tag{3.65}$$

The field amplitude at a given frequency can be calculated through the recursion relation with the given initial values based on the finite-time sequence, and then the intensity spectrum can be calculated by

$$I(f) = \left| \left[\frac{N}{2}, \frac{N}{2} \right]_{F(1,f)} \right|^2. \tag{3.66}$$

After calculating the intensity spectrum over an interested frequency range, we can obtain mode frequency f_0, and the mode Q factor from the frequency of the local maximum and the corresponding 3-dB bandwidth Δf of the intensity spectrum by $Q = f_0/\Delta f$ [3].

In Baker's algorithm, the nominator and denominator of the Padé approximation are calculated by the recursion relations at individual frequency. So the performance of the algorithm is repeated at each frequency for calculating whole-field spectrum. Because $1/\Delta t$ is much larger than the frequency of interest, the FDTD output is usually filtered and decimated to obtain a short sequence to save the computing time of the Padé approximation.

In the FFT/Padé approximation, the coefficients a_n and b_n in $P(f)$ and $R(f)$ of (3.58) are firstly determined by fitting $P(f)/R(f)$ with the limited resolution FFT output [7], and then a high-resolution field spectrum is directly calculated from $P(f)/R(f)$ with the fitted coefficients a_n and b_n. So the FFT/Padé approximation requires less computing time than Baker's algorithm in transforming the time-response signal to the frequency domain. However, the Padé approximation with Baker's algorithm can save more the computing time of FDTD simulation, especially for the cavity with nearly degenerate modes [3].

3.3.2 Calculation of Intensity Spectra for Oscillators

To examine the accuracy of Padé approximation with Baker's algorithm for intensity spectrum calculation, we first use the simple oscillator with noise as an example. Compared with the FDTD simulation data, the noise of the simple oscillator signal can be controlled artificially, and the true value without noise is also known, so the calculated value and the true value can be compared under different noise amplitudes. We consider the following simple oscillator system with noise

$$S_n = \exp[2\pi f_1 n\Delta t(-i - 1/2Q_1)] + \exp[2\pi f_2 n\Delta t(-i - 1/2Q_2)] + A\cdot\text{random} \tag{3.67}$$

In the simulation, we set the mode frequencies $f_1 = 160\,\text{THz}$ and $f_2 = 162\,\text{THz}$, and the quality factors $Q_1 = 5000$ and $Q_2 = 500$. The noise signal $A\cdot$random is a random number in the range of (−A, A), and time step $\Delta t = 4.7173 \times 10^{-17}$ seconds is assumed as the FDTD simulation. The Padé approximation is used to convert the time sequence obtained from Eq. (3.67) to frequency domain.

During the calculation of Padé approximation, the time sequence data are typically sampled. The decimation rate *deci* indicates that one datum is sampled at intervals of 2^{deci} length of the data. The sampling must meet the sampling theorem, which is that the new sampling frequency should be greater than the interest frequency range. For the time sequence given in (3.67), the maximum sampling rate can be calculated for a maximum frequency of 165 THz according to the sampling theorem. However, the actual sampling frequency cannot be obtained when the sampling rate is close to the limit of the sampling theorem, so the maximum sampling rate is taken to be *deci* = 5 in the calculations.

The intensity spectra for two-oscillator signal containing noise $A = 0.01$ with different length time sequence data are obtained by Padé approximation method and shown in Figure 3.6a. The sequence lengths are $N = 2^8$, 2^{10}, 2^{11}, 2^{12}, and 2^{13}, and the accurate frequency spectrum without noise is represented by open circles. The decimation rate is 1 in the simulation. In the case of no noise, the sequence length $N = 2^8$ can give accurate mode frequency and quality factor. However, the two peaks cannot be separated with the sequence length N of 2^8 and 2^{10}, when the noise signal $A = 0.01$. In the two cases of $N = 2^{11}$ and 2^{12}, although the two peaks can be distinguished in the spectrum, they still do not completely match

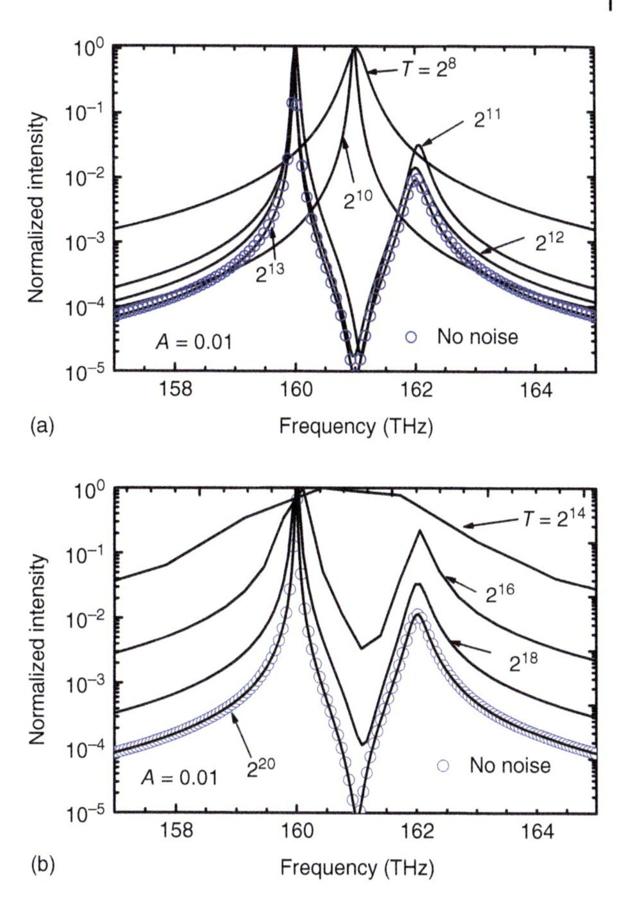

Figure 3.6 Spectra of a two-oscillator signal containing noise ($A = 0.01$) obtained by (a) Padé approximation and (b) FFT methods with different length time sequence data.

the noise-free spectrum, which indicates that the length of the time series is still insufficient. When $N = 2^{13}$, the obtained spectrum and the noise-free spectrum almost completely match. The intensity spectra obtained from the same data by FFT method are shown in Figure 3.6b. A sequence length $N = 2^{20}$ is required to obtain the precise spectrum. Although the required time sequence length for Padé approximation is increased compared with the noise-free signal, the sequence length required is still much smaller than the FFT method. For complex systems, including ultra-high Q, multimode, high noise, etc., the required time sequence should be longer. The selection of the time sequence length generally needs to ensure that the spectrum does not change significantly when the length is increased, especially for the spectrum near the desired high-Q mode.

3.4 Examples of FDTD Technique and Padé Approximation

3.4.1 Simulation for Coupled Microdisks

Two coupled microdisks surrounded by air with radius of $R = 1$ μm and refractive index of 3.2 are simulated in two-dimension x-y space by the FDTD with 10-nm

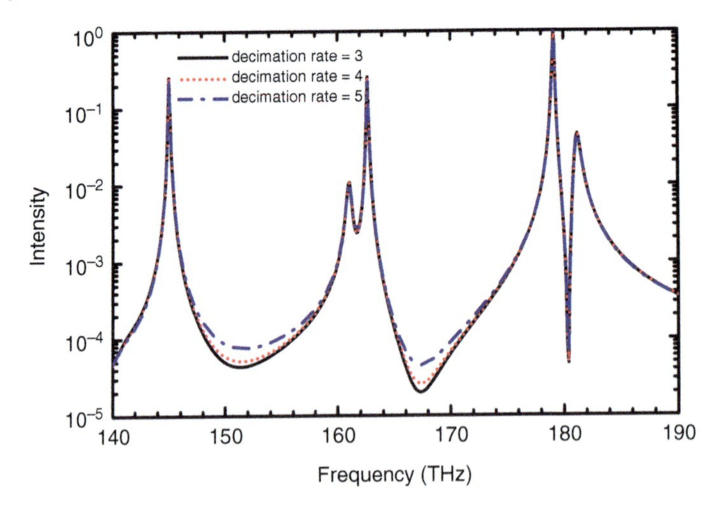

Figure 3.7 Spectra obtained by Padé approximation at different decimation rates for coupled microdisk.

square cells and the time step $\Delta t = 2.33 \times 10^{-17}$ seconds satisfying the Courant limit. The gap between the two microdisks is 50 nm. Gaussian-modulated cosine impulse (3.48) with a center frequency $f = 160$ THz, center time $t_0 = 1000\Delta t$, and pulse half-width $t_w = 400\Delta t$ is added to the E_z component at some points inside the microdisks. The sources are set to be symmetric about two symmetry axes of the coupled microdisks for simulating the modes with the corresponding symmetry conditions. In fact, the FDTD simulation can also be performed over a quarter of the coupled disks under the symmetry conditions. Fifty-thousand-step FDTD simulation is performed, and the time variation of a selected field component at some points inside the microdisks is recorded as time response signal, i.e. FDTD output. The last 2^{14} steps of FDTD output data are used to perform the transformation from the time domain to the frequency domain to reduce the influence of the initial pulse source. Because $1/\Delta t$ is much larger than the selected frequency, the FDTD output can be filtered and decimalized with the rate of $1/2^{deci}$ for saving the computing time of the transformation. The sampling theorem requires that the half of sampling frequency $1/(2^{deci}\Delta t)$ should be larger than the signal frequency range. The spectra obtained at the decimate rates $deci = 3$, 4, and 5 are plotted in Figure 3.7. The spectra obtained by Padé approximation with the different decimate rates are matched very well when *deci* satisfy sampling theorem. It can be found that the peaks completely match under different sampling conditions, and the valleys are slightly different, but the difference does not affect the calculation of the mode frequency and quality factor.

3.4.2 Simulation for Microring Channel Drop Filters

Wavelength division multiplexing is the leading technology for the practical utilization of the full optical bandwidth provided by optical fiber. The spacing of

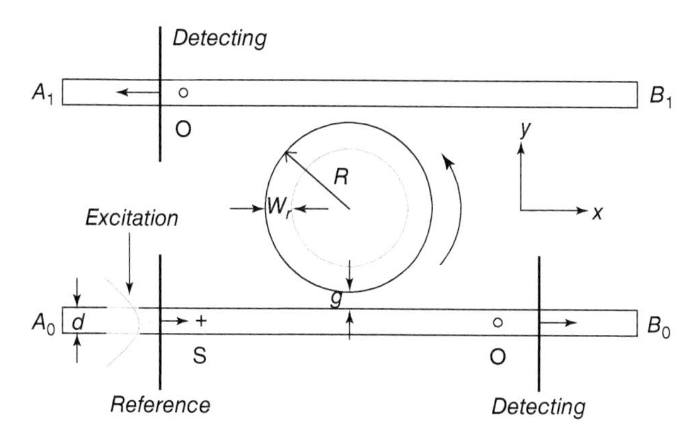

Figure 3.8 Schematic of a microring channel filter. The reference and detecting planes are indicated.

simultaneously transmitted optical data channels is closely packed to extend the whole bandwidth of optical fiber communication systems. Channel drop filters are used to obtain a single wavelength channel. FDTD technique and Padé approximation method were applied to simulate the filtering responses for microring filters, and dispersive (wavelength dependence) coupling coefficient between input waveguide and coupled region waveguide of the microring was investigated [8]. A 2D microring filter with a radius $R = 1\,\mu m$ and width $W_r = 0.2\,\mu m$ as shown in Figure 3.8 is simulated, where two parallel coupled waveguides (A_0B_0 and A_1B_1) with a width $d = 0.2\,\mu m$ separated from the ring by an air gap $g = 0.20\,\mu m$. Both the refractive indexes of the ring and waveguides are $n_{core} = 3.2$ and the external region is air. The straight waveguides support only a symmetric fundamental mode at wavelength $\lambda > 1.216\,\mu m$, which is obtained from single-mode condition. The PML absorbing boundary is used to terminate the FDTD computing window. The "bootstrapping" technique is used to set the exciting source [2], which is a 20-fs Gaussian pulse modulating a 200-THz carrier with TM polarization.

The Poynting power flux P along x direction at each point in the reference and detecting planes is obtained as the product of transverse electric field E_z and transverse magnetic field H_y as a function of frequency. The normalized spectral responses to port B_0 (T) and A_1 (D) are calculated as the ratio of the sum of Poynting power flux over the detecting plane to that over the input reference plane. By this means, T is calculated by DFT and Padé approximation. The spectra obtained by Padé approximation from 2×10^4-item FDTD output are plotted in Figure 3.9 and compared with those obtained by DFT from 2^{17}-item and 2^{18}-item FDTD outputs, which are shifted to long wavelength side by 10 and 20 nm, respectively. The Q-factors are estimated to be 2000, 1000, 600, and 320 for the four dips at 1.332, 1.433, 1.551, and 1.690 μm. From Figure 3.9, we can see that the spectra obtained by DFT at the long wavelength peak with a low Q-factor consisting of those by Padé approximation as well. However, the spectra by DFT from 2^{17}-item FDTD output at 1.551 μm and the spectra from 2^{18}-item at 1.433 μm do not give accurate result. The results show that Padé approximation gives more accurate results than

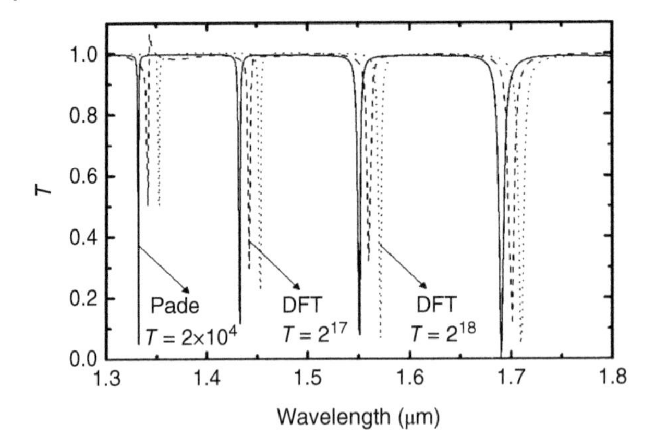

Figure 3.9 Transmission spectra T calculated by DFT from 2^{17}-item (dashed line) and 2^{18}-item (dotted line) FDTD output, and by Padé approximation from 2×10^4-item (solid line) FDTD output, where the dashed and dotted lines are shifted to long wavelength side by 10 and 20 nm, respectively.

those obtained by DFT from 13 times longer FDTD output for the resonance with $Q = 2000$.

Although the running time for FDTD simulation is cut down by Padé approximation, longer time is required for time-to-frequency transformation because Padé approximation usually takes a longer computing time than DFT for the same-length time-series data. The situation becomes worse for the spectral response calculation because the transformations for both electric and magnetic field components have to be done at each point over the plane of flux distribution. Usually, the bus and drop waveguides support only the fundamental transverse mode. The field distributions should be the same in each port of bus and drop waveguides far away from the coupling region. Furthermore, the transverse electric field component E_z and transverse magnetic field component H_y in the cross-section plane are related by a factor dependent on the x-direction propagation constant β. So the ratio of Poynting power flux over the detecting and reference planes is equal to the ratio of field component intensities at the center points "O" and "S" of the detecting and reference planes. The transmission calculated by the ratio of E_z^2 (H_y^2) at the center points "O" and "S" is shown in Figure 3.10, and that calculated by the ratio of the Poynting power flux P over the detecting and reference plane is also shown for comparison. The spectra obtained by the corresponding ratio of E_z^2 (dashed line), H_y^2 (dotted line), and P (solid line) agree very well. So the simplified calculation of T and D as the ratio of the field component intensities at the center points by Padé approximation can give enough accurate results and save memory by more than 2000 (160×13) times than that by the ratio of Poynting power flux by DFT, as the detecting and reference plane covers 1.6 μm region in y direction and the space step in the FDTD simulation is 10 nm. However, the simplified calculation is not valid for multimode bus and drop waveguides, which will have different field distribution in each port because high-order modes are excited in the waveguides.

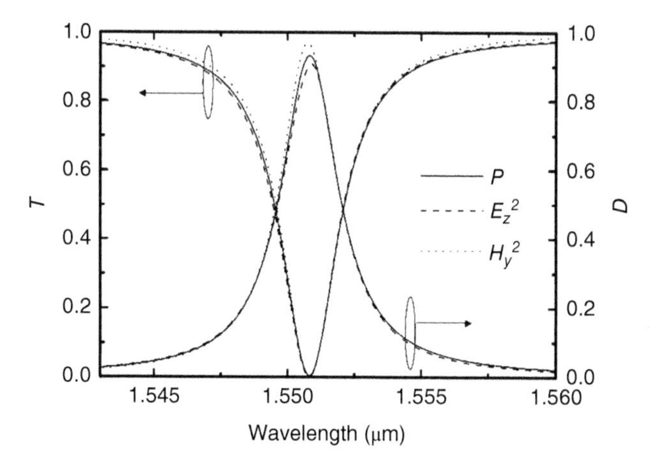

Figure 3.10 Transmission spectra T and drop spectra D calculated as the ratio of E_z^2 (dashed line), H_y^2 (dotted line) at points "O" and "S" on the axis of waveguides, and the ratio of P (solid line) in the detecting and reference planes.

3.4.3 Light Delay Simulation for Coupled Microring Resonators

In this section, we simulate the light delay of TM mode for the embedded ring resonators as shown in Figure 3.11 [9]. The external radius and the length of the straight waveguide of the racetrack ring are 3 and 2.04 μm, respectively, the radius of the embedded microring is 2.6 μm, the waveguide width is 0.2 μm with a refractive index of 3.2 surrounded by air, and the air gap between the waveguides is 0.2 μm. The structure parameters are adjusted to have the same resonance wavelength for the racetrack resonator and the inside microring with the mode number difference is an odd number for producing an electromagnetically induced transparency (EIT)-like effect. A uniform mesh with cell size of 10 nm and a 50-cell PML absorbing boundary condition are used in the FDTD simulation, and the time step is chosen to be 2.33×10^{-17} seconds satisfying the Courant condition. A Gaussian-modulated cosine impulse (3.48) with the pulse half width $t_w = 2^9 \Delta t$ and the pulse center time $t_0 = 3t_w$ are used in the following FDTD simulation.

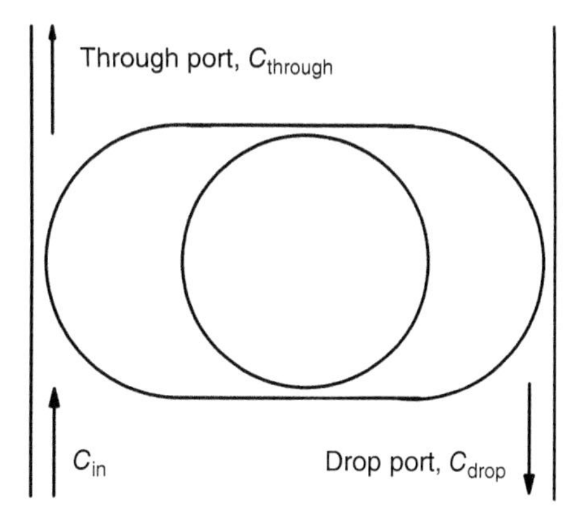

Figure 3.11 The embedded ring resonators of a racetrack and inside ring coupled to two straight waveguides. Source: Huang and Yang [9]. © 2009, Optical Society of America.

We calculate the time delay by the FDTD technique and the Padé approximation. The "bootstrapping" technique is used to set the input pulse in the FDTD simulation. The ratio of Poynting power flux over the output plane to the input plane is simply simulated as the ratio of field intensities at the center points of the output plane to the input plane. We first record the input pulse at the input plane vs. time as the input FDTD data for an isolated input waveguide, and then record output pulse at a through plane vs. time as the output FDTD data for the whole embedded microring resonators of Figure 3.11 under the same input pulse. The complex input and output field spectra are obtained from the recorded FDTD data with the pulse transmit time of 6 ps by the Padé approximation, and then the complex transmission spectra $t(\omega) = A(\omega)\exp(-i\phi)$ with the amplitude $A(\omega)$ and the phase shift ϕ are calculated as the ratio of the output field spectra to the input field spectra. The light time-delay spectrum can be calculated from the phase shift spectrum ϕ as

$$\tau = \frac{d\phi}{d\omega}, \tag{3.68}$$

where ω is the angular frequency and the time variation factor of the field is $\exp(-i\omega t)$.

The transmission spectra $A^2(\omega)$ and time-delay spectra τ of the through port obtained by the FDTD technique and the Padé approximation are plotted in Figure 3.12a,b as open circles. The peak transmission coefficient is 0.76 and time delay is 28.5 ps at the peak of 1493.65 nm. We also calculate the transmission and time-delay spectra of the through port by transfer matrix method [10] based on coupling coefficients calculated by the FDTD simulation. The coupling coefficients between the racetrack resonator and input waveguide of 0.15 and those between the two rings of 0.11 are obtained by FDTD simulation before the pulse transmission over one period inside the microrings. The mode wavelength of 1493.65 nm corresponds to mode numbers 37 and 26 for the racetrack resonator and the embedded inside microring, respectively. Assuming the transmission loss of 3 dB cm^{-1}, and the two coupling rings have the same resonance wavelength with the difference of mode numbers of an odd number, we calculate the transmission and time-delay spectra for the embedded ring resonators by the transfer matrix method. The transmission and time-delay spectra obtained by the transfer matrix method are plotted in Figure 3.12a,b as solid lines, which agree well with FDTD results. The numerical results indicate that the Padé approximation is a powerful tool to evaluate the light delay from the FDTD output for saving the FDTD computing time in modeling complex optical waveguide structures.

Finally, the propagation of a continuous exciting source at the wavelength 1493.65 nm of the transmission peak is simulated, and the transmission coefficients at the through and the drop ports vs. time are plotted in Figure 3.13. The initial response up to 7 ps is shown in Figure 3.13a with the time steps corresponding to light wave transmission one period inside the resonator. Because the coupling coefficient is 0.15 between the input waveguide and the racetrack microring, the transmission coefficient of the through port firstly increases from zero to about 0.85 as the input continuous wave arrives at time about 0.14 ps.

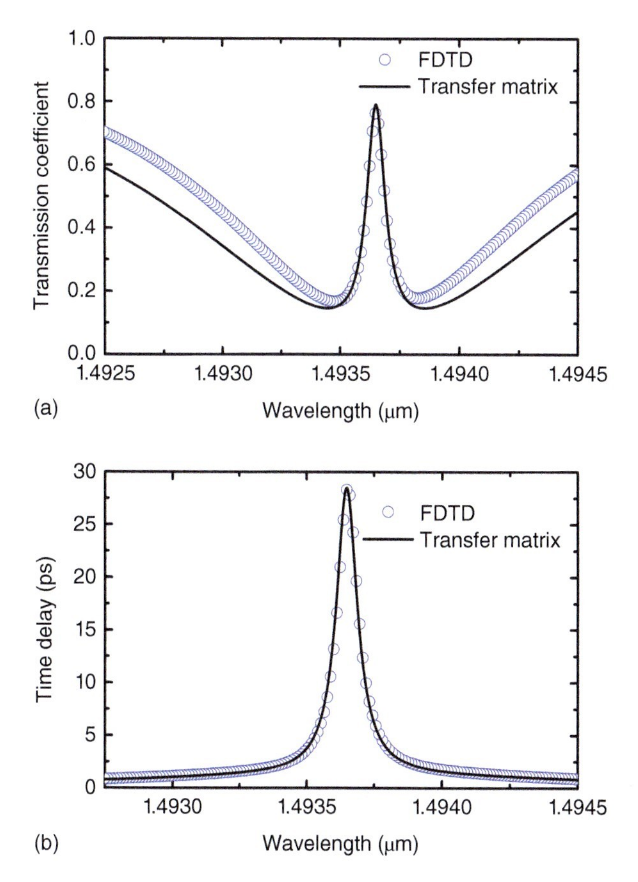

Figure 3.12 (a) Transmission and (b) time-delay spectra of the through port obtained by the FDTD technique and the Padé approximation are plotted as the open circles, and those by the transfer matrix method as the solid lines. Source: Huang and Yang [9]. © 2009, Optical Society of America.

3.4.4 Calculation of Propagation Loss in Photonic Crystal Waveguides

Finally, FDTD technique combined with the Padé approximation is applied to analyze the propagation loss in a photonic crystal waveguide. The time series of electromagnetic wave of FDTD output are transferred into frequency domain for obtaining the guide-mode frequencies, and the decaying constant of the fundamental guided mode is calculated from the Q-factor and the group velocity of the guided mode. The 2D photonic crystal with dielectric cylinders in air arrayed in a square lattice is considered with refraction index $n = 3.4$ and radius $r = 0.18a$, where a is the lattice constant. Photonic crystal waveguide is created by removing a row of dielectric cylinders and the photonic crystal structure at either side of the waveguide has finite periods [11]. Only band gaps for E polarization (electric fields lie along the cylinders) exist in this structure. So we only consider the propagation of electromagnetic wave with E polarization. The photonic crystal waveguide can be divided into a series of supercells along the waveguide axis of the photonic crystal waveguide, with a line

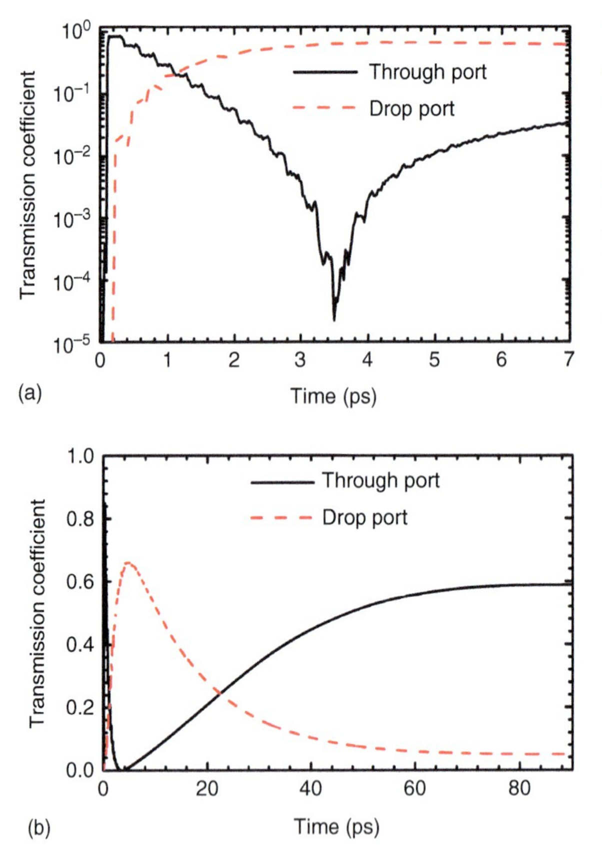

Figure 3.13 Transmission coefficients vs. time with the time from 0 to (a) 7 ps and (b) 90 ps at the through and the drop ports for injecting continuous exciting source at the wavelength 1493.65 nm of the transmission peak. Source: Huang and Yang [9]. © 2009, Optical Society of America.

of cylinders in each supercell. Because of the translational symmetry of the waveguides, we can use the periodic boundary conditions satisfying the Bloch theorem at the boundary in the waveguide axis of the supercell. The wave vector k is included in the period term of $\exp(ika)$, where a is the period in the direction of propagation. At the boundary parallel to this direction, PML-absorbing boundary conditions are applied to absorb the outgoing wave from the photonic crystal waveguide. Because the whole structure has mirror symmetry with respect to the waveguide axis, we can add even or odd symmetry conditions to the electric field components at the waveguide axis to excite the even and odd modes, respectively. At the same time, the symmetry conditions reduce the calculation domain to a half. The Gaussian function-modulated cosine impulse is added to the field at a point inside the waveguide as an exciting pulse, which covers the interesting frequency interval. The time variation of electric field component in some selected points inside the calculating window is recorded as the FDTD output.

Then we calculate the field spectrum from the FDTD output by the Padé approximation and get the mode frequency ω and Q-factor from the field spectrum. An 8000-item FDTD output is required to obtain stable field spectrum for Q-factor ranging from 100 to 1×10^7. The decaying rate $\beta = \omega/(2Qv_g)$ can be calculated from

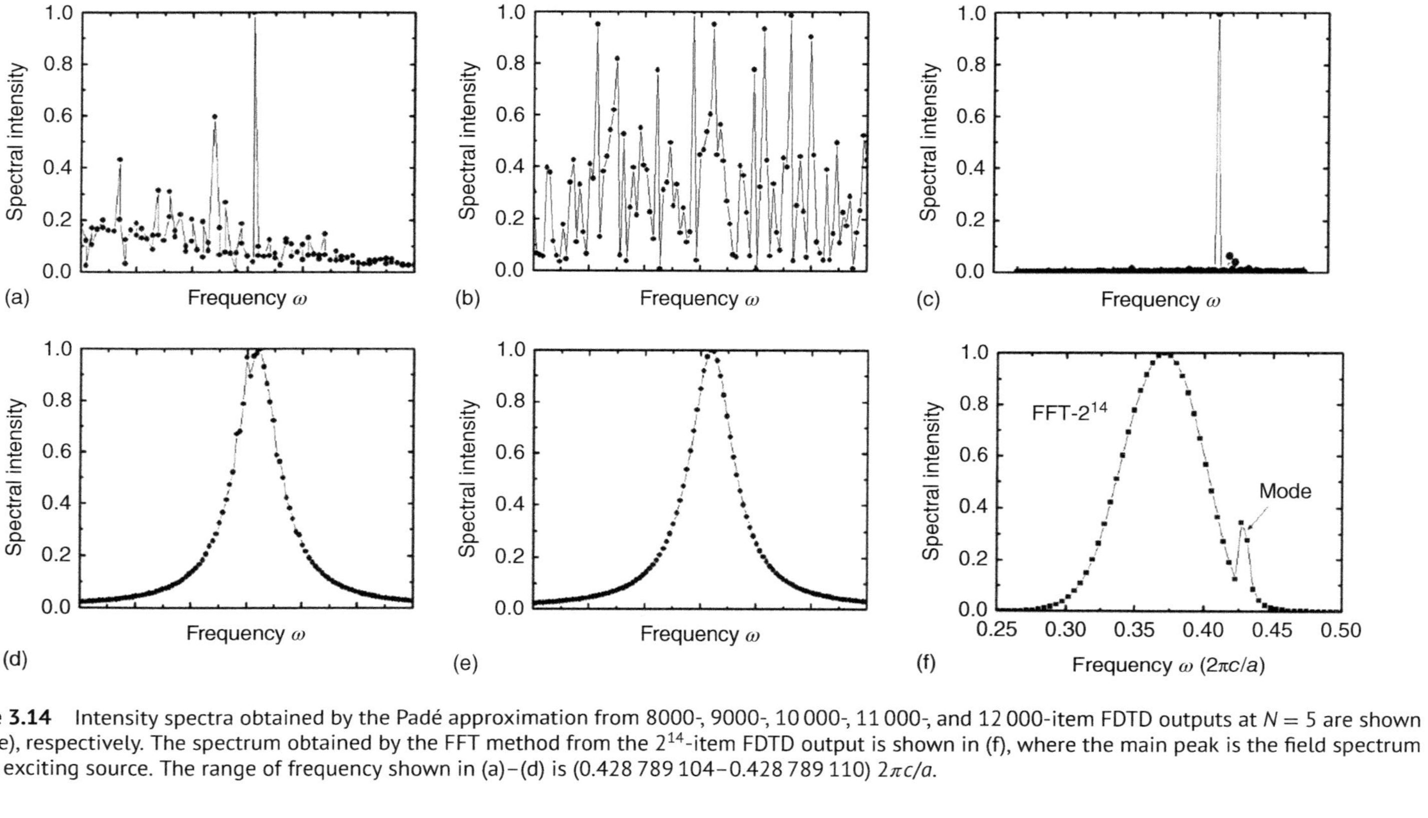

Figure 3.14 Intensity spectra obtained by the Padé approximation from 8000-, 9000-, 10 000-, 11 000-, and 12 000-item FDTD outputs at $N = 5$ are shown in (a–e), respectively. The spectrum obtained by the FFT method from the 2^{14}-item FDTD output is shown in (f), where the main peak is the field spectrum of the exciting source. The range of frequency shown in (a)–(d) is (0.428 789 104–0.428 789 110) $2\pi c/a$.

the Q-factor as the function of ω, with the group velocity v_g obtained from the dispersion curve [11]. The corresponding FDTD computing time is 0.386 ps, and the lowest decay rate is 10^{-4} cm^{-1}. In contrast, we can only get field spectrum with the resolution of 2.4×10^{11} Hz from the above FDTD output by fast Fourier transformation, and obtain the reliable value of $Q\sim10^3$ at the wavelength of 1550 nm. Guided modes with much larger Q-factor need longer FDTD output series. When the Q-factor reaches 4×10^{11} at $k=0.7\times\pi/a$ and $\omega=0.429\times2\pi c/a$ as $N=7$ (N is the number of the dielectric cylinders aside the waveguide), the corresponding decaying rate is 10^{-7} cm^{-1}, just a 20 000-item FDTD output is enough for obtaining stable field spectrum. In Figure 3.14, we show spectra intensity of guided mode at $k=0.7\times\pi/a$ and $\omega=0.429\times2\pi c/a$ with a Q-factor of 4.5×10^8 as $N=5$. The spectra obtained by the Padé approximation from the 8000-, 9000-, 10 000-, 11 000-, and 12 000-item FDTD outputs are shown in Figure 3.14a–e, respectively. We can see that FDTD outputs of 12 000-item are required for obtaining smooth field spectrum by Padé approximation. However, a wide mode peak as shown in Figure 3.14f is just observed from the spectrum calculated by the FFT method from the 2^{14}-item FDTD output.

3.5 Summary

In this chapter, the FDTD method and Padé approximation with Baker's algorithm are introduced for the simulation of electromagnetic wave in the microcavities. The difference formulas of Maxwell's equations in rectangular and cylindrical coordinates, the setting of Yee's grid, stability conditions, and boundary absorption conditions are presented for the FDTD method in 3D and 2D structures. The Padé approximation method with Baker's algorithm is used for processing simple oscillator signal with noise and FDTD output signal. It can be found that the Padé approximation method does not depend on the conditional parameters and can obtain accurate spectrum from shorter time sequence data. Therefore, the Padé approximation combined with the FDTD method can be used to solve various electromagnetic problems in complex structures, including optical microcavities.

References

1 Yee, K.S. (1966). Numerical solution of initial boundary value problems involving Maxwell's equation in isotropic media. *IEEE Trans. Antennas Propag.* 14: 302.
2 Taflove, A. and Hagness, S.C. (1998). *Computational Electrodynamics: The Finite-Difference Time-Domain Method*, 2e. Boston, MA/London: Artech House.
3 Guo, W.H., Li, W.J., and Huang, Y.Z. (2001). Computation of resonant frequencies and quality factors of cavities by FDTD technique and Padé approximation. *IEEE Microwave Wireless Compon. Lett.* 11: 223–225.
4 Li, B.J. and Liu, P.L. (1996). Numerical analysis of the whispering gallery modes by the finite-difference time-domain method. *IEEE J. Quantum Electron.* 32: 1583–1587.

5 Berenger, J.P. (1994). A perfectly matched layer for the absorption of electromagnetic waves. *J. Comput. Phys.* 114: 185–220.
6 Baker, G.A. and Gammel, J.L. (1970). *The Padé Approximant in Theoretical Physics*. Academic Press.
7 Dey, S. and Mittra, R. (1998). Efficient computation of resonant frequencies and quality factors of cavities via a combination of the finite-difference time-domain technique and Padé approximation. *IEEE Microwave Guid. Wave Lett.* 8: 415–417.
8 Chen, Q., Yang, Y.D., and Huang, Y.Z. (2006). Distributed mode coupling in microring channel drop filters. *Appl. Phys. Lett.* 89: 061118.
9 Huang, Y.Z. and Yang, Y.D. (2009). Calculation of light delay for coupled microrings by FDTD technique and Pade approximation. *J. Opt. Soc. Am. A* 26: 2419–2426.
10 Yariv, A. (2000). Universal relations for coupling of optical power between microresonators and dielectric waveguides. *Electron. Lett.* 36: 321–322.
11 Chen, Q., Huang, Y.Z., Guo, W.H., and Yu, L.J. (2005). Calculation of propagation loss in photonic crystal waveguides by FDTD technique and Padé approximation. *Opt. Commun.* 248: 309–315.

4
Deformed and Chaotic Microcavity Lasers

4.1 Introduction

Semiconductor whispering-gallery-mode (WGM) microcavity lasers are potential compact and energy-efficient light sources for photonic integration due to their small footprints, high quality (Q) factors, and planar geometry [1, 2]. Among WGM microcavity lasers with various shapes, those with a circularly symmetric geometry have attracted the most research interest owing to their high-Q WGMs [3, 4]. However, one major shortcoming for the circular microcavity lasers is that their lasing emission is typically nondirectional and homogeneous along the cavity rim due to their circularly rotational symmetry, which is undesirable for practical applications.

To realize directional lasing emission from WGM microcavity lasers, researchers have been studying various deformed microcavities to properly break the cavity symmetry, while preserving high-Q WGMs with efficient output coupling [5–7]. The WGM microcavity lasers with local boundary defects or smoothly deformed cavity shapes were proposed and demonstrated for directional emission [8, 9]. By carefully designing the cavity shapes, unidirectional lasing emissions were obtained from various asymmetric microcavity lasers, including spiral and limaçon microcavity lasers [10, 11]. Furthermore, waveguide-coupled deformed microcavity lasers have been demonstrated for on-chip optical interconnects [12, 13]. In this chapter, deformed and chaotic microcavity lasers for directional emission are reviewed. This chapter is organized as follows. In Section 4.2, a brief overview is given for the nondeformed circular microcavities. In Section 4.3, the deformed microcavity lasers with boundary defects are presented. In Section 4.4, the chaotic microcavity lasers with smoothly deformed cavity shapes are presented. Finally, a summary is given in Section 4.5.

4.2 Nondeformed Circular Microdisk Lasers

4.2.1 Whispering-Gallery Modes in Circular Microdisks

In this section, the WGMs in circular microdisks are analyzed. A practical three-dimensional (3D) microdisk with a thin slab structure can be simplified to

Microcavity Semiconductor Lasers: Principles, Design, and Applications, First Edition.
Yong-zhen Huang and Yue-de Yang.

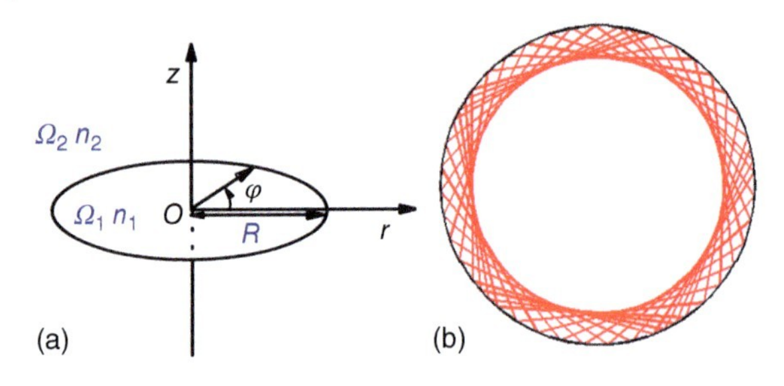

Figure 4.1 (a) An illustration of a 2D circular microdisk, and (b) light ray confined inside microdisk with total internal reflection.

a two-dimensional (2D) microdisk with an effective refractive index by assuming a given field distribution in the direction perpendicular to the slab plane. The 2D model is widely used because it can give accurate mode structure and predict high-Q WGMs more efficiently, although it may give incorrect Q factors sometimes. An illustration of a 2D circular microdisk is shown in Figure 4.1a. The center of the cavity O is placed at the coordinate origin, R is the radius of the microdisk, n_1 and n_2 are the refractive indices of the internal and external regions, respectively. Figure 4.1b shows a light ray with a conserved incident angle χ satisfying $\sin\chi > n_2/n_1$, which is confined inside the microdisk with total internal reflection and results in high Q WGMs uniformly distributing along the cavity rim region.

The WGMs in a 2D microdisk can separated into transverse electric (TE) modes and transverse magnetic (TM) modes based on their polarization. The TM (TE) WGMs are defined as the modes having magnetic (electric) fields in r–φ plane and electric (magnetic) field perpendicular to the plane. By omitting the time dependence $\exp(-i\omega t)$, the z-direction magnetic field H_z of the TE modes and the electric field E_z of the TM modes in the microdisk are expressed as $F_z(\vec{r})$. The z-direction field distribution $F_z(\vec{r})$ satisfies the 2D Helmholtz equation

$$\nabla^2 F_z(\vec{r}) + n^2(\vec{r})k^2 F_z(\vec{r}) = 0, \tag{4.1}$$

where $k = \omega(\mu_0\varepsilon_0)^{1/2} = \omega/c$ is the wavenumber in vacuum, $n(\vec{r})$ is the refractive index distribution and equals to n_1 and n_2 in the internal and external regions for the microdisk, respectively. In the polar coordinates, Eq. (4.1) is written as

$$\left(\frac{\partial^2}{\partial r^2} + \frac{1}{r}\frac{\partial}{\partial r} + \frac{1}{r^2}\frac{\partial^2}{\partial \varphi^2}\right) F_z(r,\varphi) + n^2(r,\varphi)k^2 F_z(r,\varphi) = 0, \tag{4.2}$$

where r and φ are the radius and angle in the polar coordinates, respectively.

For the microdisk with the refractive index distribution independent of position angle, the solution of Eq. (4.2) in radial direction can be expressed as Bessel function or Hankel function with a given angular mode number ν. In angular direction, the field distributions can be expressed as traveling-wave or standing-wave representations due to the circularly rotational symmetry. The traveling-wave field distributions satisfy

$$F_z = \begin{cases} AJ_\nu(n_1 kr)\exp(i\nu\varphi) & r < R \\ BH_\nu^{(1)}(n_2 kr)\exp(i\nu\varphi) & r > R \end{cases}, \tag{4.3}$$

where J_ν and $H_\nu^{(1)}$ are the Bessel function and the Hankel function of the first kind. The Bessel function is selected because it keeps convergence at $r \to 0$. The first kind of Hankel function is selected because it has the following asymptotic form at $r \to \infty$

$$H_\nu^{(1)}(n_2kr) \approx \sqrt{\frac{2}{\pi x}} \exp\left[i\left(n_2kr - \frac{\pi}{2}\nu - \frac{\pi}{4}\right)\right]. \tag{4.4}$$

With time dependence $\exp(-i\omega t)$, it represents the outgoing wave. According to the continuous condition of F_z at the microdisk boundary, Eq. (4.3) can be rewritten as

$$F_z = \begin{cases} \dfrac{J_\nu(n_1kr)}{J_\nu(n_1kR)} \exp(i\nu\varphi) & r < R \\ \dfrac{H_\nu^{(1)}(n_2kr)}{H_\nu^{(1)}(n_2kR)} \exp(i\nu\varphi) & r > R \end{cases}, \tag{4.5}$$

by normalizing the amplitude of the field F_z at the boundary to 1. Based on the z-direction field F_z, the azimuthal direction electromagnetic field E_φ and H_φ of the TE and TM modes can be obtained from Maxwell's equation as

$$\text{TE} \qquad E_\varphi = \frac{-i}{kn^2}\frac{\partial H_z}{\partial r} = \begin{cases} -\dfrac{i}{n_1}\dfrac{J_\nu'(n_1kr)}{J_\nu(n_1kR)} \exp(i\nu\varphi) & r < R \\ -\dfrac{i}{n_2}\dfrac{H_\nu^{(1)\prime}(n_2kr)}{H_\nu^{(1)}(n_2kR)} \exp(i\nu\varphi) & r > R \end{cases}, \tag{4.6}$$

$$\text{TM} \qquad H_\varphi = \frac{i}{k}\frac{\partial E_z}{\partial r} = \begin{cases} in_1\dfrac{J_\nu'(n_1kr)}{J_\nu(n_1kR)} \exp(i\nu\varphi) & r < R \\ in_2\dfrac{H_\nu^{(1)\prime}(n_2kr)}{H_\nu^{(1)}(n_2kR)} \exp(i\nu\varphi) & r > R \end{cases}, \tag{4.7}$$

where the symbol $'$ denotes the first derivative with respect to the argument. The eigenequation of the WGMs in a microdisk can be obtained according to the continuous condition of φ direction fields (4.6) and (4.7) as

$$J_\nu(n_1kR)H_\nu^{(1)\prime}(n_2kR) - \eta J_\nu'(n_1kR)H_\nu^{(1)}(n_2kR) = 0, \tag{4.8}$$

where $\eta = n_2/n_1$ and $\eta = n_1/n_2$ for the TE and TM modes, respectively. The confined energy of the WGMs in a circular microcavity decreases exponentially with time

$$W(t) = W(0)\exp(-\alpha t), \tag{4.9}$$

where W represents the energy stored in the cavity and α is the energy decay rate. The decay rate of the amplitude of the electromagnetic field should be half of that of the energy, i.e. equals $\alpha/2$. Therefore, the mode frequency will be a complex frequency considering the time dependence $\exp(-i\omega t)$ as

$$\widetilde{\omega} = \overline{\omega} - i\frac{\alpha}{2}. \tag{4.10}$$

The relationship between wave number k and angular frequency in vacuum ω is as follows

$$k = \frac{\widetilde{\omega}}{c} = \frac{\overline{\omega}}{c} - i\frac{\alpha}{2c} = Re(k) + iIm(k). \tag{4.11}$$

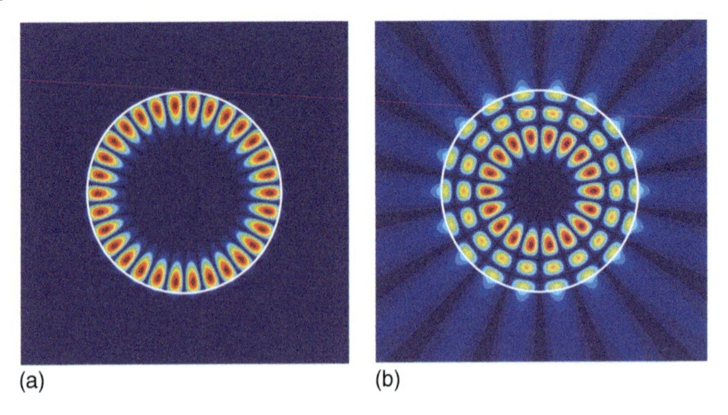

Figure 4.2 Analytical magnetic field distributions of TE standing wave modes with (a) $v = 15$, $m = 1$, and $kR = 7.4623 - i\ 9.357 \times 10^{-8}$ and (b) $v = 9$, $m = 3$, and $kR = 7.6042 - i\ 6.236 \times 10^{-2}$, in a microdisk with a refractive index of 2.65 surrounded by air.

So the root of the eigenvalue Eq. (4.8) is a complex number with a negative value of Im(k), which lies in the fourth quadrant and close to the real axis for the WGMs with low optical loss. Then the mode lifetime τ and quality factor Q of the WGMs in the microcavity can be obtained from the imaginary part of k as

$$\tau = -\frac{1}{2\,c Im(k)}, \tag{4.12}$$

$$Q = -\frac{Re(k)}{2\,Im(k)}. \tag{4.13}$$

For a given angular mode number ν, Eq. (4.8) is to be solved numerically for the discrete values of $k = k_{\nu,m}$ with the radial mode number m, and $m = 1$ refers to the fundamental WGM. The WGMs with angular mode number $\nu \neq 0$ are doubly degenerate, and can be expressed as two traveling-wave representations propagating in clockwise (CW) and counterclockwise (CCW) directions with angular mode numbers $-\nu$ and ν, or two standing-wave representations with the field distributions of $\sin\nu\varphi$ and $\cos\nu\varphi$ dependence in the azimuthal direction with ν restricted to a positive number.

Figures 4.2a,b show the analytical magnetic field distributions of the fundamental and third radial-order TE standing-wave modes in a 2D microdisk with a refractive index of 2.65, which is close to the effective index of a thin semiconductor slab with strong confinement in vertical direction, laterally confined by air. The fundamental radial-order WGM with $\nu = 15$ can be well confined inside the microdisk as shown in Figure 4.2a, and hence has ultrahigh Q factor of 3.99×10^7. The third radial-order mode with $\nu = 9$ has a strong field distribution in the external region and a low Q factor of 61.

For the WGMs with ultrahigh Q factors, the field in the external region is very weak. If the outside field can be neglected, Eq. (4.8) can be simplified under the perfect confinement approximation as

$$J_\nu(kn_1R) = 0. \tag{4.14}$$

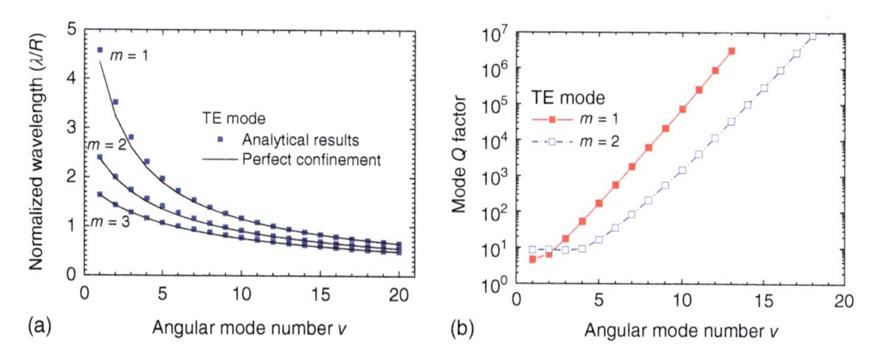

Figure 4.3 Variation of (a) wavelengths and (b) Q factors with the angular mode numbers v for different radial-order TE modes in the circular microdisk with a refractive index of 2.65 surrounded by air.

The approximate mode wavenumber can be obtained as

$$k_{vm} = \frac{x_{vm}}{n_1 R}, \tag{4.15}$$

where x_{vm} is the m_{th} nonzero root of Bessel function $J_v(x)$. The perfect confinement approximation Eq. (4.14) completely loses the information of quality factor.

For the circular microdisk with the refractive index $n_1 = 2.65$ surrounded by air, the variations of the normalized wavelength (λ/R) with the angular quantum number are shown in Figure 4.3a for the TE modes of different radial mode number m. The mode wavelengths obtained by the perfect confinement approximation are also presented as a comparison. It can be seen that the mode wavelengths of the perfect confinement approximation agree well with the exact solution. With the increase of the angular quantum number v, the effect of the perfect confinement approximation becomes better because the mode-field distribution tends to be confined to the resonator. Figure 4.3b shows the variation of the Q factor with the angular mode number v for different radial-order TE modes in the circular microdisk. It can be found that the Q factor increases exponentially with the angular mode number v for the fundamental WGMs as v is greater than 3 (it can be fitted as $Q(v) = \exp[1.2139(v-4) + 3.9567]$), and it is much larger than that of the higher-order modes. In addition, the Q factor also increases exponentially with the radius of the microdisk if keeping the resonance wavelength as a constant value. The Q factor of the WGM with $\lambda/R \approx 1$ ($v = 12$) is about 10^6, which indicates the possibility of wavelength-scale microcavity laser. It should be noted that, the Q factor of WGM is very sensitive to the refractive index of the microdisk. However, high Q modes still exist in a slightly larger microdisk with weak refractive index contrast.

Based on the analytical solution, it can be found that high Q WGMs can be confined in the circular microdisks with small mode volume V. Compared with conventional semiconductor lasers, the microdisk lasers with a large Q/V can promise lower lasing threshold and narrower linewidth with small device sizes, and also provide a platform for studying light–matter interaction.

4.2.2 Circular Microdisk Semiconductor Lasers

Lasing in semiconductor microdisks was first observed by Mccall et al. in InP/InGaAsP disks on a pedestal with InGaAs quantum wells as active medium under optical pumping [3]. Room-temperature pulsed injection lasing with sub-milliamp threshold was demonstrated by Levi et al. with a double-disk structure, where the top disk was used as p-contact layer, and the current was injected through the top and bottom pedestals [14]. However, the microdisks on a pedestal prevent the heat dissipation and limit the current injection efficiency as the WGMs mostly distribute in the rim region. Levi et al. reported the electrically pumped InGaP/InGaAs microdisk laser vertically confined by semiconductor materials (also called microcylinder laser) [15]. The weak vertical confinement leads to additional radiation loss and may limit the laser size, which is discussed in Chapter 9 in detail. Owing to their high Q WGMs, circular microdisk semiconductor lasers with different material platforms have been extensively investigated in the following decades after the first demonstration of the microdisk lasers, for their potential applications in fundamental physics study and photonic integration. However, one major shortcoming appears as the isotropic emission along the cavity rim, which poses significant difficulty in the efficient collection of the output light from the circular microdisk lasers.

4.3 Deformed Microcavity Lasers with Discontinuous Boundary

To realize directional or unidirectional emission from the WGM microcavity lasers, it is necessary to break the circularly rotational symmetry. In this section, the deformed microcavity lasers with a discontinuity at the cavity boundary are considered. Local perturbation is a natural idea for realizing directional emission by scattering or refracting light out from the discontinuous boundary. The local boundary defects can break the cavity symmetry and result in directional or unidirectional emission in free space with specific designs.

4.3.1 Microdisk Lasers with a Local Boundary Defect

Soon after the first demonstration of semiconductor circular microdisk laser, Levi et al. experimentally demonstrated that microdisk laser with a local boundary defect can improve the emission directionality [8]. An ideal microdisk laser emitted light in a symmetric pattern in the disk's plane. A practical microdisk laser typically exhibited some edge roughness in the form of variations in radius, and then the far-field emission did not maintain circular symmetry. The laser output emission was nondirectional and uncontrollable. Designing patterned asymmetries in the cavity shape could provide control of both direction and intensity of light output without dramatically increasing laser thresholds. The asymmetries could be formed by introducing a linear grating to the circumference of the disk or a defect to the

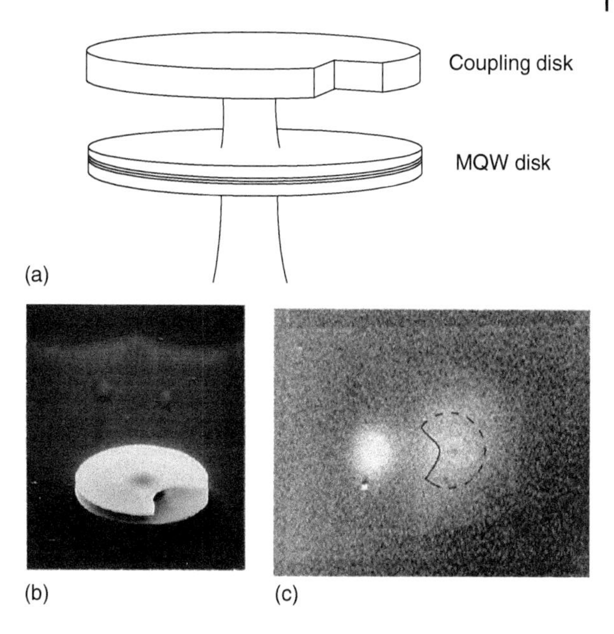

Figure 4.4 (a) Schematic of the double-disk structure of the InGaAs/InGaAsP microcavity laser. (b) SEM images of a fabricated double-disk microcavity laser with 10 μm in diameter. (c) An edge-emitting lasing output from the microcavity laser at the wavelength of 1.5 μm. Source: Reproduced from Chu et al. [16] with permission of AIP publishing.

boundary with the shape of a "tab." These deformed microcavities lased with thresholds approximately twice that of their ideal circular disks while the emission directionality was improved.

To couple light output from a microdisk laser and maintain a high Q value simultaneously, Chu et al. experimentally demonstrated double-disk structure microcavity laser [16]. The double-disk microcavity laser consisted of a lower lasing disk with multiple quantum well (MQW) active layer and a top waveguiding disk with a "notch" providing a leaking source for output coupling, as shown in Figure 4.4a. The lasing light coupled out horizontally from the top disk, while the nondeformed bottom lasing disk maintained a high cavity Q value. Figure 4.4b shows the scanning electron microscope (SEM) images of a fabricated double-disk microcavity laser with 10 μm in diameter. Figure 4.4c shows the edge-emitting lasing output from the opening of a 10-μm top disk at the wavelength of 1.5 μm. The location of the top disk is retraced using a dashed line. The figure clearly shows that the notch on the top disk provides unidirectional lasing output from the double-disk laser.

Microdisk lasers with a notch were demonstrated for realizing directional emission and controlling the lasing modes simultaneously [17]. The Q factor and emission pattern control of the WGMs in notched microdisk resonators were studied numerically by Boriskina et al. [18]. The results showed that the proposed notched microcavity design could provide efficient control of both frequency separation and Q factors of two symmetrical types of originally double-degenerate WGMs, as well as directional light output. Using midinfrared quantum cascade lasers as a model system, Wang et al. demonstrated unidirectional emission WGM elliptical microcavity quantum-cascade laser (QCL) with a wavelength scale notch at the boundary [19]. An in-plane beam divergence as small as 6 ° and a peak optical power of ~5 mW at room temperature were achieved. The beam divergence was insensitive to the pumping current and to the notch geometry, demonstrating the robustness of the cavity design. The underlying mechanism is that the notch can act as a small scatterer that scatters light rays toward the opposite side of the elliptical disk, where the outgoing rays are collimated as a parallel beam by the elliptical boundary.

4.3.2 Spiral-Shaped Microcavity Lasers

A linear variance of the radius with the polar angle gives the spiral-shaped microcavity with a localized defect, and the boundary is defined as

$$r = R(1 - \varepsilon\varphi/2\pi), \tag{4.16}$$

where r is the azimuthally varying radius of the spiral microcavity, φ is the azimuthal angle, R is the radius at $\varphi = 0$, and ε is the deformation parameter. The radius jumps back to R at $\varphi = 2\pi$ creating a notch, and laser light output can be obtained from this notch as shown in Figure 4.5a. Chern et al. reported the experimental demonstration of the unidirectional emission InGaN MQW spiral-shaped microcavity lasers [10]. Unidirectionality lasing emission is shown as the far-field pattern (FFP) in Figure 4.5b. The experimental data (open circles) agree well with the FFP obtained from the numerical solution of the Helmholtz equation (solid line) and model calculations (dashed). The modes of the spiral-shaped microcavity were simulated using a generalization of the scattering quantization approach at $n = 2.6$, $\varepsilon = -0.1$, and $nkR \approx 200$, and are shown in Figure 4.5a. The FFP shows that there is indeed notch-emitting mode in the spiral microcavity with high intensity near the perimeter.

Due to the broken chiral symmetry, the modes in the spiral microcavity exhibit a pronounced chirality and are predominantly composed of CW rotating components

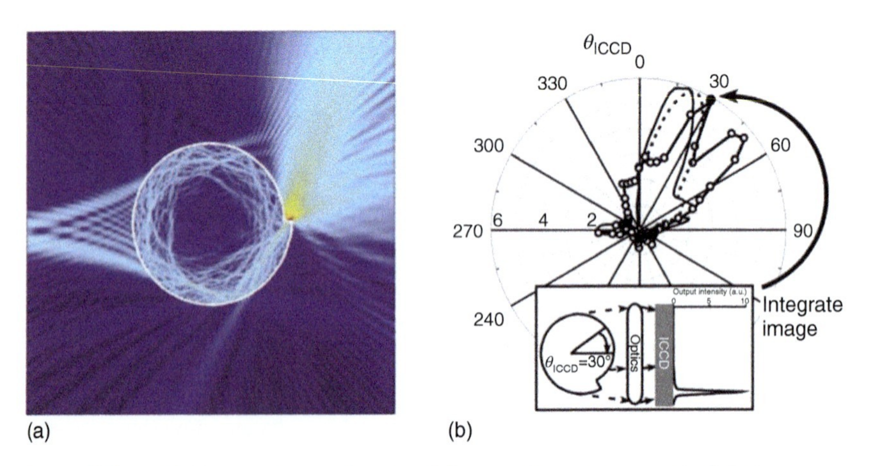

Figure 4.5 (a) Real-space false color plot of the modulus of the electric field for a simulated mode at $n = 2.6$, $\varepsilon = -0.1$, and $nkR \approx 200$. (b) Experimental (open circles) and simulated (solid and dashed) FFPs. Source: Reproduced from Chern et al. [10] with permission of AIP publishing.

corresponding to ray motion, which could not escape at the notch, and will limit its unidirectional emission characteristics. Furthermore, Lee et al. observed quasiscarred modes, i.e. modes localized along simple geometric structures that are not supported by any periodic ray in the spiral microcavity [20]. The quasiscarred modes have weak field distribution around the notch and prevent the light output through the notch.

Wiersig et al. had systematically studied the optical modes in the spiral microcavity, and found that the broken chirality leads to the appearance of copropagating pairs of nearly degenerate and highly nonorthogonal modes [21]. Both the quasiscarred modes and WG-like modes were found in the spiral microcavity. Figures 4.6a,b show the simulated intensity distributions of two nearly degenerate quasiscar TE modes with similar mode frequencies ($kR = 41.4676$ and 41.4627) and Q factors ($Q = 606$ and 597) in the spiral microcavity with $\varepsilon = 0.04$. The mode patterns are very similar for the two modes, and the corresponding angular momentum distributions also indicate the consistency as shown in Figure 4.6c. Both modes have the dominant components in the CCW direction (here the deformation parameter is positive, and the notch direction is opposite to Figure 4.5), and are highly nonorthogonal. Moreover, the WG-like modes show similar properties as the nearly degenerate modes have the similar field patterns and angular momentum distributions. The physical mechanism is the asymmetric scattering between CW and CCW propagating waves induced by the notch.

In the circular microcavity lasers, the mode competition between the doubly degenerate CW and CCW modes leads to the instability of the output power, which

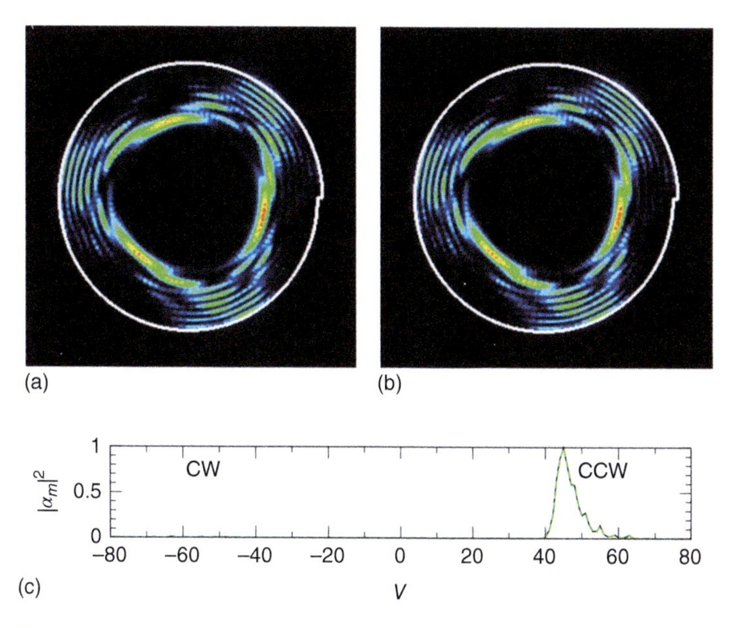

Figure 4.6 (a and b)Simulated intensity distributions of nearly degenerate quasiscar TE modes and (c) the corresponding angular momentum distributions. Source: Reproduced from Wiersig et al. [21] with permission of American Physical Society.

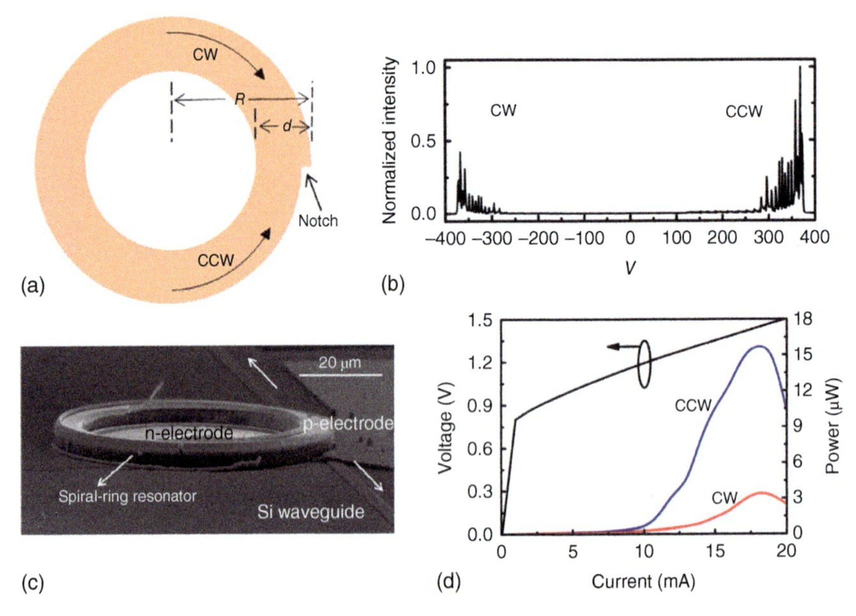

Figure 4.7 (a) Top-view schematic of a spiral-ring microcavity. (b) Simulated angular momentum distribution of one typical mode in the spiral-ring microcavity. (c) SEM image of the fabricated spiral-ring microcavity laser coupled to an underlying silicon waveguide. (d) Measured output powers vs. continuous-wave injection currents for spiral-ring microcavity lasers. Source: Sui et al. [22].

is undesirable for the practical applications. In the spiral-shaped microcavity, the asymmetric scattering results in copropagating pairs with dominant components in the CCW direction [21], which can be used to realize unidirectional emission by laterally evanescent coupling a waveguide to the cavity. Because the WGMs mostly distribute in the rim region, a spiral-ring microcavity structure was used to improve the current injection efficiency [22]. Figure 4.7a shows the top-view schematic of a spiral-ring microcavity used in experiment. Figure 4.7b shows the simulated angular momentum distribution of one high-Q mode at the wavelength of 1561 nm in the spiral-ring cavity with $R = 30\,\mu\text{m}$, a notch width of 500 nm, and a ring width of 5 μm. The spiral-ring cavity has a refractive index of 3.2 surrounded by silicon oxide similar to the experiment. Due to the asymmetric scattering at the notch, all the high-Q modes will have dominant components in the CCW direction. Figure 4.7c shows the SEM image of the hybrid spiral-ring cavity coupled to an underlying silicon waveguide fabricated with the bonding technology. After cleaving the silicon waveguide, the hybrid spiral-ring lasers with the p-side up were bonded on a thermoelectric cooler for measurement. The output powers from both sides of the silicon waveguide are measured by butt-coupling a multimode optical fiber to the cleaved silicon waveguide end faces. Figure 4.7d shows the output powers of the CW and CCW directions for a 30-μm-radius hybrid spiral-ring laser. The power abrupt fluctuations caused by the CW and CCW mode competition are not observed and the power from CCW direction is almost five times as great as that from CW direction for the hybrid spiral-ring microcavity laser. Therefore, the output power instability existing in the hybrid microdisk and ring lasers is eliminated for the hybrid spiral-ring laser, and unidirectional emission from CCW direction is achieved [22].

4.3.3 Waveguide-Connected Spiral Microcavity Lasers

In-plane waveguide-coupled microcavity lasers with unidirectional emission should be a potential solution for the applications in on-chip optical interconnects. One candidate for such a waveguide-integrated on-chip light source is the spiral microcavity laser, which offers the key merit of direct gapless coupling from a characteristic notch that is only a fraction of the microcavity radius to an integrated waveguide [13]. Furthermore, the connected waveguide can almost eliminate the asymmetric scattering induced by the notch ensuring high efficient output through the waveguide.

For the waveguide-connected spiral microcavity with $r_0 = 30\,\mu m$, $\varepsilon = 0.05$, and a refractive index of 3.2 surrounded by air, two kinds of high Q TE modes are found with the field distribution as shown in Figure 4.8a. The simulation results suggest a high waveguide-coupling efficiency via the spiral notch. The waveguide-coupling efficiency is defined as the ratio of the energy flux through the output-waveguide to the total energy flux coupled out of the cavity. The simulated waveguide-coupling efficiency exceeds 70% for the simulated WG-like modes. The field-amplitude distributions inside the waveguide reveal multiple high-order transverse modes, suggesting a multimode waveguide, as shown in the zoom-in views of Figure 4.8a. To spatially overlap the current injection with the high-Q modes of the spiral microcavity, ring-shaped p-contacts are used on top of the spiral rim region. Figure 4.8b shows the optical microscope image of a fabricated waveguide-coupled AlGaInAs/InP spiral microcavity laser with patterned electrodes. Room-temperature continuous-wave electrically injected lasing was realized for the AlGaInAs/InP waveguide-coupled spiral microcavity lasers with a disk radius of $30\,\mu m$. Figure 4.8c shows the fiber-coupled laser output power vs. injection current of the waveguide-coupled spiral microcavity lasers with the outer-ring (solid line), middle-ring (dashed line), and disk injection (dotted line) designs. The ring-shaped current injection can reduce the lasing threshold efficiently. In addition, high-speed direct modulation was also demonstrated for the spiral microcavity lasers [13].

4.4 Chaotic Microcavity Lasers with Smoothly Deformed Boundary

Apart from deformed microcavities with discontinuous boundary, the chaotic microcavity lasers with smoothly deformed boundary have been studied intensively for realizing directional or unidirectional emission shortly after the first demonstration of microdisk lasers [9]. Since unidirectional emission is essential to many applications, the microcavity lasers with many shapes, which typically can be written as simple functions in polar coordinate, have been proposed for realizing unidirectional

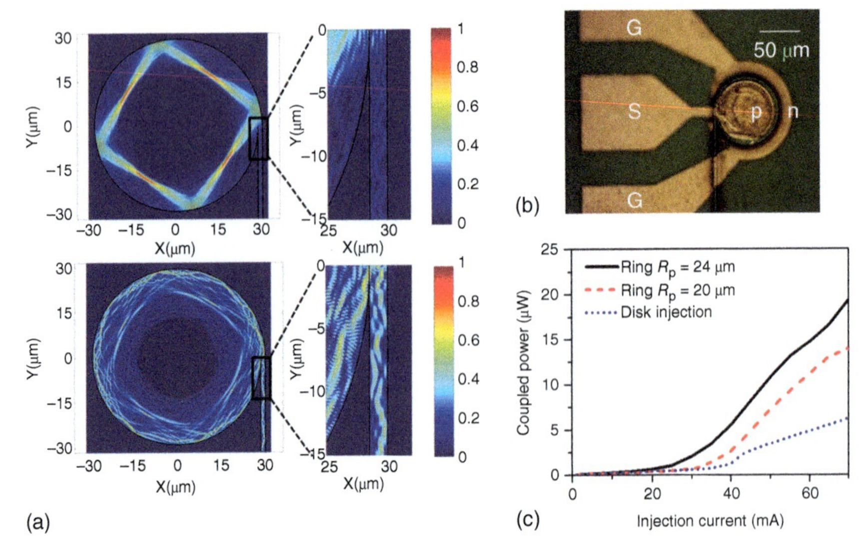

Figure 4.8 (a) Internal-field amplitude distributions normalized to the peak amplitude inside the cavity for the quasiscar mode (top) and WG-like mode (bottom). (b) Optical microscope image of a fabricated spiral microcavity laser. (c) Fiber-coupled laser output power vs. injection current of the waveguide-coupled spiral microcavity lasers. Source: Yang et al. [13].

emission over the years. However, the deformation is typically accompanied by the Q spoiling; thus, the tradeoff between the Q factor and directionality is a critical issue for deformed microcavities. In this section, several kinds of deformed microcavity with directional or unidirectional emission are reviewed.

4.4.1 Quadrupolar-Shaped Microcavity Lasers with Directional Emission

A deformed microcavity with a quadrupolar shape is first considered with the boundary defined as

$$r = R(1 + \varepsilon \cos 2\varphi), \tag{4.17}$$

where r is the azimuthally varying radius of the quadrupolar-shaped microcavity, φ is the azimuthal angle, R is the average radius, and ε is the deformation parameter. The WGMs of circular microdisks have high Q because the rays impinge upon the boundary with a conserved incident angle $\sin\chi > n_2/n_1$ and are thus trapped by total internal reflection. In contrast, the incident angle of the light rays in deformed microcavities is not conserved, and the refractive escape could happen. Starting from a light ray with an incident angle above the critical angle of total reflection, after a large number of reflections, the ray may impinge upon the boundary below the critical angle and escape with high probability. Figure 4.9a shows the regular and chaotic ray trajectories in the circular microdisk (left) and the quadrupolar-shaped microcavity with $\varepsilon = 0.1$ (right).

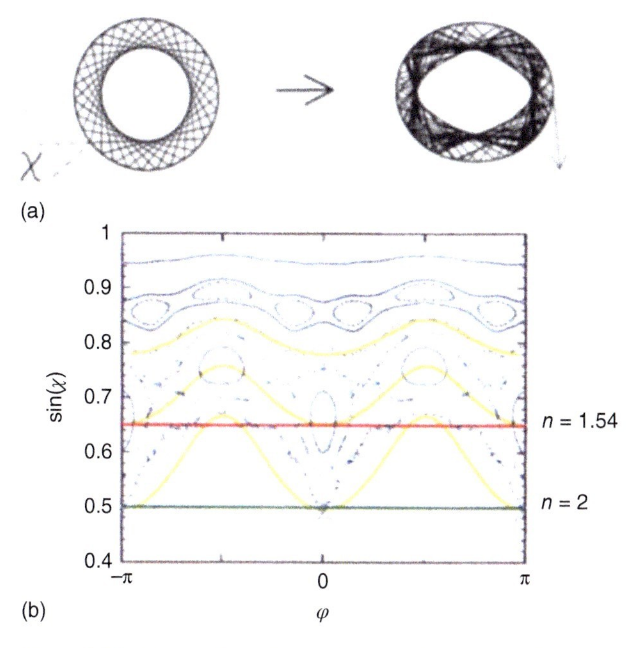

Figure 4.9 (a) Regular and chaotic ray trajectories in the circular microdisk (left) and the quadrupolar-shaped microcavity with $\varepsilon = 0.1$ (right). (b) Poincaré SOS showing phase-space ray dynamics for quadrupolar-shaped microcavity with $\varepsilon = 0.072$. Source: Reproduced from Nöckel and Stone [5] with permission of Springer Nature.

For smooth convex deformations from a circle, the ray dynamics becomes increasingly chaotic with increasing deformation according to the Kolmogorov–Arnold–Moser (KAM) theorem. For a smooth perturbation of an integrable system, some of the invariant tori survive, while others are destroyed giving rise to partially chaotic dynamics. In terms of the chaoticity of the internal ray dynamics, the degree of deformation was classified into weak, moderate, and large, with nearly integrable ray dynamics, mixed-phase space, and predominantly chaotic dynamics, respectively [7]. Figure 4.9b shows the partial Poincaré surface of section (SOS) of the ray dynamics in a quadrupolar-shaped microcavity with a moderate deformation $\varepsilon = 0.072$, and a refractive index of n surrounded by air. The $\sin\chi$ and the angular position φ are recorded at each collision. The two horizontal lines represent the critical lines at $\sin\chi = 1/n$ for $n = 2, 1.54$. "Islands" indicate trajectories oscillating around stable periodic orbits. Unbroken KAM curves appear above $\sin\chi > 0.8$, and no light ray can cross such a curve. Trajectories below $\sin\chi < 0.8$ lying outside islands are all chaotic, but follows for some time the adiabatic curve predicted by the equation

$$\sin\chi(\varphi) = \sqrt{1 - (1 - S^2)\kappa(\varphi)^{2/3}}, \tag{4.18}$$

where $0 < S < 1$ is an adiabatic constant and $\kappa(\varphi)$ is the curvature of the boundary curve at azimuthal angle φ [5]. The dynamics of light ray trajectories consists of a rapid motion along adiabatic curves and a slow chaotic diffusion transverse to them. When the adiabatic curve touches the critical angle, the ray can escape tangentially.

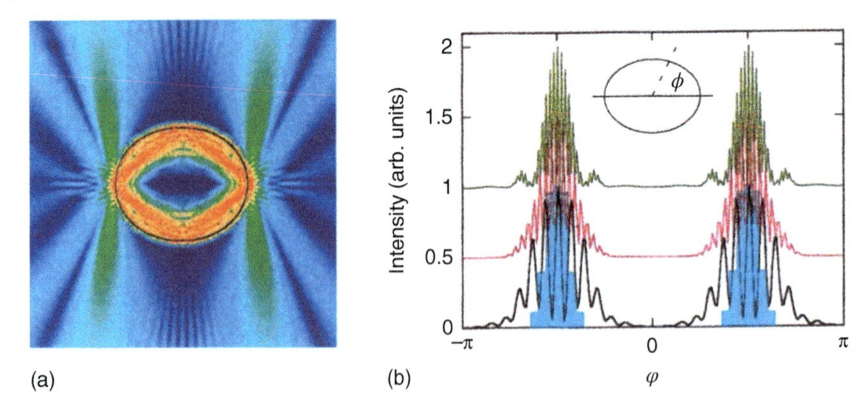

Figure 4.10 (a) Simulated electric field intensity distribution of the TM mode with $kR = 45.14$ in the quadrupolar-shaped microcavity with $\varepsilon = 0.11$. (b) FFPs of the intensity of three modes with $kR = 12.1$, 27.9, and 45.4. Blue histogram is the prediction of the ray-optics model. Source: Reproduced from Nöckel and Stone [5] with permission of Springer Nature.

It can be seen that the minima of the adiabatic curves occur at the points of maximum curvature according to Eq. (4.18). The localization in the spatial coordinate (at the maximum of the curvature) and in the angle (light is emitted tangentially) results in strong emission maxima in the far field in directions tangent to the highest curvature points [5]. Figure 4.10a shows the numerically simulated field distribution of a TM mode in the quadrupolar-shaped microcavity with $\varepsilon = 0.11$. It can be clearly seen that the high-intensity regions in the near field just outside the surface at the highest curvature points $\varphi = 0, \pi$, and the high-emission-intensity lines (green) emanating from these points in the tangent directions. Figure 4.10b shows the FFPs of the intensity of three modes with $kR = 12.1, 27.9$, and 45.4 in the quadrupolar-shaped microcavity with a refractive index of 2. Two emission peaks at $\varphi = \pm\pi/2$ corresponding to the tangent directions to the points of highest curvature are found.

The prediction based on Eq. (4.18) concerning the tangential emission from the highest-curvature points fails if regular islands are located around the minima of the relevant adiabatic curve, as shown in Figure 4.9a for $n = 1.54$. As the rays cannot enter the regular islands, they do not escape at the maximum of the curvature but mainly at two points separated by roughly the size of the islands. In this case, the peaks split and maximum emission does not occur at the points of highest curvature [9].

Gmachl et al. reported the experimental demonstration high-power directional emission from a "flattened" quadrupolar-shaped QCL with an effective refractive index of 3.3, where the cavity shape was described by $r = R(1 + 2\varepsilon \cos 2\varphi)^{1/2}$ [6]. For the flattened quadrupolar-shaped QCL with small deformation ε, the basic picture of chaotic orbits escaping refractively still held. However, multimode lasing with low-output emission power was observed. At larger deformations, the experimental results showed that the "bow-tie" resonances around the stable periodic orbits in phase spaces emerged and were responsible for highly directional and high-power emission.

4.4.2 Limaçon Microcavity Lasers with Unidirectional Emission

Owing to the partially reserved rotational symmetry in quadrupolar-shaped microcavity, unidirectional emission is impossible. To realize unidirectional emission, the rotational symmetry should be broken totally. Wiersig and Hentschel proposed the limaçon microcavity for realizing robust unidirectional emission while preserving high Q factors [11]. The cavity boundary is defined as

$$r = R(1 + \varepsilon \cos \varphi), \tag{4.19}$$

where r is the azimuthally varying radius of the limaçon microcavity, φ is the azimuthal angle, R is the average radius, and ε is the deformation parameter. To obtain high efficient unidirectional emission, the degree of deformation is large and the microcavity has predominantly chaotic dynamics. High Q factor modes are scarred modes confined with low optical loss due to the wave localization along unstable periodic ray trajectories. The key idea for achieving unidirectional emission is to exploit light emission along unstable manifolds of the chaotic saddle of the ray dynamics. The output directionality is universal for all the high Q modes in the limaçon cavity with the size much larger than the resonance wavelength, because the corresponding escape routes of rays are similar. This property enables a robust laser cavity design for realizing unidirectional emission without selective excitation of specific modes in experiments [7].

Figures 4.11a,b show a ray trajectory the limaçon cavity with $\varepsilon = 0.43$ and refractive index $n = 3.3$, where the dynamics is predominantly chaotic. The arclength s is normalized to the cavity's perimeter s_{max}. Starting a test ray with an initial value well above the critical line (square), it rapidly approaches the leaky region (triangle). Without refractive escape, the ray trajectory would fill the phase space in a random fashion (small dots in Figure 4.11b). Periodic ray trajectories do exist but are always unstable, except for the two islands in the leaky region. Stable ray trajectories are also confined to the tiny region with $\sin\chi > 0.99$. The FFP of the limaçon cavity is determined by the motion of ray trajectories approaching the critical angle, also called the unstable manifold in the leaky region ($\sin\chi < 1/n$), which can be computed from the survival probability distribution (SPD) for an ensemble of rays starting uniformly in phase space. Figure 4.11c shows the calculated SPD of the limaçon cavity for TE polarization. The Fresnel-weighted unstable manifold in the leaky region is concentrated on very few high-intensity spots, giving highly directional output. For TE polarization, directionality around $\varphi = 0$ is observed, whereas in the TM case additional, smaller peaks occur, i.e. the unidirectionality in the TE case is better than that in the TM case. The magnification of SPD shown in Figure 4.11d reveals the differences between TE (top) and TM (bottom) polarization in the leaky region. Such a difference originates from Fresnel's law due to the existence of the Brewster angle for the TE polarization, which leads to a much lower reflectivity with the incident angle slightly smaller than critical angle compared to the TM polarization.

To prove the unidirectionality, numerical simulation was performed for limaçon cavity, and the results confirmed that the limaçon cavity supports high Q modes of both TE and TM polarization. As an example, Figure 4.12 shows the simulation

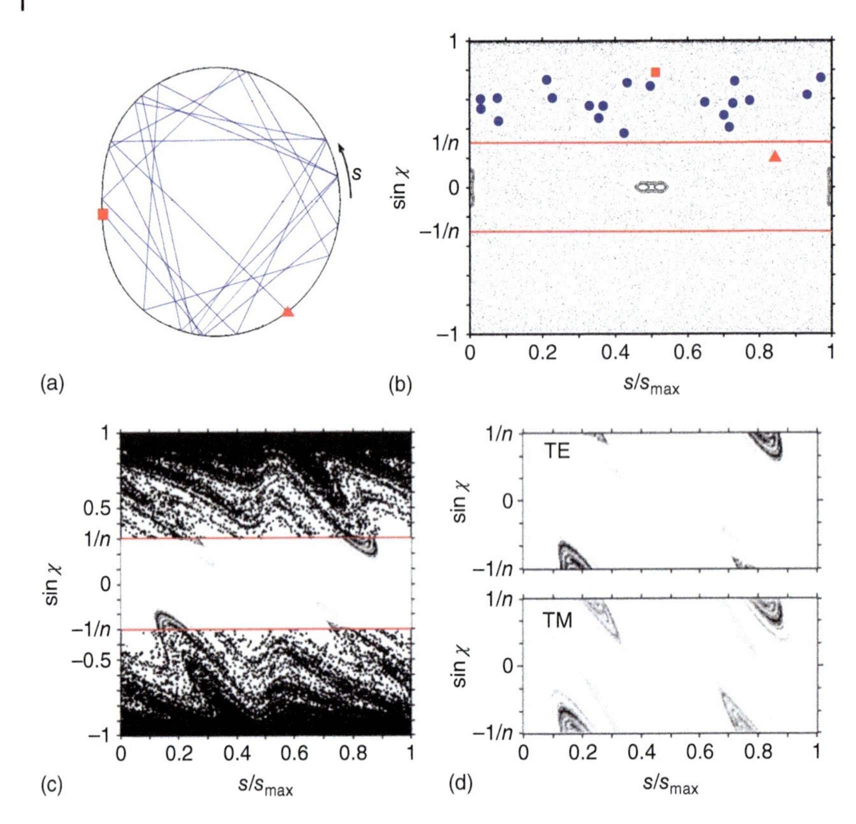

Figure 4.11 (a) Chaotic ray trajectory in the limaçon cavity with $n = 3.3$ and $\varepsilon = 0.43$, and (b) the corresponding Poincaré SOS for the trajectory (thick dots) starting above the critical line (square) and refractively escaping after only 20 bounces (triangle). The small dots show a typical trajectory in the corresponding closed billiard system. (c) SPD of the limaçon cavity for TE polarization. (d) Magnification of SPD for TE (top) and TM (bottom) polarization in the leaky region. Source: Reproduced from Wiersig and Hentschel [11] with permission of American Physical Society.

results for a TE mode at $kR = 26.0933$, which corresponds to a free-space wavelength of about 900 nm for the limaçon cavity with $R = 3.75\ \mu\text{m}$. The mode is spatially confined well near the boundary of the cavity as shown in Figure 4.12a with a Q factor of 1.85×10^5. To further investigate the character of these optical modes, the Husimi projection obtained from the overlap of the wave function with a coherent state was considered. The Husimi projection in Figure 4.12c shows the mode intensity is enhanced around an unstable periodic ray trajectory located around $|\sin\chi| \approx 0.86$ well above the critical line. Hence, the confined mode has only exponentially small intensity in the leaky region guarantees high-Q factor. Although the Husimi projection has a small contribution in the leaky region, it is precisely this outgoing light that determines the FFP. Figure 4.12d shows that the Husimi projection in the leaky region agrees well with the unstable manifold in Figure 4.11d, confirming its responsibility for the directional emission. In addition, the simulation results also proved

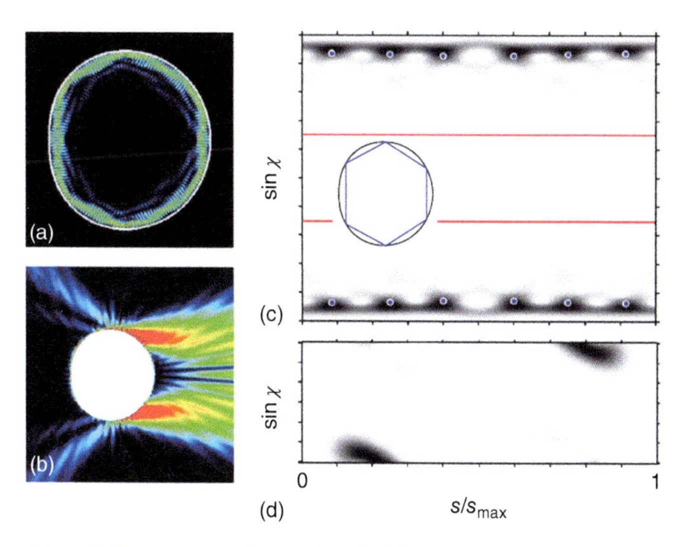

Figure 4.12 Calculated intensity distribution of an even-parity TE mode (a) inside and (b) outside a limaçon cavity with $n = 3.3$ and $\varepsilon = 0.43$. (c) Husimi projection of the TE mode in (a), and (d) magnified Husimi function in the leaky region. Source: Reproduced from Wiersig and Hentschel [11] with permission of American Physical Society.

that the TM modes have worse unidirectionality compared to the TE modes, which is consistent with the ray dynamics analysis.

Using the QCLs emitting at ~10 μm as a model system, Yan et al. reported the experimental demonstration of directional light emission from limaçon-shaped microcavity semiconductor lasers [23]. The dominant polarization of QCL is TM due to the selection rules of the optical transition of the active material. The effective refractive index of TM modes for the QCL is 3.2. Figure 4.13a shows the FFPs obtained by wave simulation and ray optics simulations for one high-Q TM WGMs ($Q > 10^7$) in the limaçon-shaped microcavity with $\varepsilon = 0.4$ and $R = 80\,\mu$m. Both wave and ray optics simulations show the unidirectional emission with the far-field divergence angle of about 30°, which is defined as the full width at half maximum (FWHM) of the main far-field emission lobe. The inset of Figure 4.13a shows the external intensity distribution of the mode obtained by wave simulation, where the main lobe is resulted from the refractive escape from the points A′ and B′. The fabricated limaçon-shaped QCLs were tested in pulsed mode at room temperature with a pulsed injection current of a width of 125 ns and a repetition rate of 80 kHz. Figure 4.13b shows the measured light output power vs. current (L–I) and voltage vs. current (V–I) curves for a representative QCL with the same parameter used in simulation. A peak output power of 4 mW, a threshold current density around 2.0 kA cm^{-2}, and a maximum slope efficiency of about 12 mW A^{-1} are obtained from the L-–I curve for the QCL. The laser has a threshold current of about 380 mA. Figure 4.13c shows the lasing spectra of the limaçon microcavity QCL measured at different currents above threshold. Single-mode lasing is observed at 380 mA around threshold, whereas multimode lasing is observed at higher injection current. Figure 4.13d shows the measured FFPs at pumping currents of 500

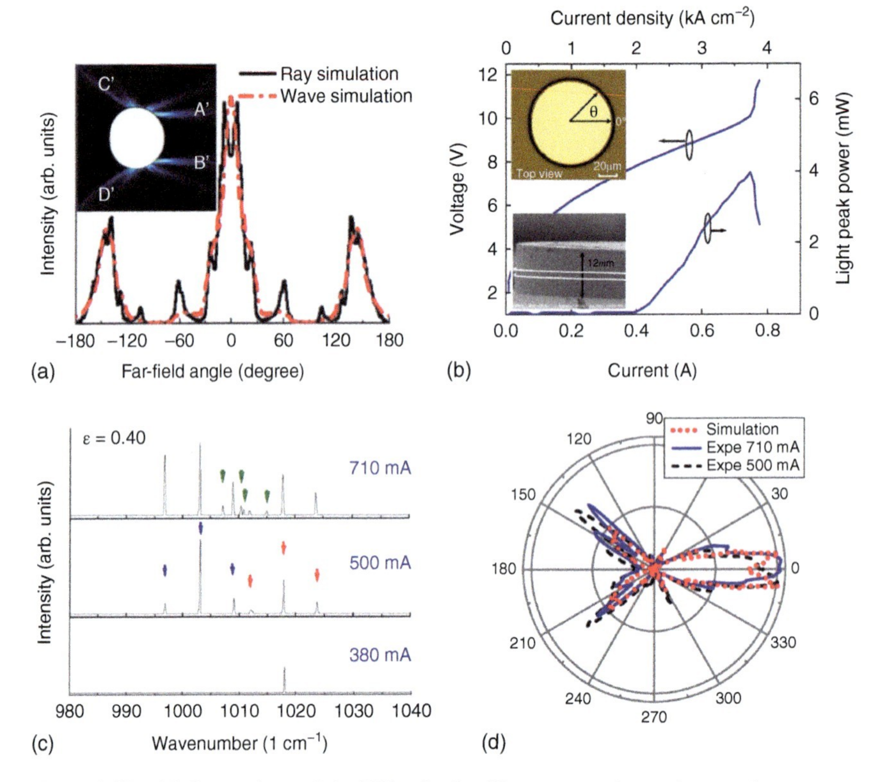

Figure 4.13 (a) Comparison of the FFPs obtained by wave-optics and ray-optics simulations for the limaçon-shaped microcavity with $n = 3.2$, $\varepsilon = 0.4$, and $R = 80\,\mu\text{m}$; inset: the external intensity distribution. (b) Voltage and peak output power as a function of injection current for the limaçon-shaped microcavity QCL; the upper left inset: the top-view microscope image; and the lower left inset: the side-view SEM image. (c) Laser spectra at different pumping currents. (d) Comparison of FFPs obtained by ray optics simulation and experiment at pumping currents of 500 and 710 mA. Source: Reproduced from Yan et al. [23] with permission of AIP publishing.

and 710 mA compared to the simulation results in polar coordinates, which are in good agreements even through multiple modes lasing. The device performance was insensitive to the deformation in the range of $0.37 < \varepsilon < 0.43$ for $R = 80\,\mu\text{m}$ limaçon microcavity QCLs indicating limaçon microcavity as a robust laser cavity design for realizing unidirectional emission.

4.4.3 Wavelength-Scale Microcavity Lasers with Unidirectional Emission

In the limaçon microcavity with size much larger than the optical wavelength ($kR \gg 1$), unidirectional emission has been obtained by manipulating the ray dynamics via optimize the deformation parameter. However, the classical ray model breaks down as the cavity size approaches resonance wavelength, as wave phenomena become significant resulting in high Q mode with poor directional emission properties [24]. Moreover, the output directionality among different modes is no longer universal, in contrast to the prediction of the ray model. Song et al. reported unidirectional emission from wavelength-scale microcavity laser by coupling an isotropic high Q mode to an anisotropic low Q mode [25].

The inset of Figure 4.14a shows the top-view SEM image of a GaAs limaçon microcavity with embedded InAs quantum dots as the gain media. The microcavity has a modified limaçon shape with the cavity boundary defined as $r = R(1 + \varepsilon \cos \varphi)(1 - \varepsilon_1 \cos 2\varphi) + d$, where $R = 890\,\text{nm}$, $\varepsilon = 0.28$, $\varepsilon_1 = 0.06$, and $d = 60\,\text{nm}$. Figure 4.14a shows the lasing spectrum of a limaçon microcavity under optical pumping with three evident peaks at wavelengths of 908, 942, and 978 nm. Figure 4.14b shows the measured FFPs for the lasing modes 1 and 3 in Figure 4.14a, which indicate a dramatic difference between them. The lasing mode 1 at 978 nm has bidirectional emission in forward and backward directions, while the lasing mode 3 at 908 nm has output mostly in forward direction. This phenomenon cannot be explained by the ray model, numerical methods were used to study the modes in the limaçon microcavity. Figure 4.14c shows the Q factors obtained by numerical method for the high Q modes (dots) and the low Q modes (squares) in the same cavity as those in Figure 4.14a. Modes 1, 2, and 3 correspond to the three lasing peaks obtained in Figure 4.14a. Two sets of modes with different frequency spacing are observed, and the high Q modes have the field distributions similar to the common WGMs in most cases. A dip is found at $kR = 7.1$ around the frequency cross of the two sets of modes, which is attributed to the mode coupling between these modes due to the cavity deformation. The directionality is defined as $U = \int I_\varphi \cos \varphi d\varphi / \int I_\varphi d\varphi$, where I_φ represents the angle dependent far-field emission intensity. $U = 0$ corresponds to isotropic or bidirectional emission, whereas a positive (negative) value corresponds to emission mainly in forward (backward) direction. Figure 4.14d shows the calculated U for both the high Q and low Q modes, where crosses are U values of three lasing modes in Figure 4.14a, which are consistent with the experimental results. Around the mode cross point, the high Q mode 4 and low Q mode 6 have similar FFPs as shown in the inset of Figure 4.14d. The mode coupling results in decrease of mode Q factor and improvement of directionality for the high Q modes, as the low Q modes have anisotropic emission. Such a phenomenon can also be understood by the field distributions and Husimi projections of these modes. The directional emission of the low Q modes concentrated around period-3 orbits is resulted from the beam shifts, which can be neglected for the high Q modes concentrated around period-4 orbits as the incident angle is far above the critical angle [7].

The bounce number of the underlying orbits for high Q modes decreases with a reduction of the cavity size, and the incident angles approach the critical angle of total reflection. Redding et al. reported the unidirectional emission resulted from the local chirality of the modes in ultrasmall microcavities [26]. Figure 4.15a shows the calculated intensity distribution of the CW (left) and CCW (right) waves for a high Q mode at $kR \approx 3.2$ in a limaçon cavity with the boundary defined by (4.19) at $\varepsilon = 0.41$. The intensity maxima for the CW and CCW waves are spatially separated. The dashed line describes the classical period-3 ray orbit, and the solid lines depict the path for the CW and CCW waves. The split in the CW and CCW orbits is due

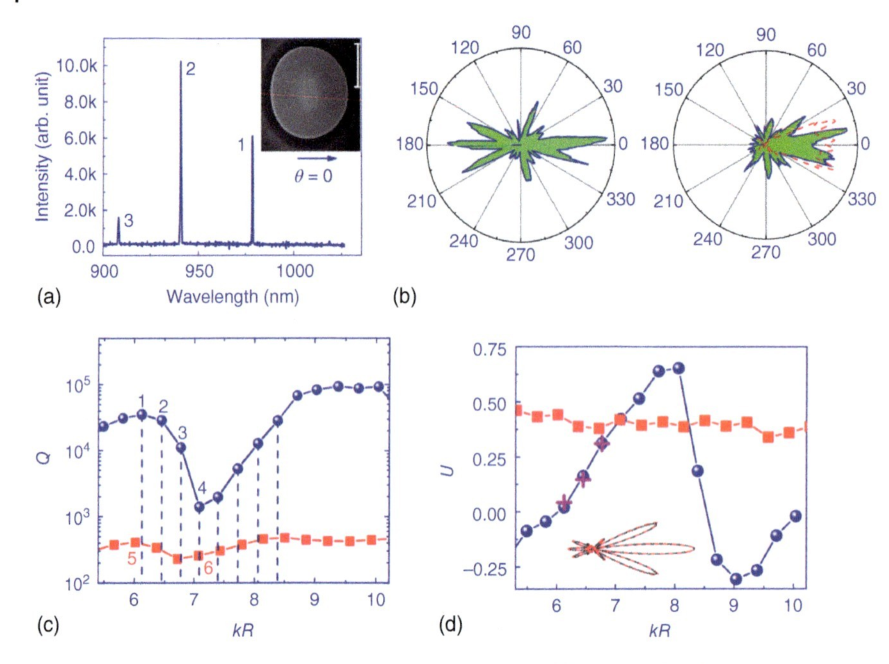

Figure 4.14 (a) Measured laser emission spectrum of a GaAs-modified limaçon microcavity; Inset: SEM image of a device with $R = 890$ nm. (b) Measured FFPs for the lasing modes 1 (left), and 3 (right) shown in (a); dashed curve in (b): the FFP obtained by a ray-tracing model. (c) Q factors and (d) directionality U obtained by numerical method for the high Q modes (dots) and the low Q modes (squares) in the same cavity as that in (a). Inset in (d): Calculated FFPs for modes labeled 4 (solid line) and 6 (dashed line) in (c). Source: Reproduced from Song et al. [25] with permission of American Physical Society.

to the Goos–Hänchen shift and the Fresnel filtering. For the CW wave, the incident angle at point 1 is smaller than that at point 3, light leakage from point 1 exceeds that from point 3 as shown in Figure 4.15b, which results in directional emission predominantly from point 1 in the $\varphi = 0$ direction. Similarly, the CCW wave has the directional emission predominantly from point 3 in the $\varphi = 0$ direction. The simulation results indicate the unidirectional emission for the high Q mode at $kR \approx 3.2$. The spatially separated distribution of the CW and CCW waves results in the local chirality, which is defined as the intensity difference between the CW and CCW waves over the total intensity and plotted in Figure 4.15c as a function of angle φ. For high Q mode at $kR \approx 3.2$, a strong local chirality is observed, while the local chirality of the mode at $kR \approx 5.2$ corresponding to the period-4 orbits is near zero. The intensity of the lasing modes at 875 nm in a limaçon cavity with $\varepsilon = 0.41$ and $R = 460$ nm is plotted in Figure 4.15d as a function of the incident pump power, which show a distinct threshold behavior. The lasing peak corresponds to the high Q mode at $kR \approx 3.2$ in the simulation. The inset of Figure 4.15d shows the measured FFP for the lasing mode, which is in good agreement with the simulation results. Moreover, the local chirality of the high Q modes enables the directional evanescent coupling by introducing a waveguide to a specific position.

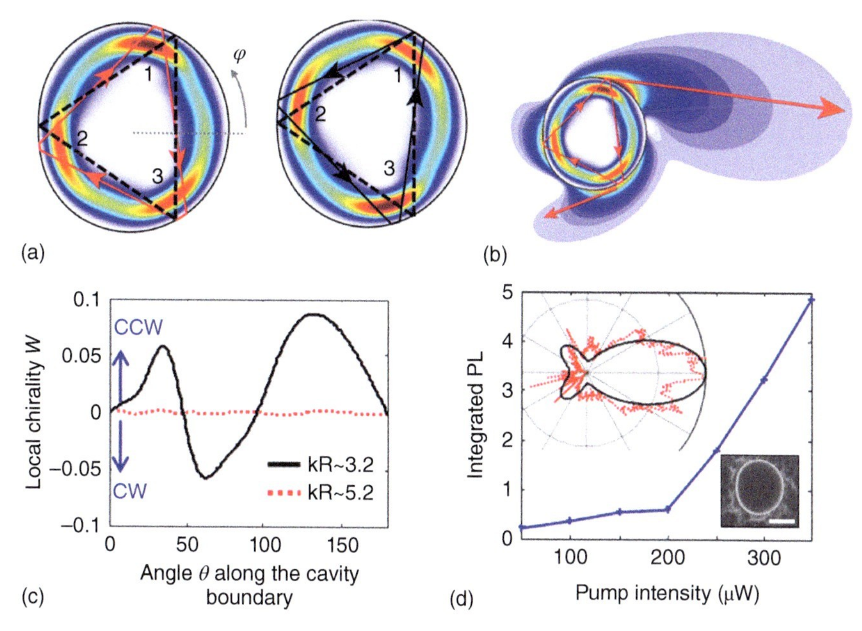

Figure 4.15 (a) Calculated spatial intensity distribution of the CW (left) and CCW (right) waves for a high Q mode at $kR \approx 3.2$ in a limaçon cavity with $\varepsilon = 0.41$. (b) Enhanced CW wave intensity in the outside for showing directional emission. (c) Local chirality W for the modes at $kR \approx 3.2$ (solid line) and $kR \approx 5.2$ (dashed line). (d) Intensity of the emission peak at 875 nm as a function of the incident pump power; Inset: (top)FFP of laser emission (thin dotted line) agreeing well with the calculated results (thick solid line); (bottom) SEM image of limaçon cavity of $R = 460$ nm. Source: Reproduced from Redding et al. [26] with permission of American Physical Society.

In the short wavelength limit ($kR \gg 1$), the ray model is appropriate for designing deformed microcavity with directional or unidirectional emission, but is failure for the wavelength-scale deformed microcavity with $kR \approx 1$, where the mode spacing is large and the deformation can be regarded as a weak perturbation. The perturbation theory was proposed as a powerful tool to analyze the optical modes in deformed microcavities [27, 28]. The deformed microcavity with a boundary defined as

$$r = R(1 + \delta f(\varphi)), \tag{4.20}$$

in polar coordinates, where R is the radius of the unperturbed circular microcavity, and δ is perturbation parameter. Dubertrand et al. conducted the perturbation theory for the microcavity with mirror symmetry of $f(\varphi) = f(-\varphi)$, and the doubly degenerate WGMs will split into two standing-wave modes with even and odd symmetry to the $\varphi = 0$ plane [27]. The perturbation theory has been applied successfully to compute mode frequencies, Q factors, and FFPs of different kinds of deformed microcavities. Ge et al. proved that the perturbation theory was very useful to understand the extreme sensitivity of the FFP with subwavelength boundary deformations, which was caused by a strong mixing of nearly degenerate cavity resonances with different angular momenta [28]. With the perturbation theory, Kraft and Wiersig solved the inverse problem for realizing unidirectional light emission from weakly deformed microcavities [29]. With a given data of far-field profiles, the best-fitting shape of the microcavity could be determined and the optimal solution was unique for specified mode numbers and refractive index. Thus, the optical microcavity devices with a certain needed FFP can be designed based on the perturbation theory.

4.4.4 Waveguide-Coupled Chaotic Microcavity Lasers

For the on-chip application of chaotic microcavity lasers, waveguide coupling is necessary. The traditional evanescent wave coupling requires precise alignment. Attaching a waveguide directly to a nondeformed circular microcavity can realize efficient output with high Q reconstructed modes due to the mode coupling between different radial-order modes, as the waveguide also working as a perturbation dramatically affects the mode-field distributions [30]. In the polygonal microcavities, the mode fields distribute nonuniformly along the cavity boundary, and a waveguide attached to the position with weak mode field can couple the laser light out without significant Q spoiling [31].

For the moderately deformed microcavities with mixed-phase space, high Q modes are typically confined in the regular islands around the stable periodic orbits. In classical ray optics, the light rays of these modes are confined perfectly by total internal reflection and cannot leak out. The light emission is explained by the wave optics as the evanescent escaping due to the curvature of the boundary and the refractive escaping originated from regular-to-chaos tunneling. Song et al. reported a robust mechanism termed "chaos-assisted channeling" to achieve waveguide output from chaotic microcavities with high Q resonances [12].

Figure 4.16a shows the Poincaré SOS for a quadrupolar-shaped microcavity with $\varepsilon = 0.08$, where an attached waveguide introduces a vertical exit window in the phase space. This exit window seriously spoils the Q factor of WGMs with the ray trajectories approaching the waveguide, but only mildly influences the Q factor of the modes confined in the islands around the period-4 orbits. The first 800 bounces of a chaotic trajectory outside the period-4 islands are indicated by the blue dots, and its real space representation is shown in Figure 4.16b. Light in a high Q mode undergoes chaotic motion once it tunnels from the regular islands into the neighboring chaotic region. And then the emission light is efficiently collected by the attached waveguide due to the "chaos-assisted channeling," as the chaotic rays diffuse laterally to the exit window as illustrated in Figure 4.16a instead of vertically down to the critical line. The inset of Figure 4.16c shows the top-view SEM image of a waveguide-coupled GaAs quadrupolar microcavity with $R = 3\,\mu\text{m}$. Figure 4.16c shows the measured laser spectrum, where three lasing peaks at wavelengths of 951, 963, and 977.5 nm are observed. The FFPs of these lasing modes are recorded by fabricating a large ring structure enclosing the device. Figure 4.16d shows the measured FFPs of the corresponding lasing modes in Figure 4.16c. The inset of Figure 4.16d shows the top-view microscope image of mode 1, where the dark regions indicate high intensity. The results show that all lasing modes produce similar outputs along the waveguide.

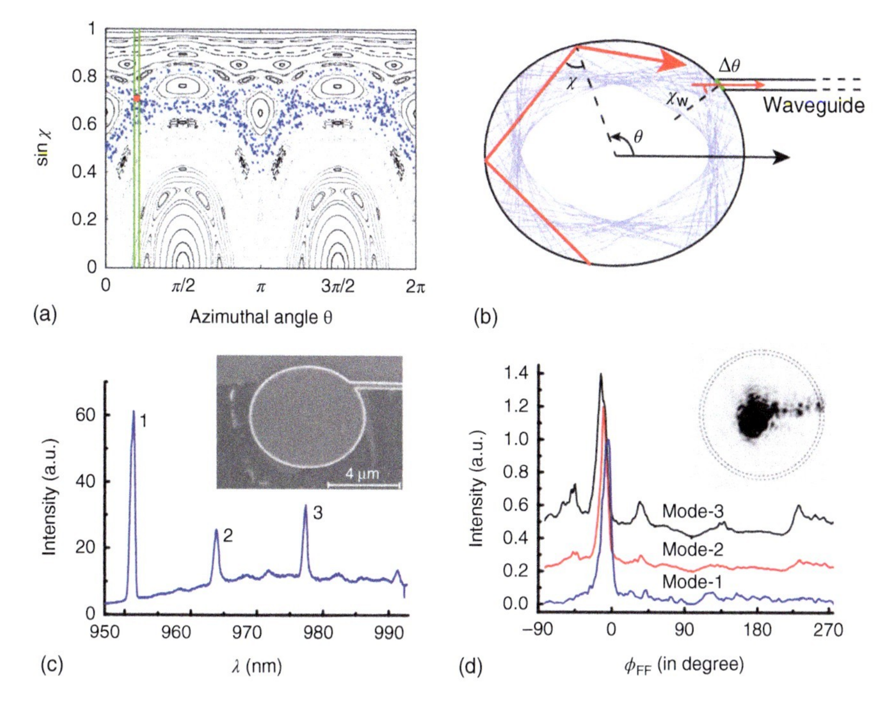

Figure 4.16 (a) Poincaré SOS for a quadrupolar-shaped microcavity with $\varepsilon = 0.08$; vertical lines mark the exit window due to the attached waveguide; blue dots indicate a typical chaotic trajectory out of the period-4 islands. (b) Schematic of a waveguide-coupled quadrupolar-shaped microcavity, and the real space representation of the chaotic ray trajectory around the period-4 islands. (c) Measured laser spectrum; inset: top-view SEM image. (d) Measured FFPs of the three lasing modes shown in (c); inset: top-view microscope image of mode 1. Source: Reproduced from Song et al. [12] with permission of American Physical Society.

4.5 Summary

In the circular microdisks, WGMs are confined by the total internal reflection on the cavity boundary with ultrahigh Q factors. However, the circularly rotational symmetry results in isotropic emission of the laser light limiting the application of the microcavity lasers. Thus, deformed and chaotic microcavity lasers have been proposed and demonstrated for realizing directional emission. In this chapter, some results of deformed and chaotic microcavity semiconductor lasers have been reviewed.

Introducing a local defect to the cavity boundary is a natural idea to improve the output directionality. It has been demonstrated that a deformed elliptical microcavity with a carefully designed notch allows for unidirectional emission with low divergence angle. In addition to the local perturbation, the microcavities with smoothly deformed boundary have been studied intensively for directional or unidirectional emission.

The modes characteristics of the deformed microcavities can be well predicted by the ray optics under the short wavelength limit ($kR \gg 1$). In these microcavities, the directionality is usually universal among different modes as the directional emission is obtained from manifolds in the leakage region by the statistical model. The limaçon microcavity has been demonstrated as a robust laser cavity design for realizing unidirectional emission even in the case of multiple modes lasing. For the wavelength-scale microcavities, the ray model is failure as the wave optics dominates the mode properties. Mode coupling between different modes is then essential to control the Q factors and directionality of the modes inside microcavity. Unidirectional emission can be obtained from the wavelength-scale microcavity laser by designing the cavity parameters.

For the application of microcavity lasers in on-chip photonic integration, waveguide coupling is important. By attaching an output waveguide to the deformed microcavity, high-efficiency waveguide coupling has been demonstrated. The waveguide direct connection enables a robust design for realizing waveguide unidirectional emission, which is essential for the high-density on-chip photonic integration of multiple functional devices.

References

1 Miller, D.A.B. (2009). Device requirements for optical interconnects to silicon chips. *Proc. IEEE* 97: 1166–1185.

2 Roelkens, G., Liu, L., Liang, D. et al. (2010). III-V/silicon photonics for on-chip and intra-chip optical interconnects. *Laser Photonics Rev.* 4: 751–779.

3 Mccall, S.L., Levi, A.F.J., Slusher, R.E. et al. (1992). Whispering-gallery mode microdisk lasers. *Appl. Phys. Lett.* 60: 289–291.

4 He, L., Özdemir, Ş.K., and Yang, L. (2013). Whispering gallery microcavity lasers. *Laser Photonics Rev.* 7: 60–82.

5 Nöckel, J.U. and Stone, A.D. (1997). Ray and wave chaos in asymmetric resonant cavities. *Nature* 385: 45–47.

6 Gmachl, C., Capasso, F., Narimanov, E.E. et al. (1998). High-power directional emission from microlasers with chaotic resonators. *Science* 280: 1556–1564.

7 Cao, H. and Wiersig, J. (2015). Dielectric microcavities: model systems for wave chaos and non-Hermitian physics. *Rev. Mod. Phys.* 87: 61–111.

8 Levi, A.F.J., Slusher, R.E., Mccall, S.L. et al. (1993). Directional light coupling from microdisk lasers. *Appl. Phys. Lett.* 62: 561–563.

9 Nockel, J.U., Stone, A.D., Chen, G. et al. (1996). Directional emission from asymmetric resonant cavities. *Opt. Lett.* 21: 1609–1611.

10 Chern, G.D., Tureci, H.E., Stone, A.D. et al. (2003). Unidirectional lasing from InGaN multiple-quantum-well spiral-shaped micropillars. *Appl. Phys. Lett.* 83: 1710–1712.

11 Wiersig, J. and Hentschel, M. (2008). Combining directional light output and ultralow loss in deformed microdisks. *Phys. Rev. Lett.* 100: 033901.

12 Song, Q.H., Ge, L., Redding, B., and Cao, H. (2012). Channeling chaotic rays into waveguides for efficient collection of microcavity emission. *Phys. Rev. Lett.* 108: 243902.
13 Yang, Y.D., Zhang, Y., Huang, Y.Z., and Poon, A.W. (2014). Direct-modulated waveguide-coupled microspiral disk lasers with spatially selective injection for on-chip optical interconnects. *Opt. Express* 22: 824–838.
14 Levi, A.F.J., Slusher, R.E., Mccall, S.L. et al. (1992). Room-temperature operation of microdisk lasers with submilliamp threshold current. *Electron. Lett.* 28: 1010–1012.
15 Levi, A.F.J., Slusher, R.E., Mccall, S.L. et al. (1993). Room-temperature lasing action in $In0_{.51}Ga_{0.49}P/In_{0.2}Ga_{0.8}As$ microcylinder laser-diodes. *Appl. Phys. Lett.* 62: 2021–2023.
16 Chu, D.Y., Chin, M.K., Bi, W.G. et al. (1994). Double-disk structure for output coupling in microdisk lasers. *Appl. Phys. Lett.* 65: 3167–3169.
17 Backes, S.A., Cleaver, J.R.A., Heberle, A.P., and Kohler, K. (1998). Microdisk laser structures for mode control and directional emission. *J. Vac. Sci. Technol., B* 16: 3817–3820.
18 Boriskina, S.V., Benson, T.M., Sewell, P., and Nosich, A.I. (2006). Q factor and emission pattern control of the WG modes in notched microdisk resonators. *IEEE J. Sel. Top. Quantum Electron.* 12: 52–58.
19 Wang, Q.J., Yan, C.L., Yu, N.F. et al. (2010). Whispering-gallery mode resonators for highly unidirectional laser action. *Proc. Natl. Acad. Sci. U.S.A.* 107: 22407–22412.
20 Lee, S.Y., Rim, S., Ryu, J.W. et al. (2004). Quasiscarred resonances in a spiral-shaped microcavity. *Phys. Rev. Lett.* 93: 164102.
21 Wiersig, J., Kim, S.W., and Hentschel, M. (2008). Asymmetric scattering and nonorthogonal mode patterns in optical microspirals. *Phys. Rev. A* 78: 053509.
22 Sui, S.S., Tang, M.Y., Yang, Y.D. et al. (2015). Hybrid spiral-ring microlaser vertically coupled to silicon waveguide for stable and unidirectional output. *Opt. Lett.* 40: 4995–4998.
23 Yan, C.L., Wang, Q.J., Diehl, L. et al. (2009). Directional emission and universal far-field behavior from semiconductor lasers with limaccedilon-shaped microcavity. *Appl. Phys. Lett.* 94: 251101.
24 Shim, J.B., Wiersig, J., and Cao, H. (2011). Whispering gallery modes formed by partial barriers in ultrasmall deformed microdisks. *Phys. Rev. E* 84: 035202.
25 Song, Q.H., Ge, L., Stone, A.D. et al. (2010). Directional laser emission from a wavelength-scale chaotic microcavity. *Phys. Rev. Lett.* 105: 103902.
26 Redding, B., Ge, L., Song, Q.H. et al. (2012). Local chirality of optical resonances in ultrasmall resonators. *Phys. Rev. Lett.* 108: 253902.
27 Dubertrand, R., Bogomolny, E., Djellali, N. et al. (2008). Circular dielectric cavity and its deformations. *Phys. Rev. A* 77: 013804.
28 Ge, L., Song, Q.H., Redding, B., and Cao, H. (2013). Extreme output sensitivity to subwavelength boundary deformation in microcavities. *Phys. Rev. A* 87: 023833.
29 Kraft, M. and Wiersig, J. (2016). Inverse problem for light emission from weakly deformed microdisk cavities. *Phys. Rev. A* 94: 013851.

30 Yang, Y.D., Wang, S.J., and Huang, Y.Z. (2009). Investigation of mode coupling in a microdisk resonator for realizing directional emission. *Opt. Express* 17: 23010–23015.

31 Yang, Y.D., Huang, Y.Z., Che, K.J. et al. (2009). Equilateral-triangle and square resonator semiconductor microlasers. *IEEE J. Sel. Top. Quantum Electron.* 15: 879–884.

5

Unidirectional Emission Microdisk Lasers

5.1 Introduction

Semiconductor whispering-gallery-mode (WGM) microdisk lasers are suitable for realizing low-threshold operation due to their compact sizes and high quality (Q) factors [1]. However, directional emission and output power are greatly limited by the rotational symmetries of the circular microdisks. Various microdisk lasers with deformed geometries [2–6] or evanescently coupled waveguide [7, 8] have been demonstrated to realize efficient directional emission. In addition, polygonal microcavities can also support high Q WGMs with nonuniform mode-field patterns along the cavity perimeter, and directly connecting a waveguide to the position with weak mode fields was used for realizing waveguide output [9, 10].

The mode fields of WGMs in a circular microdisk distribute uniformly along the cavity perimeter. One can expect that the mode Q factors should degrade dramatically by directly connecting a waveguide to the cavity boundary, because the light rays of the WGMs will approach the exit window of the output waveguide once a period results in a strong leakage loss. However, the connected waveguide destroys the circular symmetry of the microdisk and leads to mode coupling between different radial-order WGMs with close wavelengths, which results in high Q-coupled modes with field distributions localized along simple geometric structures preventing the leakage through the connected waveguide [11]. Semiconductor microdisk lasers connecting one or multiple waveguides have been demonstrated experimentally for realizing directional or unidirectional emission [12, 13]. In this chapter, unidirectional emission semiconductor microdisk lasers with waveguide coupling are reviewed. This chapter is organized as follows. In Section 5.2, a brief overview of mode coupling in waveguide-connected microdisks is given. In Section 5.3, the characteristics of waveguide-connected semiconductor microdisk lasers are presented. In Section 5.4, unidirectional-emission microring lasers are presented and compared with microdisk lasers. In Section 5.5, a locally deformed microring resonator is proposed and applied to realize unidirectional emission hybrid microlaser on silicon. In Section 5.6, wide-angle emission microdisk lasers with a flat side and microdisk lasers with multiport are discussed. Finally, a summary is given in Section 5.7.

Microcavity Semiconductor Lasers: Principles, Design, and Applications, First Edition.
Yong-zhen Huang and Yue-de Yang.

5.2 Mode Coupling in Waveguide-Connected Microdisks

5.2.1 Whispering-Gallery Modes in Circular Microdisks

The analytical model for the WGMs in two-dimensional (2D) circular microdisk is given in Chapter 4. For the microdisk with a radius of R and a refractive index of n_1 surrounded by the media with a refractive index of n_2, the eigenequation for the WGMs is obtained based on the continuous condition at the cavity boundary as

$$J_\nu(n_1 kR)H_\nu^{(1)\prime}(n_2 kR) - \eta J_\nu'(n_1 kR)H_\nu^{(1)}(n_2 kR) = 0, \tag{5.1}$$

where the time dependence $\exp(-i\omega t)$ is omitted, ν is the angular mode number, J_ν and $H_\nu^{(1)}$ are the Bessel function and the Hankel function of the first kind, $k = \omega(\mu_0\varepsilon_0)^{1/2} = \omega/c$ is the wavenumber in vacuum, and η equals n_2/n_1 and n_1/n_2 for the transverse-electric (TE) and transverse-magnetic (TM) modes, respectively. The TE and TM WGMs are denoted as $\mathrm{TE}_{\nu,m}$ and $\mathrm{TM}_{\nu,m}$, and corresponding z-direction magnetic field H_z and electric field E_z are expressed as $J_\nu\,(n_1 kr)$ and $H_\nu^{(1)}(n_2 kr)$ in the internal and external regions of the microdisk. The Hankel function $H_\nu^{(1)}(kn_2 r)$ represents the outgoing wave with the time dependence $\exp(-i\omega t)$, and is divergent as $r \to \infty$ due to the imaginary part of k. By solving the eigen Eq. (5.1) numerically with given angular mode number ν and radial mode number m, the wavenumber k of the WGMs can be obtained as a complex number. The resonance wavelengths and Q factors are obtained from the real and imaginary parts of k as

$$\lambda = \frac{2\pi}{\mathrm{Re}(k)}, \tag{5.2}$$

$$Q = -\frac{\mathrm{Re}(k)}{2\mathrm{Im}(k)}. \tag{5.3}$$

In the nondeformed circular microdisk, the WGMs are doubly degenerate modes when $\nu \neq 0$, and can be represented as clockwise (CW) and counter clockwise (CCW) traveling-wave modes with the angular dependence of $\exp(-i\nu\varphi)$ and $\exp(i\nu\varphi)$ owing to the circularly rotational symmetry, or symmetric and antisymmetric standing-wave modes with the angular dependence of $\cos(\nu\varphi)$ and $\sin(\nu\varphi)$ due to the mirror symmetry, where ν is restricted to be a positive value and φ is the azimuthal angle. To realize unidirectional emission, the circular symmetry of the microdisk should be broken and the mode degeneracy will also be removed. If the microdisk still has a mirror-symmetric plane, the WGMs will split into two nearly degenerate standing-wave modes with symmetric and antisymmetric F_z field distributions relative to the plane. In this section, the standing-wave representations of the WGMs are considered for the microdisk with mirror symmetry. The field distributions of the symmetric WGMs in the polar coordinate can be expressed as

$$F_z = \begin{cases} \dfrac{J_\nu(n_1 kr)}{J_\nu(n_1 kR)}\cos(\nu\varphi) & r < R \\[2ex] \dfrac{H_\nu^{(1)}(n_2 kr)}{H_\nu^{(1)}(n_2 kR)}\cos(\nu\varphi) & r > R \end{cases}, \tag{5.4}$$

with the amplitude of the field F_z normalized to 1 at the boundary. There are many WGMs confined in the microdisk, and some of them may have close wavelength. For two WGMs with the angular mode numbers of ν_1 and ν_2, the superposition of the field along the boundary can be written as

$$\cos(\nu_1\varphi) + \cos(\nu_2\varphi) = 2\cos\left(\frac{\nu_1+\nu_2}{2}\varphi\right)\cos\left(\frac{\nu_1-\nu_2}{2}\varphi\right), \tag{5.5}$$

by simply assuming the same amplitude for the two WGMs. If $\nu_1, \nu_2 \gg |\nu_1 - \nu_2|/2$ is satisfied, the field described in Eq. (5.5) is modulated by slowly varying function of $\cos[(\nu_1 - \nu_2)\varphi/2]$, and the amplitude shows a cosine function–modulated envelope with a node (antinode) number of $|\nu_1 - \nu_2|$. Similar envelope is expected for the antisymmetric mode.

For a 2D microdisk with a diameter of 4.5 μm and a refractive index of 3.2 surrounded by air ($n_1 = 3.2$, $n_2 = 1$), the mode wavelengths (Q factors) of $TM_{18,3}$ and $TM_{15,4}$ modes obtained from Eq. (5.1) are 1489.5 and 1489.4 nm (6.05×10^7 and 8.29×10^4), respectively. When two WGMs have nearly the same mode wavelength, their mode superposition can result in localization of mode field along simple geometric structures such as regular polygons (or stars if the difference between the radial mode numbers is greater than 1). The side number of the polygon is equal to the difference between the angular mode numbers of the two WGMs. The triangular-shaped field patterns constructed by the superposition of $TM_{18,3}$ and $TM_{15,4}$ with the same amplitude and the phase differences of π and 0 are plotted in Figure 5.1a,b, respectively, for the symmetric modes relative to the horizontal middle line. However, the phase difference between the two WGMs is not a constant value as they do not have exactly the same mode wavelength, so the superposition-mode distributions are not invariable. Furthermore, mode superposition only describes the superposition of the field distributions, and the two modes still have different wavelengths and Q factors.

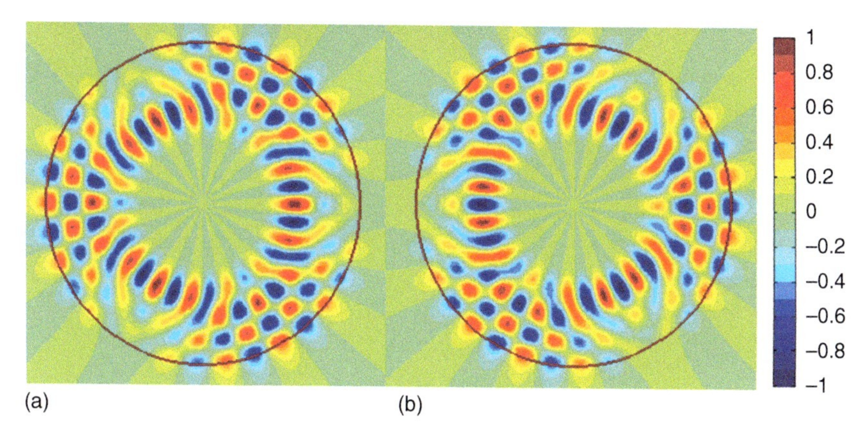

Figure 5.1 The triangular-shaped mode-field patterns constructed by the superposition of $TM_{18,3}$ and $TM_{15,4}$ with phase difference of (a) π and (b) 0 in the perfect microdisk. Source: Yang et al. [11]. © 2009, Optical Society of America.

5.2.2 Mode Coupling in Waveguide-Connected Microdisks

The mode coupling between the WGMs with different angular mode number is forbidden in a nondeformed circular microdisk. However, for the microdisk with an output waveguide directly connected to the cavity boundary in radial direction as shown in the inset of Figure 5.2a, the mode coupling between two modes with close wavelengths occurs due to the break of the symmetry [11]. Different from the mode superposition, the mode coupling will result in two new coupled modes. A 4.5-μm-diameter microdisk with a refractive index of 3.2 and a 0.6-μm-wide output waveguide surrounded by air is simulated by finite-difference time-domain (FDTD) method [14]. In the simulation, the uniform mesh cell size is 10 nm and time step is Courant limit. A cosine impulse modulated by a Gaussian function $P(x_0, y_0, t) = \exp[-(t - t_0)^2/t_w^2] \cos(2\pi ft)$ is used as an exciting source, where t_0 and t_w are the times of the pulse center and the pulse half width, respectively, and f is the center frequency of the pulse. In the waveguide-connected microdisk, mirror symmetry relative to the middle line of the output waveguide still maintains; thus, the modes can be divided into two groups based on their symmetries. Symmetric or antisymmetric exciting sources are used to simulate the modes of different symmetries independently. The perfect matched layer (PML)-absorbing boundary condition is used as the boundaries to terminate the FDTD computation window. 2^{18}-step FDTD simulation is performed with an impulse at $f = 200$ THz, $t_w = 29\Delta t$, and $t_0 = 3t_w$, and the time variation of field is recorded as an FDTD output. The Padé approximation with Baker's algorithm [15] is used to transform the last 2^{15}-step FDTD output from the time domain to the frequency domain.

The obtained intensity spectra for TM modes are plotted in Figure 5.2a as the solid and dashed lines for the symmetric and antisymmetric modes relative to the x-axis, respectively. The intensity spectrum for the TM modes in the corresponding microdisk without the output waveguide is also calculated and plotted in Figure 5.2b with the detail of spectrum from 1490 to 1490.4 nm in the inset. The modes with wavelength difference less than 5 nm are marked by circles in Figure 5.2b, which result in the coupled modes in Figure 5.2a. All of the WGMs with radial mode number $\nu < 4$ appear in the spectrum of Figure 5.2b, and their Q factors are larger than 10^4. However, only the coupled modes can keep high Q factors in the microdisk connecting an output waveguide and appear in Figure 5.2a. The Q factors of 9.1×10^3 and 2.8×10^4 are obtained for the symmetric and antisymmetric modes at the wavelength of 1490 nm in Figure 5.2a, which correspond to the coupled modes between $TM_{18,3}$ and $TM_{15,4}$ with the wavelength difference of 0.1 nm in the nondeformed circular microdisk. The peaks at 1433 and 1552 nm in Figure 5.2a have the Q factors from 1.7×10^3 to 3×10^3, which correspond to the coupled modes between $TM_{17,3}$ and $TM_{14,4}$, and $TM_{19,3}$ and $TM_{16,4}$, respectively. The mode wavelength differences between $TM_{17,3}$ and $TM_{14,4}$, and $TM_{19,3}$ and $TM_{16,4}$ in the nondeformed circular microdisk are 2.9 and 2.6 nm, respectively, which are larger than those between $TM_{18,3}$ and $TM_{15,4}$. The Q factors of 6.5×10^2 and 1.1×10^3 are obtained for the symmetric and antisymmetric modes at the wavelength of 1576 nm, which correspond to the coupled modes between $TM_{20,2}$ and $TM_{11,5}$ with the wavelength difference of

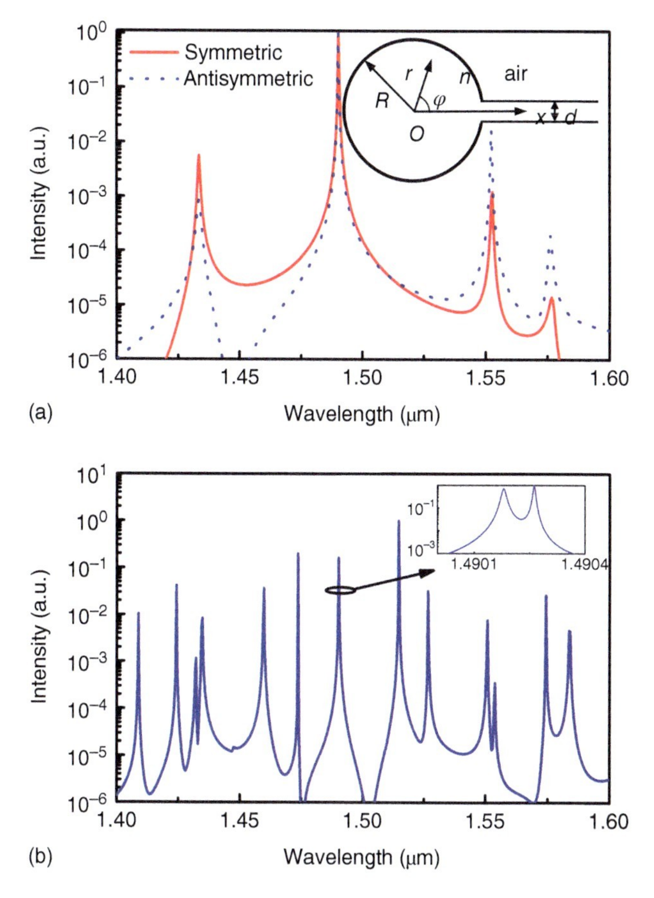

Figure 5.2 The intensity spectra for TM modes obtained by FDTD simulation and Padé approximation for the 4.5-μm-diameter microdisk (a) with a 0.6-μm-wide output waveguide and (b) without the output waveguide, the circles mark the modes for coupling. Source: Yang et al. [11]. © 2009, Optical Society of America.

4.4 nm in the nondeformed circular microdisk. The Q factor of $TM_{11,5}$ is less than 10^3; thus, $TM_{11,5}$ does not appear as a peak in Figure 5.2b.

Using a long-time optical pulse with a very narrow frequency bandwidth covering only one high-Q mode, the field distribution of a specific mode can be obtained by the FDTD simulation. The electric field distributions of the symmetric TM coupled modes are plotted in Figure 5.3a,b for the high Q and low Q modes at the wavelength of 1490 nm with equilateral-triangular mode-field patterns, and in Figure 5.3c,d for the high Q and low Q modes at the wavelength of 1786 nm with square mode-field patterns. The field at the right-side output waveguide is magnified five times for high Q-coupled modes in Figure 5.3a,b. The mode-field patterns in Figure 5.3a,b are similar to the superposition field distribution in Figure 5.1a,b, respectively. The mode coupling of two WGMs results in the equilateral-polygonal shaped mode pattern with the side number equal to the difference of angular mode numbers of the two WGMs, which prevents a strong leak through the waveguide and leads to high Q

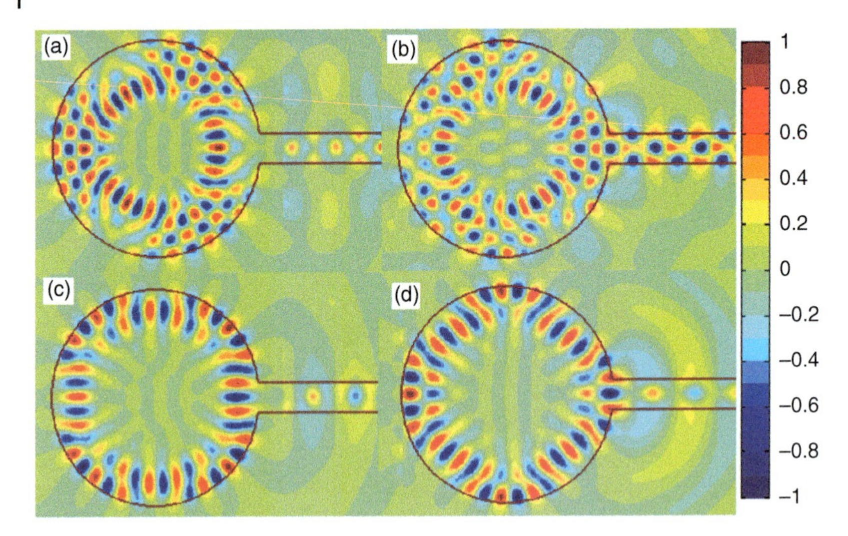

Figure 5.3 The field distributions of (a) high Q and (b) low Q TM coupled modes at wavelength of 1490 nm, and (c) high Q and (d) low Q TM-coupled mode at wavelength of 1786 nm in the 4.5-μm-diameter microdisk with a 0.6-μm-wide output waveguide. The field at the right-side output waveguide is magnified five times for high Q-coupled modes in (a) and (c). Source: Yang et al. [11]. © 2009, Optical Society of America.

factors. The Q factor of the coupled mode in Figure 5.3c at 1786 nm is 2.4×10^4, corresponding to the mode coupling between $TM_{21,1}$ and $TM_{17,2}$ with the wavelength difference of 1.4 nm.

The Q factors obtained by the FDTD simulation can be simply expressed as $1/Q = 1/Q_r + 1/Q_c$ with Q_r and Q_c related to the radiation loss and the output coupling loss, respectively. The output coupling loss is contributed to the directional emission in the output waveguide. The radiation loss typically is much lower than the coupling loss for the WGMs, which indicate high-output coupling efficiency from the output waveguide. However, the mode Q factor will be limited by material absorption coefficient α_i as $Q_A = n_g k_0/\alpha_i$ in a practical microcavity. With the mode group index $n_g = 3.5$ for the semiconductor material, the material absorption limited Q factors are 1.4×10^5 and 1.4×10^4, respectively, as the absorption $\alpha_i = 1$ and $10\,\text{cm}^{-1}$ at the wavelength of 1550 nm. The output coupling efficiency of microcavity lasers can be expressed as $Q_c^{-1}/(Q^{-1} + Q_A^{-1})$. When Q is much smaller than Q_A, the output coupling efficiency is Q/Q_c, which can be calculated as the ratio of the energy flux through the output waveguide to the total emission energy flux of the cavity by the FDTD simulation.

The mode Q factors and the output efficiencies as Q/Q_c vs. the output waveguide width are plotted in Figure 5.4a,b for the symmetric and antisymmetric modes in the 4.5-μm-diameter microdisk. Two coupled modes are marked by symbols S (short wavelength) and L (long wavelength) according to their mode wavelengths. As the width of the output waveguide is zero, the L- and S-coupled modes correspond to $TM_{18,3}$ and $TM_{15,4}$ with the wavelengths of 1489.5 and 1489.4 nm, respectively, in the nondeformed circular microdisk. But the Q factor of the L mode 5×10^5 obtained

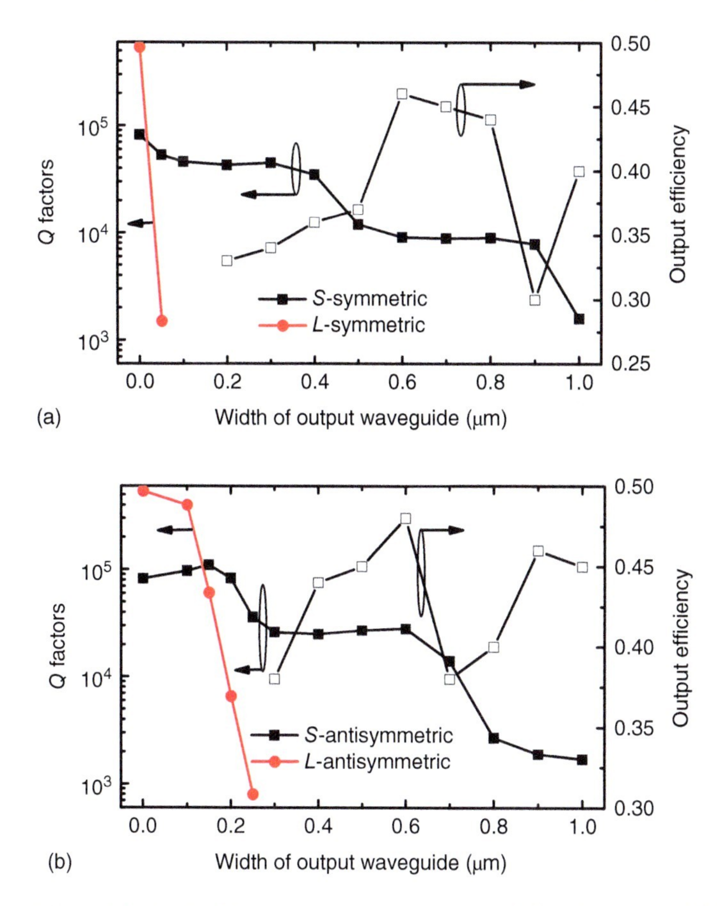

Figure 5.4 Mode Q factors and output coupling efficiencies vs. the width of the output waveguide for (a) symmetric and (b) antisymmetric modes in the 4.5-μm-diameter microdisk, where symbols S and L mark the coupling modes with short and long wavelengths around 1490 nm. Source: Yang et al. [11]. © 2009, Optical Society of America.

by FDTD simulation is of the order of a hundredth of that obtained by Eq. (5.1) for $TM_{18,3}$, and the Q factor of the S mode has the same value as that of $TM_{15,4}$. In fact, exact value of Q factor less than 10^5 can be obtained by the FDTD simulation with the mesh cell size of 10 nm and single precision numbers. The L-coupled modes have a low Q factor and the field distribution similar to Figure 5.3b, which strongly couples with the output waveguide. Similarly, uncoupled WGMs have uniform envelop of the field pattern along the perimeter of the microdisk and low Q factor due to strong coupling with the output waveguide.

To verify the generality of the above results, the TE WGMs in the microdisk with an output waveguide are also investigated. The intensity spectra are calculated by FDTD technique and Padé approximation with the same condition as TM modes. For a 5-μm-diameter microdisk with a 0.8-μm-wide output waveguide, the obtained intensity spectra are plotted in Figure 5.5a as the solid and dotted lines for the symmetric and antisymmetric modes relative to the x-axis, respectively. The intensity spectrum

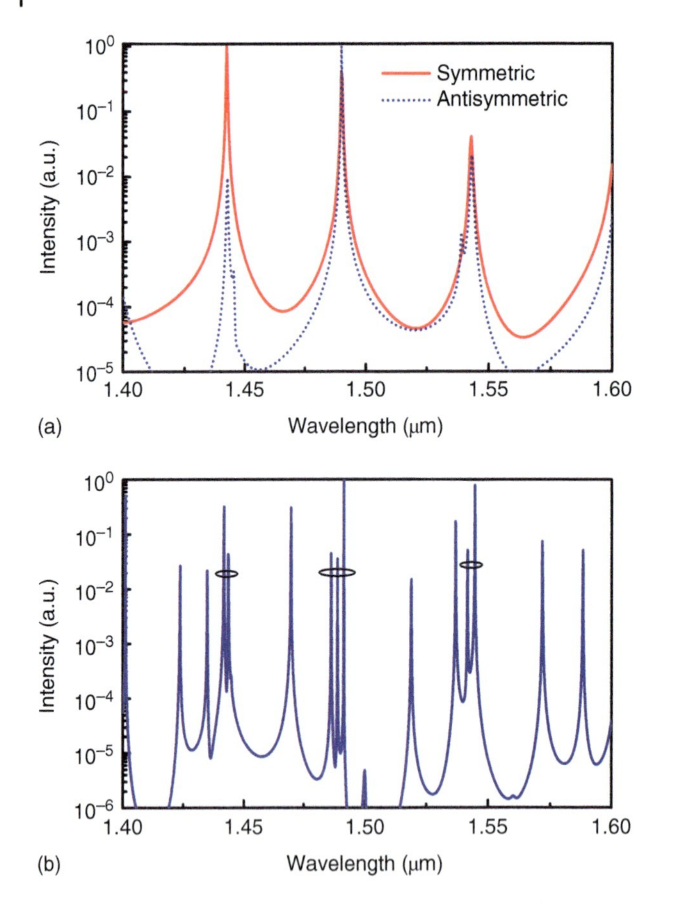

Figure 5.5 The spectra of TE modes obtained by FDTD simulation and Padé approximation for a 5 μm diameter microdisk (a) with a 0.8-μm-wide output waveguide and (b) without the output waveguide, the circles mark the modes for coupling. Source: Yang et al. [11]. © 2009, Optical Society of America.

for TE modes in the corresponding microdisk without the output waveguide is also calculated and plotted in Figure 5.5b, with the modes with wavelength difference less than 5 nm marked by circles. There are many WGMs with radial mode number $\nu < 4$ appearing in Figure 5.5b. Similar to TM modes, only the coupled modes, which are induced by the mode coupling between two modes with almost the same wavelength, can keep high Q factors in the microdisk with the output waveguide and appear in Figure 5.5a.

Mode Q factors of 3.5×10^3 and 8.6×10^3 are obtained for the symmetric and antisymmetric modes at the wavelength of 1490 nm in Figure 5.5a, which correspond to the coupled modes between $TE_{28,1}$ and $TE_{20,3}$. The peaks at 1443 and 1542 nm in Figure 5.5a correspond to the coupled modes between $TE_{29,1}$ and $TE_{21,3}$, and $TE_{19,3}$ and $TE_{16,4}$, respectively, and have the Q factors from 1.6×10^3 to 4.3×10^3. In Figure 5.5b, there is $TE_{17,4}$ with mode wavelength near 1490 nm besides $TE_{28,1}$ and $TE_{20,3}$. The coupled modes between $TE_{28,1}$ and $TE_{17,4}$ can have a Q factor of 10^3

when the width of the output waveguide is 0.6 μm, but do not appear in the intensity spectra of Figure 5.5a as the width of the output waveguide is 0.8 μm. Similar to TM modes, the TE WGMs without mode coupling have a very small Q factor and do not appear in Figure 5.5a.

The mechanism of the enhancement of mode Q factors is the external mode coupling induced by the waveguide loss [16]. The coupling of two-resonance system can be generally understood in terms of a 2×2 Hamiltonian matrix:

$$H = \begin{pmatrix} E_1 & V \\ W & E_2 \end{pmatrix}, \tag{5.6}$$

where E_1 and E_2 are complex energies of the uncoupled system with negative imaginary parts corresponding to finite lifetimes, V and W are coupling coefficients. The eigenvalues of the coupled system $E_\pm$ can be obtained as

$$E_\pm = \frac{E_1 + E_2}{2} \pm \sqrt{\frac{(E_1 - E_2)^2}{4} + VW}. \tag{5.7}$$

The amplitudes of the imaginary parts of $E_\pm$ represent the losses of the coupled resonances, which are inversely proportional to the lifetimes of the eigenstates. In the case of internal coupling where $V = W^*$ and VW is a real positive number, each state is individually coupled to the continuum, and the imaginary parts of the two eigenvalues become closer from that of the uncoupled system. In the more general case of external coupling with $V \neq W^*$, the resonance are coupled through continuum. For the optical modes, Eq. (5.7) can be expressed as

$$\omega_\pm = \frac{\omega_1 + \omega_2}{2} \pm \sqrt{\frac{(\omega_1 - \omega_2)^2}{4} + \kappa_{12}\kappa_{21}}, \tag{5.8}$$

where $\omega_\pm$ and $\omega_{1,2}$ represent the complex mode frequencies after and before mode coupling, respectively, and the product of the coupling coefficients $\kappa_{12}\kappa_{21}$ is a real number for the internal coupling and is a complex number for the more general case of external coupling.

By varying the mode frequencies, different coupling phenomena can be observed with different coupling coefficients. If the uncoupled two modes have exactly the same mode frequency ($\omega_1 = \omega_2$), the external mode coupling leads to splits in both real and imaginary parts of the mode frequency, which means that the two coupled modes will have enhanced and reduced mode Q factors, respectively. It has been shown that external coupling can produce long-lived scar-like modes in the optical microcavities [17]. In the waveguide-connected microdisk, the dominant loss is the waveguide output coupling and scattering localized around the waveguide-connecting area. The interference between two nearby modes in the loss area leads to external mode coupling, and the coupling strength is relative to the losses of the two modes. When the coupling strength is comparable to the frequency difference between the two WGMs, the mode coupling effect will be obvious. High Q-coupled modes correspond to those with destructive interference in the loss area.

In the microdisk with relatively large size, high radial-order WGMs can have high Q factors ($>10^4$), which means there will be many high Q modes in one free spectral range (FSR). For a circular microdisk with a radius of 10 μm, a refractive index of

3.2, and a 2-μm-wide output waveguide, the intensity spectrum covering one FSR is obtained by the FDTD simulation and Padé approximation for the TE modes and plotted in Figure 5.6a. The wavelengths of high Q modes are 1542.3, 1546.6, 1550.3, and 1552.9, and the corresponding mode Q factors are obtained as 1.5×10^4, 1.1×10^5, 6.5×10^4, and 2.3×10^4, respectively [13]. Choosing a narrow-band exciting source centered at the resonant frequency, single mode-field patterns for the high Q modes can be obtained by the FDTD simulation. The field patterns of the symmetric modes at the wavelengths of 1542.3, 1546.6, 1550.3, and 1552.9 nm are simulated and plotted in Figure 5.6b. The field in the output waveguide is magnified 10 times for clarity. The field distributions have the polygonal patterns with the side numbers of 3, 7, 5, and 4, and the corresponding output coupling efficiencies, defined as the ratio of output power confined in the output waveguide to the total radiation power from the resonators, are 0.89, 0.77, 0.92, and 0.79 based on the simulated mode-field patterns.

The analytical wavelengths and Q factors of the WGMs can be obtained from the eigen Eq. (5.1). The mode wavelengths of the TE WGMs in the 10-μm-radius circular microdisk vs. the radial mode numbers are plotted in Figure 5.7 as the open circles for the modes with mode Q factors larger than 10^4 from 1538 to 1552 nm covering one FSR. Four group modes of WGMs $TE_{67,14}$ and $TE_{64,15}$, $TE_{121,1}$ and $TE_{114,2}$, $TE_{108,3}$ and $TE_{103,4}$, and $TE_{94,6}$ and $TE_{90,7}$ with small mode wavelength differences form the high Q-coupled modes in Figure 5.6. The numbers indicated in Figure 5.7 are the corresponding angular mode numbers. And the side number of field pattern in Figure 5.6b is equal to the difference of the angular mode numbers of the corresponding WGMs involved in the mode coupling. As shown in Figure 5.6b, the coupled modes have polygonal field patterns. It can be expected that the coupled modes with triangular and square field patterns can still have high Q factor if three or four ports are symmetrically connected to the microdisk [13].

5.3 Waveguide-Connected Unidirectional Emission Microdisk Lasers

According to the above numerical simulations and theoretical analyses, waveguide-connected microdisks can have high Q coupled modes with the field patterns formed by the superposition of two WGMs with close wavelengths. Mode coupling is a universal phenomenon in the microdisk with broken circular symmetry, especially for the microdisk with relatively larger size as high radial-order WGMs can also have high Q factors. Based on the high Q-coupled modes, one can expect that unidirectional microdisk lasers can be realized by directly connecting an output waveguide to the microdisk.

5.3.1 Lasing Characteristics of Unidirectional Emission Microdisk Lasers

Waveguide-connected microdisk lasers were first fabricated on an AlGaInAs/InP multiple quantum well (MQW) epitaxial wafer with the laser cavity laterally

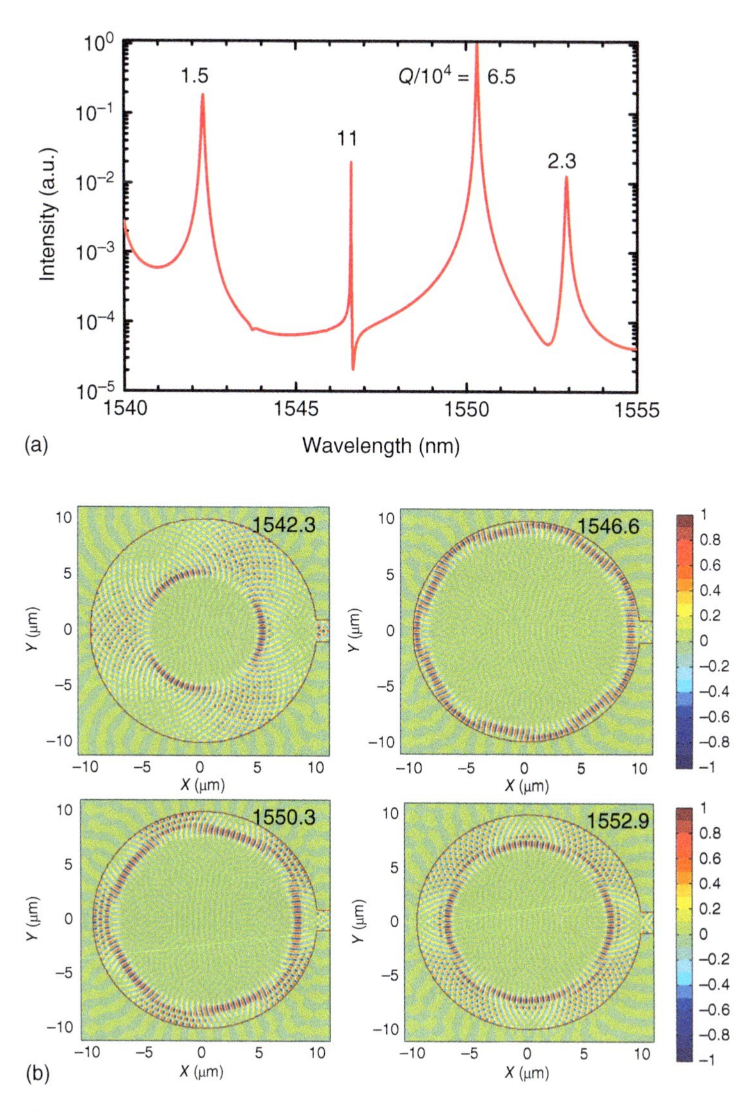

Figure 5.6 (a) The intensity spectrum of TE modes obtained by FDTD simulation and Padé approximation for a 10-μm-radius circular microdisk with a 2-μm-wide output waveguide. (b) The magnetic field distributions of TE modes at wavelengths of 1542.3, 1546.6, 1550.3, and 1552.9 nm obtained by FDTD simulation, with the field at the output waveguide magnified 10 times. Source: Huang et al. [13]. © 2010, IOP Publishing.

confined by the p-electrode metal [12]. An 800-nm SiO_2 was deposited by plasma-enhanced chemical vapor deposition (PECVD) on the laser wafer. The laser cavity patterns were transferred onto the SiO_2 layer using standard contact photolithography and inductive-coupled plasma (ICP) techniques. The patterned SiO_2 layer was used as a hard mask for the following ICP dry-etching process with a total etched depth of about 5 μm. A chemical etching process was then used to improve the smoothness of the etched side walls, and then the residual SiO_2 hard

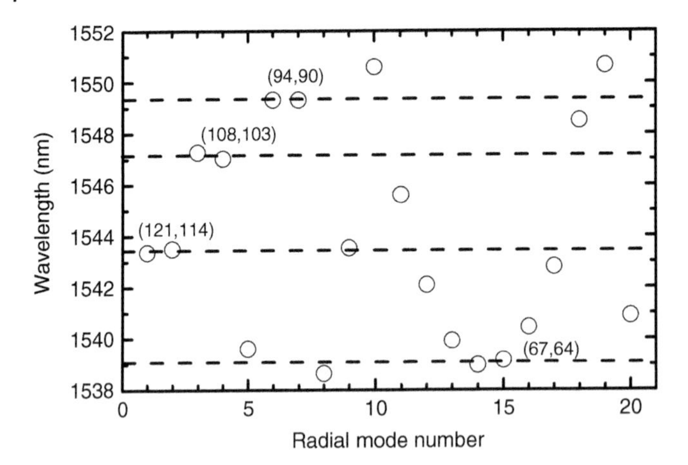

Figure 5.7 Mode wavelengths vs. the radial mode number for high Q TE WGMs in the microdisk with a radius of 10 μm and a refractive index of 3.2. The dashed lines mark the WGMs suitable for mode coupling. Source: Huang et al. [13]. © 2010, IOP Publishing.

masks were removed using diluted HF solution. Finally, a 450-nm SiO_2-insulating layer was deposited on the whole wafer and a contact window was opened on the top of laser cavity using ICP-etching process again, and Ti–Au and Au–Ge–Ni were deposited as p-contact and n-contact metals, respectively.

After cleaving over the output waveguide, the microdisk laser was placed on a Cu sink for the following experiment. The output power and laser spectrum were measured by coupling an optical fiber to the cleaved facet of the output waveguide. The detailed laser spectrum around 1557 nm near the threshold at 8 mA is plotted in Figure 5.8. Fitting the laser spectrum by a three-peak Lorentzian function, we have the mode wavelengths of 1556.07, 1556.84, and 1556.98 nm with the full width at half maximum (FWHM) of 0.172, 0.127, and 0.139 nm, respectively. Mode Q factors of 9.0×10^3, 1.2×10^4, and 1.1×10^4 can be estimated as the ratio of the peak wavelength to the width. The two modes at the wavelengths of 1556.84 and 1556.98 nm can be attributed to the degenerate modes of the symmetric and antisymmetric coupled modes relative to the output waveguide.

For the microdisk lasers laterally confined by p-electrode metal, the large-area electrode results in relatively large capacitance, which limits direct modulation performance of the microdisk laser. In addition, the metal confinement also leads to multiple modes lasing as the light rays with the incident angle below the critical angle can also have high reflectivity on the metal surface. To control the lasing modes efficiently, AlGaInAs/InP microdisk lasers buried in dielectric material of benzocyclobutene (BCB) cladding were fabricated [18, 19]. In the fabrication, a ~200-nm silicon nitride (SiN_x) layer was deposited by PECVD after the ICP etching of InP laser cavity for better adhering to the following spin-coated BCB layer for planarization. The hard-cured BCB film was etched without any mask to exposure the top of the microdisks. Then, a 400-nm SiO_2 layer is deposited on the whole wafer. After that, a 450-nm SiO_2 layer is deposited on the wafer and a contact window is opened

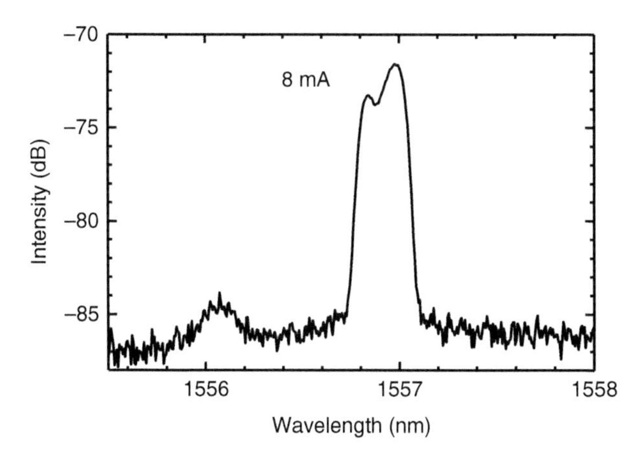

Figure 5.8 Laser spectrum for a microdisk laser with a radius of 10 μm and a 2-μm-wide output waveguide at the injection currents of 8 mA and room temperature. Source: Wang et al. [12]. © 2010, IEEE.

by photolithography and ICP etching for current injection. A Ti/Pt/Au p-electrode is then deposited by e-beam evaporation and lift-off process, forming a patterned p-electrode. The patterned p-electrode can reduce the device capacitance efficiently. The laser wafer is mechanically lapped down to a thickness of ~120 μm, and an Au–Ge–Ni metallization layer is deposited as n-electrode.

After the output waveguide is cleaved to a length of about 10~15 μm, the microdisk laser is bonded p-side up on an AlN submount with a thin-film resistance of about 35 Ω in series and mounted on a thermoelectric cooler (TEC). For the BCB-confined microdisk laser with a radius of 7 μm and a 1.2-μm-wide waveguide, the output powers coupled into a single mode optical fiber at the TEC temperatures of 287, 298, and 313 K are plotted in Figure 5.9a, where the inset shows the schematic diagram of a microdisk laser laterally confined by BCB. The threshold current is about 2.1, 3.0, and 3.9 mA estimated from the first kink position of the curves, and the output powers at 15 mA are 4.1, 3.0, and 1.8 μW measured at heat sink temperatures of 287, 298, and 313 K, respectively. The sharp decrease of the output power around 7 mA at 287 K is caused by mode competition. The lasing spectra measured at injection currents of 2 and 15 mA at the temperature of 287 K are shown in Figure 5.9b. The redshift of the lasing peaks results from the injection current-induced temperature rising. Single-mode operation with a side mode suppression ratio (SMSR) of more than 45 dB at 1542.8 nm is realized at 15 mA. Three peaks with the longitudinal mode intervals of 16.28 and 16.65 nm correspond to the FSR of WGMs in the 7-μm-radius microdisk with a group refractive index of 3.3.

5.3.2 Direct Modulation Characteristics of Unidirectional Emission Microdisk Lasers

With the pattern p-electrode, high-speed direct modulation can be achieved for the single-mode microdisk lasers. The dynamic characteristics are performed for

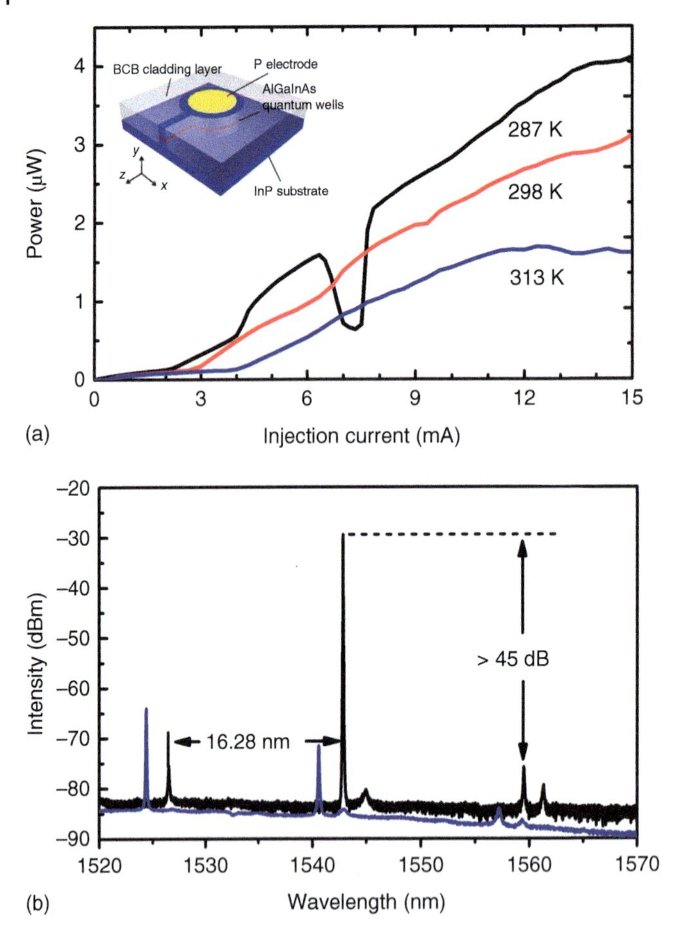

Figure 5.9 (a) Single-mode optical fiber–coupled powers vs. continuous injection current for a microdisk laser with a radius of 7 μm and a 1.2-μm-wide waveguide at 287, 298, and 313 K. Inset: Schematic diagram of a microdisk laser laterally confined by BCB. (b) Laser spectra for the microdisk laser at the injection currents of 2 and 15 mA at 287 K. Source: Lv et al. [20]. © 2013, John Wiley & Sons.

the microdisk lasers with the experimental setup shown in Figure 5.10, where parts (a) and (b) marked by the dashed boxes are used for small-signal response and large-signal modulation measurements, respectively [20]. The bias current was combined with the modulation signal using a high-frequency bias-T and fed to the microdisk laser through a radiofrequency probe. The microlaser output was coupled into a tapered single-mode fiber (SMF) and amplified by about 30 dB using an erbium-doped fiber amplifier (EDFA). Afterwards, the amplified output was spectrally filtered using a tunable bandpass filter, and the main lasing mode was monitored by an ac-coupled high-speed photodetector.

Using the apparatus as shown in Figure 5.10, the small signal response curves for the microdisk laser at a fixed temperature and different biasing currents were measured. The lasing mode around 1543 nm as shown in Figure 5.9b was chosen by

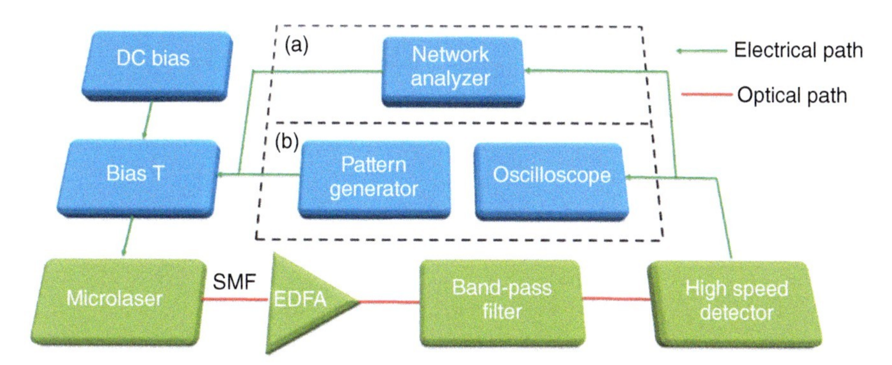

Figure 5.10 Schematic diagram of the apparatus used for small-signal response (a) and large signal response (b) measurements. Source: Lv et al. [20]. © 2013, John Wiley & Sons.

the bandpass filter as the operation mode after amplified by 30 dB using an EDFA. The small signal response curves measured at 10, 13, and 18 mA are plotted as dotted, dashed, and solid lines in Figure 5.11a, respectively. With the increase of biasing current, the resonance frequency gradually increases while the resonance peak is suppressed. Due to the limitation of network analyzer measuring range, the 3-dB bandwidths are obtained as 15.6, 18.7, and over 20 GHz at the bias currents of 10, 13, and 18 mA, respectively. The variations of resonance frequency with the injection current are plotted as the open circles in Figure 5.11b, where the horizontal axis represents the square root of the bias current minus the threshold current and the vertical axis represents the resonance frequency. The results could be fitted by the following linear equation as shown by the solid line in the figure:

$$f_R = \frac{1}{2\pi}\left(\frac{v_g a s_0}{\tau_P}\right)^{1/2} = \frac{1}{2\pi}\left(\frac{v_g a \eta_i}{qV_P}\cdot(I - I_{\text{th}})\right)^{1/2} = D\cdot(I - I_{\text{th}})^{1/2}, \tag{5.9}$$

where I_{th} is the threshold current, a is equal to $\partial g/\partial n$, v_g is the group velocity, τ_p is the mode lifetime, η_i is the injection efficiency, V_p is the mode volume, and D is a quality factor that indicates the efficiency of the modulation speed of an intrinsic laser. A D factor of 4.05 GHz/mA$^{1/2}$ is obtained by fitting Eq. (5.9) as the solid line in Figure 5.11b.

The eye diagrams of large signal modulation for the microdisk laser were measured using the apparatus as shown in Figure 5.10. A 32-Gbit/s pulse pattern generator was used to supply non-return-zero (NRZ) signals with amplitude of 1.8 V, and an EDFA and a tunable band-pass filter were utilized to amplify and choose the dominant mode around 1543 nm before the light was transmitted into a 20-GHz digital sampling oscilloscope. With the bias current of 21 mA and the TEC temperature of 287 K, the measured eye diagrams at 20, 25, and 30 Gbit/s are shown in Figure 5.12. Extinction ratios of 5.08, 4.79, and 4.05 dB are obtained for the eye diagram of 20, 25, and 30 Gbit/s, respectively. The clear eye diagrams could still be observed at 25 and 30 Gbit/s.

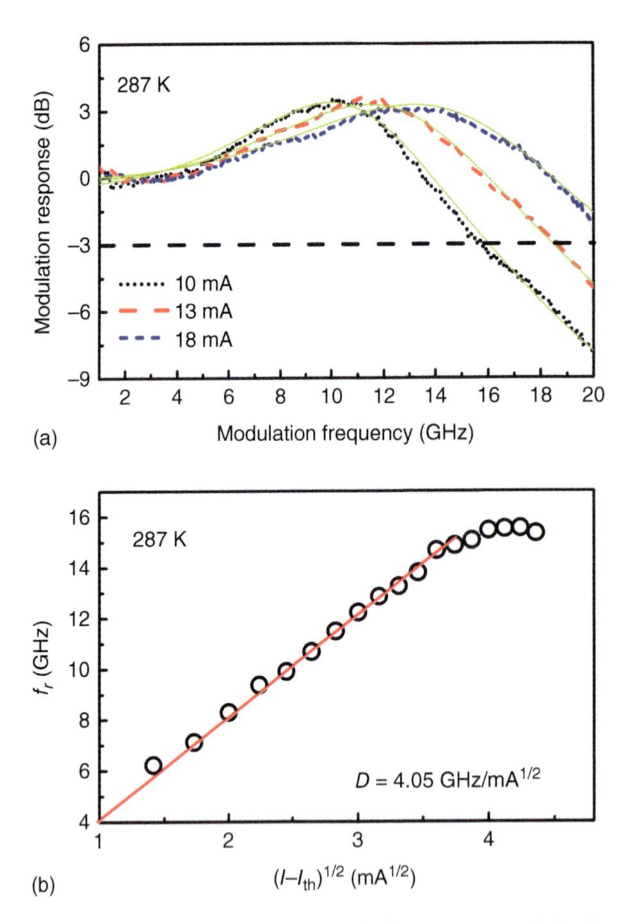

Figure 5.11 (a) Small-signal modulation responses of the 7-μm radius microdisk laser at 287 K and biasing currents of 10, 13, and 18 mA. (b) Resonance frequency vs. square root of the bias current minus the threshold current. Source: Zou et al. [19]. © 2015, Optical Society of America.

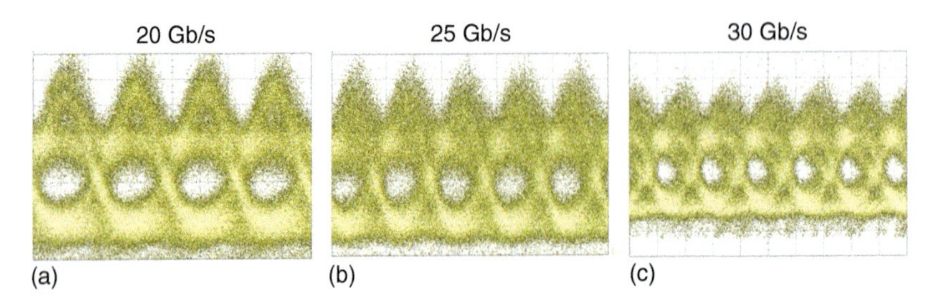

Figure 5.12 Eye diagrams at the modulation bit rates of 20, 25, and 30 Gbit/s for the microdisk laser with a side length of 7 μm and a 1.2-μm-width vertex output waveguide at 287 K. Source: Zou et al. [19]. © 2015, Optical Society of America.

5.4 Unidirectional Emission Microring Lasers

Compared with the microdisk, microring has more compact volume of the active region and better current injection efficiency as the WGMs mainly distribute in the rim region. The inner wall of the microring can bring in an additional loss and even suppress higher radial-order modes. Thus, the unidirectional emission microring lasers with an output waveguide were designed and demonstrated experimentally [21, 22].

In the waveguide-connected microdisk with a refractive index of 3.2, a radius of 10 μm, and a waveguide width of 2 μm laterally confined by the BCB with a refractive index of 1.54, there are several groups of high Q-coupled modes in one FSR. The mode intensity distributions of $|H_z|^2$ for the antisymmetric TE modes at 1548.9, 1551.4, 1553.0, 1556.4, and 1559.1 nm obtained by FDTD simulation are shown in Figure 5.13a–e, respectively. The distributions on the right side of the external region, including the output waveguide, are magnified to clarify. Compared with the air-confined microdisk shown in Figure 5.6, the triangular-shaped coupled mode cannot maintain high Q factor in the BCB-confined microdisk as the corresponding incident angle of the coupled mode (~30°) is close to the critical angle of total reflection (~28.8°). In addition, high-order coupled modes with star shapes and Q factors of several thousand are presented in Figure 5.13.

The coupled modes are formed between WGMs with close-mode wavelengths. Based on the definition for periodic orbits in microdisks, two parameters (p, q) are applied to describe the coupled modes, where p is the number of bouncing points on the boundary of a closed trajectory, and q is determined by the chord angle Θ of successive bounces, as shown in Figure 5.13f. The p-bouncing points divide the circumferential angle into p parts of $2\pi/p$, and q is the quotient of Θ divided by $2\pi/p$, i.e. $q = \Theta/(2\pi/p)$. Therefore, the mode patterns in Figure 5.13a–e are defined as (7, 1), (10, 3), (5, 1), (4, 1), and (7, 2), respectively, with the mode Q factors of 3.6×10^4, 9.0×10^3, 5.4×10^4, 1.2×10^4, and 8.3×10^3. The two parameters p and q can also be understood by the differences in the angular and radial mode numbers for the two WGMs involving mode coupling. The output efficiencies η, defined as the ratio of the energy flux through the cross section of the output waveguide to the total external energy flux, are 52.3%, 52.4%, 45.9%, 37.9%, and 48.3%. Furthermore, the mode Q factors of 6.0×10^4, 1.0×10^4, 1.1×10^5, 1.2×10^4, and 7.7×10^3 and the output efficiencies of 44.1%, 53.6%, 44.9%, 46.1%, and 55.9% are obtained for the corresponding symmetric coupled modes at the wavelengths of 1548.9, 1551.4, 1553.0, 1556.4, and 1559.1 nm, respectively.

The mode Q factors of the coupled modes in Figure 5.13 vs. the inner radius r of microrings are plotted in Figure 5.14. The radius of outer radius R is kept at 10 μm. The mode Q factors almost keep constant as r is less than 4.5 μm due to their nearly zero field distributions in the central region. Then, the fluctuations of Q factors are observed and the five categories of transverse modes show cut-off phenomenon separately as r is larger than 8.3, 7.25, 6.65, 6.25, and 5.5 μm as shown by the dashed lines in Figure 5.14. Based on the distance of the mode light ray from the center of the microring $R\cos(\Theta/2)$, the cutoff values of r are calculated to be 9.01, 8.09 7.07,

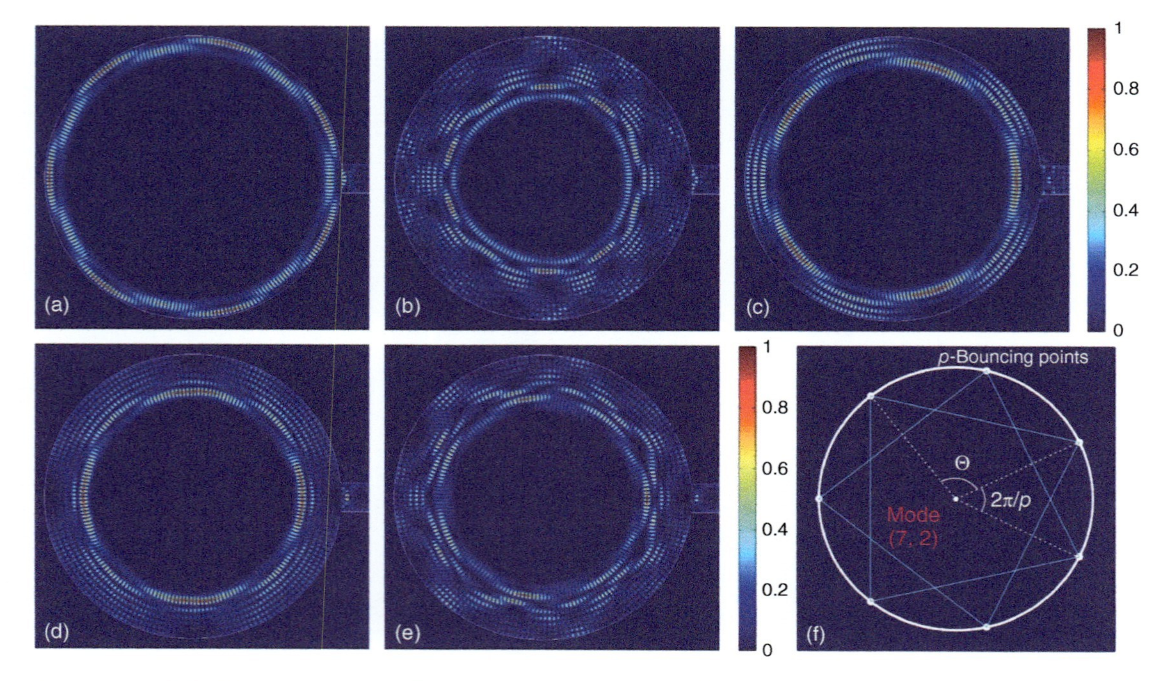

Figure 5.13 Intensity patterns of $|H_z|^2$ for antisymmetric TE modes with mode indices of (a) (7, 1), (b) (10, 3), (c) (5, 1), (d) (4, 1), and (e) (7, 2) in the microdisk with a radius of 10 μm connecting a 2-μm-wide output waveguide. (f) Schematic diagram of the definition of mode indices (p, q). Source: Ma et al. [22]. © 2014, Optical Society of America.

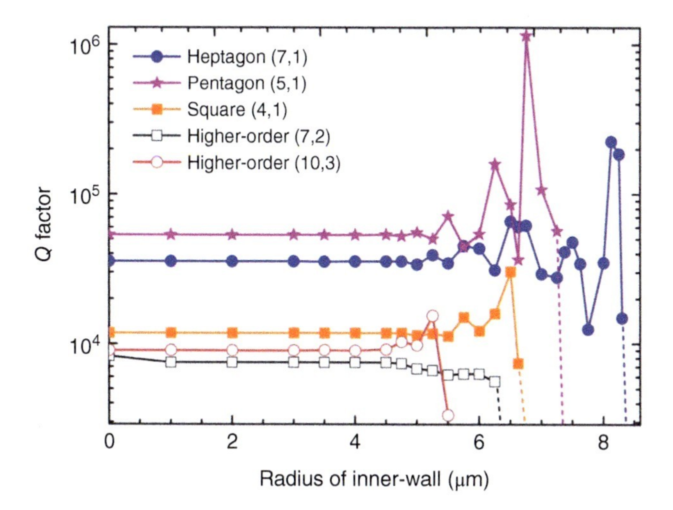

Figure 5.14 Mode Q factors vs. the radius of inner wall for the coupled modes of Figure 5.13 in corresponding microrings. Source: Ma et al. [22]. © 2014, Optical Society of America.

6.23, and 5.88 μm for modes (7, 1), (5, 1), (4, 1), (7, 2), and (10, 3), respectively. The results show that the cut-off values of r are a little larger than those estimated by FDTD simulation due to a certain width of the mode-field pattern. Further increase of r will lead to the deformation and destruction of the mode-field distributions, thus making the Q factors decrease sharply. For the microring with $R = 10$ μm, r from 7 to 8.25 μm ($r/R = 0.7$–0.825) is an optimal value range (which may be a different value for resonators with the other value of R), for the better effect of the transverse mode control and achieve single-mode unidirectional emission operation.

The waveguide-connected microdisk and microring lasers laterally confined by BCB were fabricated with the same processes in Section 5.3.1. The microdisk and microring lasers with outer radius of 10 μm were measured by coupling a multimode fiber to the cleaved waveguide facet and maintaining a TEC temperature of 290 K. Continuous wave lasing was realized with threshold currents of 5.5 and 2 mA for the microdisk laser and the microring laser with $r = 7$ μm, respectively. The lasing spectra for the microdisk laser with injection currents of 5 and 45 mA and the microring laser with injection current of 23 mA are plotted in Figure 5.15a,b, respectively, where the lasing spectra show longitudinal mode intervals from 10.65 to 11.08 nm and 11.09 to 11.63 nm, and several transverse modes appear for each longitudinal mode. Single-mode operations at 1548.42 and 1566.08 nm are obtained as shown in Figure 5.15a,b with SMSR of 28.4 and 34.6 dB, respectively. The lasing peaks are marked by the same symbols as in Figure 5.14 for the same-order coupled modes. The dominant mode in Figure 5.15a is heptagonal-shaped modes (7, 1) at the wavelength of 1548.43 nm, accompanied by pentagonal-shaped (5, 1), square-shaped (4, 1), and star-shaped (10, 3) and (7, 2) modes at wavelengths of 1550.71, 1552.55, 1550.22, and 1555.24 nm, respectively. For the microring laser with $r = 7$ μm, only two groups of peaks of the heptagonal-shaped (7, 1) and

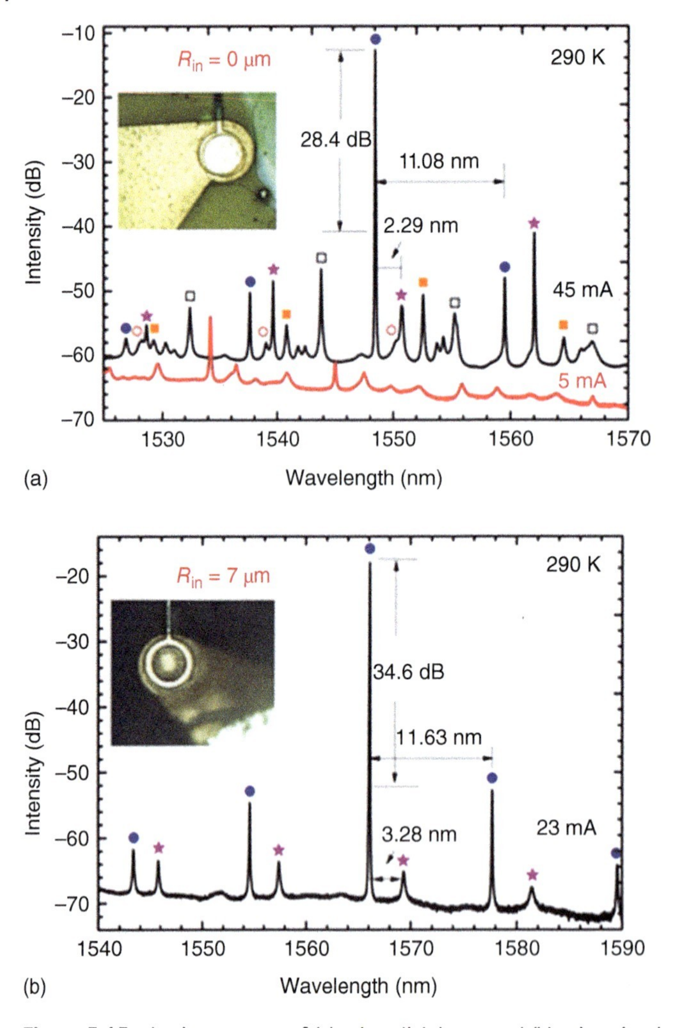

Figure 5.15 Lasing spectra of (a) microdisk laser and (b) microring laser with $r = 7\ \mu m$, measured at injection currents of (a) 5 and 45 mA, and (b) 23 mA, respectively. Inset: Optical microscope images of fabricated microring lasers. Source: Ma et al. [22]. © 2014, Optical Society of America.

pentagonal-shaped (5, 1) modes are observed in Figure 5.15b, which agree well with the simulation results in Figure 5.14. One can expect a further reduction of the width of microrings will eventually lead to real single transverse-mode lasing based on the simulation results. However, the SMSR is mainly limited by multiple longitudinal modes for the corresponding microring lasers. In addition, the dominant lasing mode is the heptagonal-shaped mode, although its Q factor is slightly lower than the pentagonal-shaped mode in simulation. The possible reason is that the low radial-order modes have lower vertical radiation loss as presented in Chapter 9, which is not considered in 2D numerical simulation.

Furthermore, the fields of WGMs in a microdisk are mainly localized in the rim of the resonators, which results in strong carrier spatial hole-burning and carrier diffusion effect affecting the dynamic characteristics greatly. The small-signal modulation responses of microdisk lasers exhibit a significant low-frequency roll-off around 3~4 GHz similar as the numerical results of rate equations. The microring lasers have better small signal modulation curves, due to the larger overlap between the injection current and the mode-field pattern, and less effect of the carrier diffusion [23].

5.5 Unidirectional Emission Hybrid Deformed-Microring Lasers

In the circular microdisk or microring, the mode competition between the doubly degenerate CW and CCW modes leads to the instability of the output power if an output waveguide is evanescent-field coupled to the cavity in angular direction. The waveguide connected to the microcavity in radial direction is one choice for obtaining unidirectional emission as shown in above. Furthermore, chiral modes were obtained by deforming the cavity to break the circular symmetry, and unidirectional emission was achieved with light emission from evanescent-field-coupled silicon waveguide [24].

Figure 5.16a shows the schematic diagram of a deformed microring for realizing chiral modes, where R is the outer radius and d is the width of the microring. The deformed microring includes a flat part with a length of L to form a notch with a width of Δ. The asymmetric scattering induced by the notch will result in copropagating pairs with dominant components in the CCW direction similar to that in a spiral microcavity, while the deformed microring can maintain higher Q factors due to the mode coupling [24, 25]. For the deformed microring with $R = 20\,\mu\text{m}$, $d = 4\,\mu\text{m}$, and $\Delta = 500\,\text{nm}$, the field intensity $|H_z|^2$ distribution for TE mode at the wavelength of 1555.48 nm is plotted in Figure 5.16b, with a mode Q-factor of 9×10^4.

To quantify the chirality of the mode, the magnetic field distribution in angular direction is expanded by $\exp(i\nu\varphi)$. The obtained normalized intensity vs. the angular momentum index ν is shown in Figure 5.16c, where positive and negative values of ν represent the CCW and CW traveling-wave components, respectively. Moreover, the unidirectional ratio $\eta_{\text{cw/ccw}}$ is 0.055, which is defined as the ratio of the highest intensities of the CW and CCW directions. The Q-factor and unidirectional ratio $\eta_{\text{cw/ccw}}$ vs. the notch width Δ are calculated and shown in Figure 5.16d for the high Q mode. Both the Q-factor and $\eta_{\text{cw/ccw}}$ decrease with the increase of Δ, which indicate a tradeoff between the mode Q factor and unidirectionality. As $\Delta > 500\,\text{nm}$, nearly total unidirectional output from the CCW direction can be expected, but the mode Q factor is relatively lower. For $\Delta < 400\,\text{nm}$, the mode Q-factor is higher than 1.3×10^5 because of lower radiation loss, but the unidirectional ratio is deteriorated.

In the deformed microring resonator with $R = 20\,\mu\text{m}$ and $\Delta = 500\,\text{nm}$, two sets of high-order transverse modes appear as the ring width is increased. Figure 5.17a,b give the mode-field intensity distributions for the high-order transverse modes at

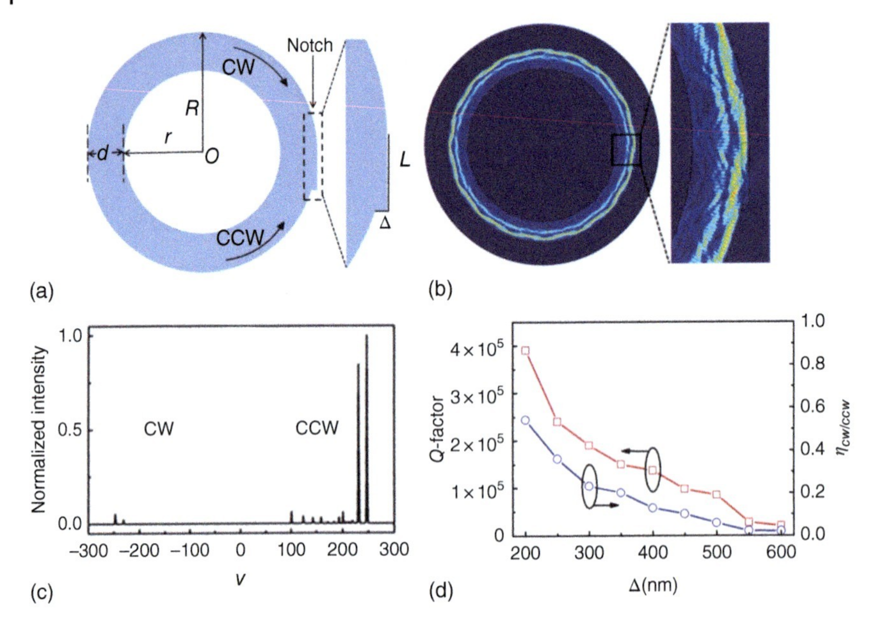

Figure 5.16 (a) Schematic diagram of a deformed microring. (b) Internal field intensity distribution and (c) normalized intensity vs. the angular momentum index v for the fundamental mode in the deformed microring with $R = 20\,\mu\text{m}$, $d = 4\,\mu\text{m}$, and $\Delta = 500\,\text{nm}$. (d) Q-factor and unidirectional ratio $\eta_{cw/ccw}$ vs. notch width Δ. Source: Sui et al. [24]. © 2016, Optical Society of America.

$d = 11\,\mu\text{m}$ and the corresponding normalized intensity vs. the angular momentum index v, respectively. The corresponding resonant wavelengths are 1555.86 and 1554.71 nm, and the Q-factors are 3.6×10^5 and 1.1×10^5, higher than that of the fundamental modes with the unidirectional ratios of 0.13 and 0.11. The field distributions are similar to the coupled modes with polygonal-shaped patterns in Figure 5.6. The numerical results indicate that there is a cutoff inner radius above which a higher-order mode cannot be supported. The cutoff inner radii are 13, 10, and 19 μm for the high-order transverse modes in Figure 5.17a,b and the fundamental mode in Figure 5.16b, respectively. Therefore, the single-transverse mode lasing can be expected for $14\,\mu\text{m} < r < 19\,\mu\text{m}$ corresponding to $1\,\mu\text{m} < d < 6\,\mu\text{m}$ in deformed microring lasers with $R = 20\,\mu\text{m}$ and $\Delta = 500\,\text{nm}$.

The hybrid deformed microring lasers vertically coupled to a Si waveguide were fabricated using wafer-bonding technology. The schematic diagram of the microlaser is shown in Figure 5.18a. The CW and CCW light can couple into the Si waveguide in opposite directions. The cross-sectional view scanning electron microscope (SEM) image of a hybrid deformed microring laser is shown in Figure 5.18b. After cleaving the silicon waveguide, the microlasers are bonded on a TEC and measured under pulsed electrical injection with a pulse duty of 10% and a pulse width of 100 ns at 285 K. The output powers from both sides of the Si waveguide are measured by butt coupling a multimode optical fiber to the cleaved waveguide facets. For the deformed microring laser with $R = 20\,\mu\text{m}$, $d = 4\,\mu\text{m}$, and $\Delta = 500\,\text{nm}$, the output powers of the CW and CCW directions from the Si waveguide are plotted in Figure 5.18. The threshold current is 7 mA. The maximum powers are

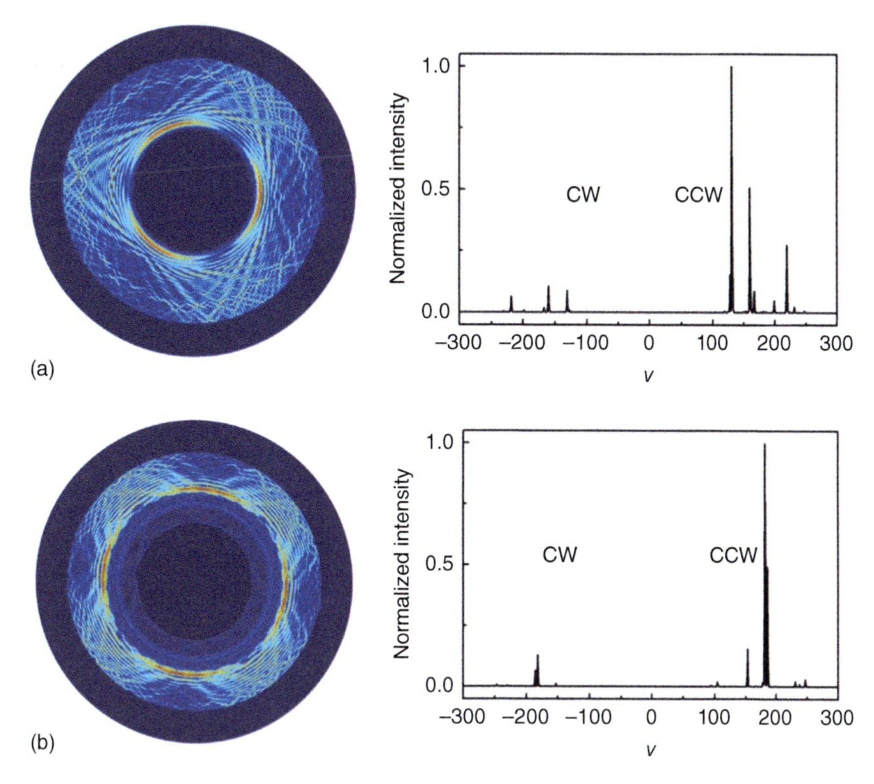

Figure 5.17 Internal field intensity distribution and normalized intensity vs. the angular momentum index v for the high-order transverse modes at (a) 1555.86 nm and (b) 1554.71 nm in the deformed microring with $R = 20\,\mu\text{m}$, $d = 11\,\mu\text{m}$, and $\Delta = 500\,\text{nm}$. Source: Sui et al. [24]. © 2016, Optical Society of America.

18.6 and 0.98 μW at the pulsed injection current of 33 mA for the CCW and CW directions, respectively. The unidirectional ratio of 0.053 is in accordance with the simulated value of 0.055. The lasing spectrum from the CCW direction at the pulsed injection current of 15 mA is shown in Figure 5.18d, and the lasing wavelength is 1554.8 nm. The lasing peak cannot be observed from the silicon waveguide for the CW direction due to its low-output power. Besides, high-order transverse modes are not observed, and single-mode lasing is realized with the SMSR of 27 dB. The mode coupling eliminates mode competition between CW and CCW propagating modes and results in better stable unidirectional mission for locally deformed microring lasers with a notch than the spiral-ring microlaser [26].

5.6 Wide-Angle Emission and Multiport Microdisk Lasers

5.6.1 Wide-Angle Emission-Deformed Microdisk Lasers

In addition to unidirectional emission microlasers, wide-angle emission light sources are important for some special applications, such as microscale illumination, monitor system, and wide-angle sensor. Wide-angle emission-deformed

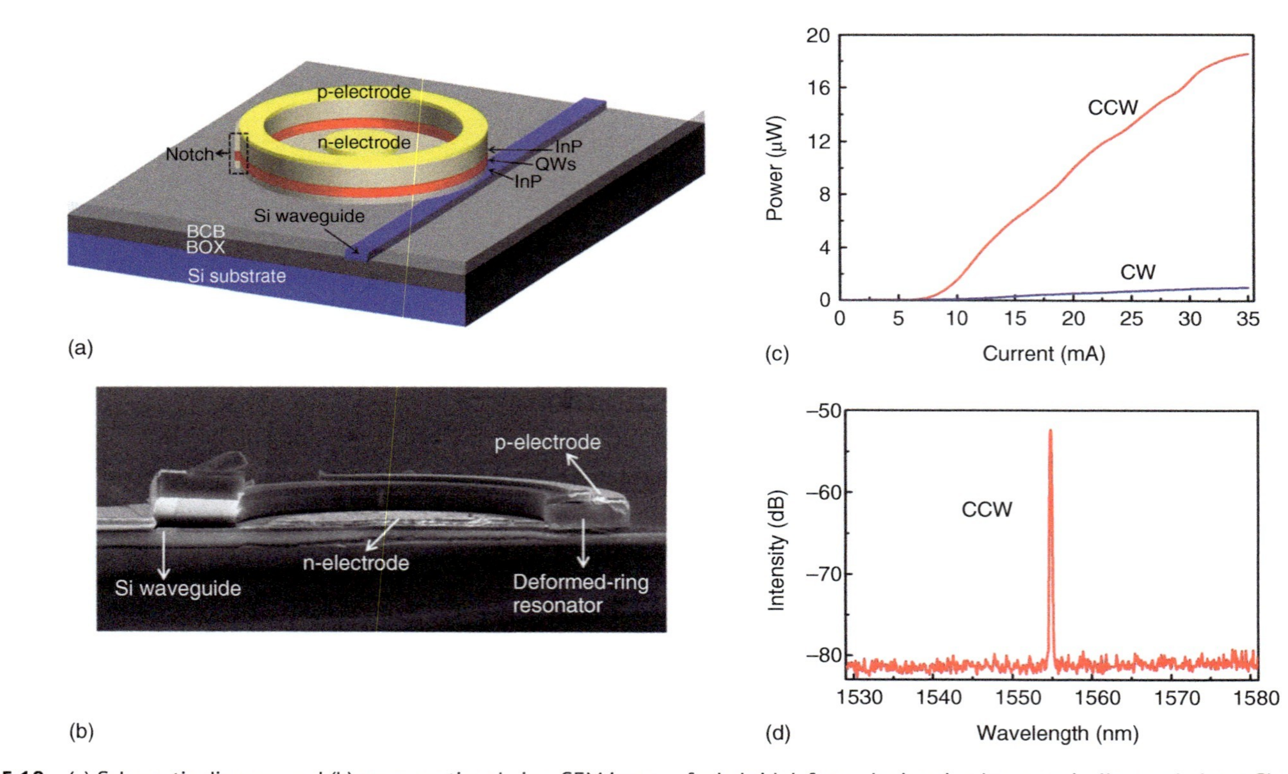

Figure 5.18 (a) Schematic diagram and (b) cross-sectional view SEM image of a hybrid deformed microring laser vertically coupled to a Si waveguide. (c) Measured CW and CCW-coupled output powers and (d) lasing spectrum from the CCW direction at 15 mA for the deformed microring laser with $R = 20\,\mu m$, $d = 4\,\mu m$, and $\Delta = 500\,nm$. Source: Sui et al. [24]. © 2016, Optical Society of America.

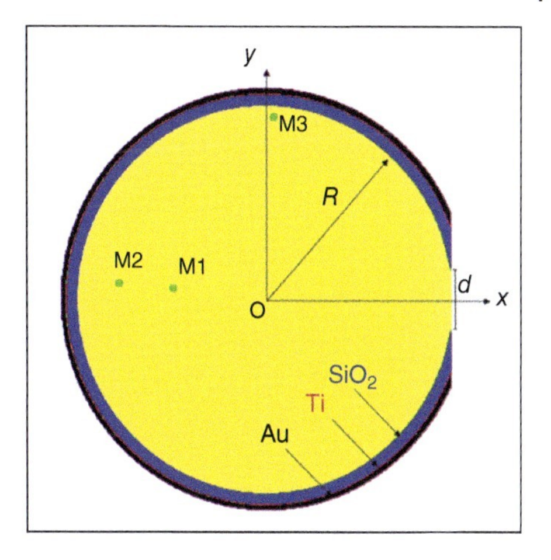

Figure 5.19 Schematic diagram of a 2D deformed-circular resonator with a flat side *d*. Source: Lin et al. [27]. © 2012, Optical Society of America.

circular microlasers can be realized by forming a flat side on the perimeter of a circular microresonator [27].

The mode characteristics of TE modes are simulated for a deformed circular microresonator confined by SiO_2 and metallic layers with a flat side as shown in Figure 5.19, by 2D FDTD technique. The radius of the circular microresonator is $R = 8\,\mu m$, the SiO_2 layer thickness is 0.4 μm, and the metallic layer consisted of 0.04 μm titanium and 0.2 μm gold as the p-electrode. The complex refractive indices of the SiO_2, titanium, and gold are 1.45, $3.7 + 4.5i$, and $0.18 + 10.2i$, respectively. The effective refractive index of an MQW laser wafer is taken to be 3.2. The exciting sources are laid in three low symmetry points M1, M2, and M3 inside the resonator. The Courant time step of 0.0467 fs and a uniform mesh with the cell size of 20 nm are used in the FDTD simulation. The simulated mode wavelengths and Q factors are plotted as the functions of the width of the flat side d for the symmetric and antisymmetric modes with the initial mode wavelength of 1528 nm in Figure 5.20.

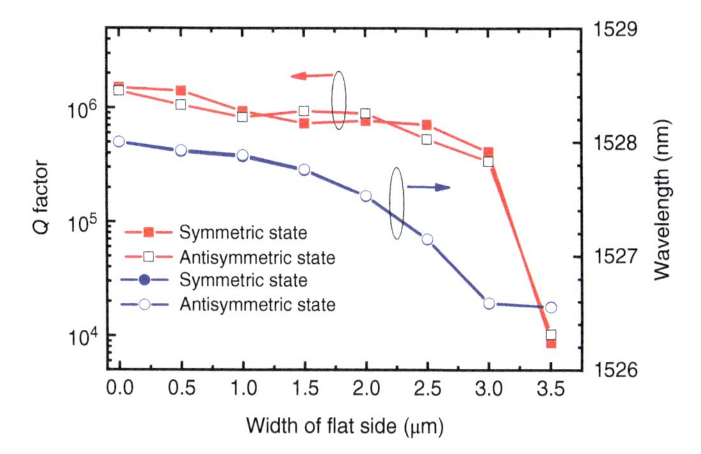

Figure 5.20 Mode Q factors and wavelengths vs. the width of the flat side width d for the symmetric and antisymmetric modes in the metallic confined deformed-circular resonator with the radius of 8 μm. Source: Lin et al. [27]. © 2012, Optical Society of America.

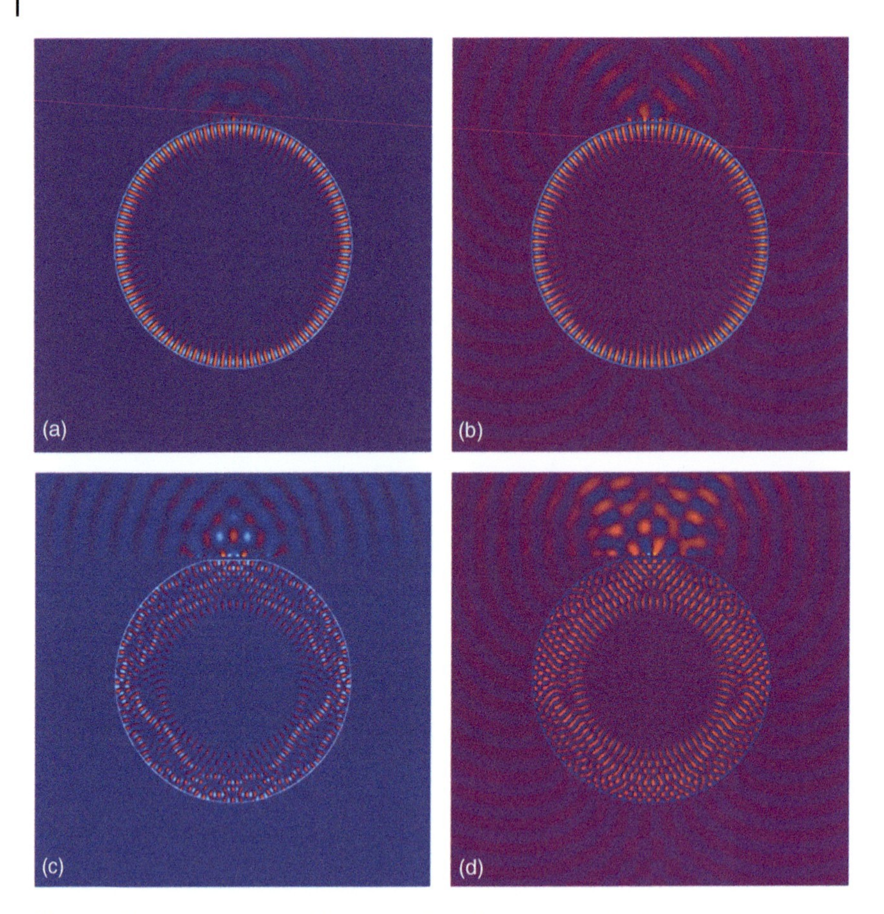

Figure 5.21 Field patterns of the magnetic field components for (a) symmetric and (b) antisymmetric modes at the wavelengths of 1527.52 nm in the deformed-circular resonator with the flat side width of 2.0 μm, and for (c) symmetric and (d) antisymmetric modes at the wavelength of 1526.55 nm with the flat side width of 3.5 μm, where the output field amplitude in the upper external region right side is magnified 100 times. Source: Lin et al. [27]. © 2012, Optical Society of America.

For the symmetric mode, the mode wavelength shifts from 1528.0 to 1526.56 nm as the width d increases from 0 to 3.5 μm, and the mode Q factor is about 1.6×10^6, 7.7×10^5, and 8.7×10^3 as the width d is 0, 2, and 3.5 μm. The corresponding antisymmetric mode has almost the same mode wavelength and mode Q factor as the symmetric mode. The results show that the deformed circular microresonator with a flat side can still have high Q confined modes.

Taking a narrow exciting source centered at a resonant wavelength, we simulate the mode-field pattern and plot magnetic field component H_z in Figure 5.21 for (a) symmetry and (b) asymmetry modes at 1527.52 nm as $d = 2.0$ μm and (c) symmetry, and (d) asymmetry modes at 1526.55 nm as $d = 3.5$ μm. The mode-field patterns in Figure 5.21a,b are similar to those of the first-order WGMs in a perfect circular resonator, with the main field intensity around the perimeter, which means that the

2-μm-wide flat side is only a perturbation. However, the mode-field patterns become very complicated as shown in Figure 5.21c,d as $d = 3.5\,\mu\text{m}$, which should be the superposition of several WGMs with a large Q factor spoiling. The field patterns in the external region are magnified 100 times for clarity, which show wide far-field pattern for deformed microdisks with a flat side.

5.6.2 Multiport Output Microdisk Lasers

Finally, the spatial coherence of a single-mode microdisk laser with two output waveguides is verified by measuring the far-field pattern of the microlaser. The spatial coherence characteristics of the two-port cylinder microlaser can be explained as the phenomena of the Young interferometer [28]. The AlGaInAs/InP microlasers, with a radius of 15 μm and two 2-μm-wide output waveguides, were fabricated using conventional photolithography and ICP etching technique. As shown in Figure 5.22a, the output waveguides corresponding to output ports a and b are normal and with a crossing angle of 45° to the cleaved output facet. The microlaser output spectra measured from the ports a and b are plotted in Figure 5.23 at an injection current of 50 mA. Single-mode operation at a wavelength of 1581.8 nm is obtained with the SMSR of 30 dB.

The far-field interference pattern of the microlaser was measured based on the schematic diagram in Figure 5.22b, where the microlaser is placed as near as possible to the rotation center O of a rotating plate. The distance h between the output ports A and B is about 17 μm for the tested microlaser; the distance from the center O to the cleaved facet R is about 50 μm; and the distance from the cleaved facet to the photodetector D is 8.2 cm. The ports A and B symmetry to the detector corresponds to $\phi = 0$. The normalized far-field patterns are measured and plotted in Figure 5.24 for the microlaser at the injection currents of 20, 40, and 50 mA. A wide peak with a FWHM larger than 100° is observed at 20 mA, as the current of 20 mA is less than the threshold current. The interference patterns are observed as the injection current is larger than the threshold current. The more evident interference pattern at 50 mA is agreement with the observed SMSR increasing with the injection current. The interference patterns have two sets of interference fringes, which can be attributed to the

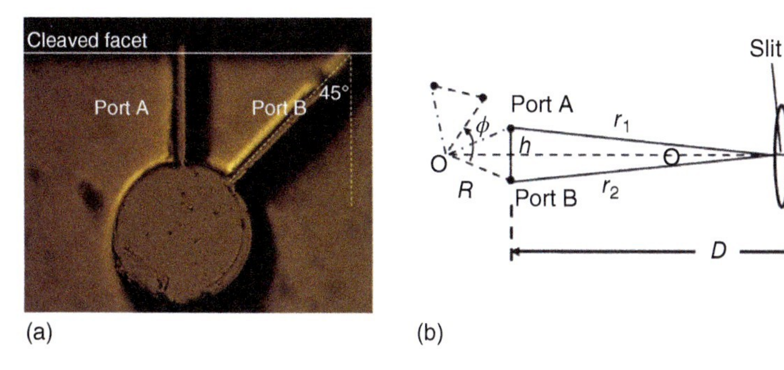

Figure 5.22 (a) The microscope image of a microdisk laser with two output waveguides [29]. (b) Schematic diagrams for measuring far-field pattern of the microdisk laser with two output ports A and B. Source: Lin et al. [28]. © 2012, Optical Society of America.

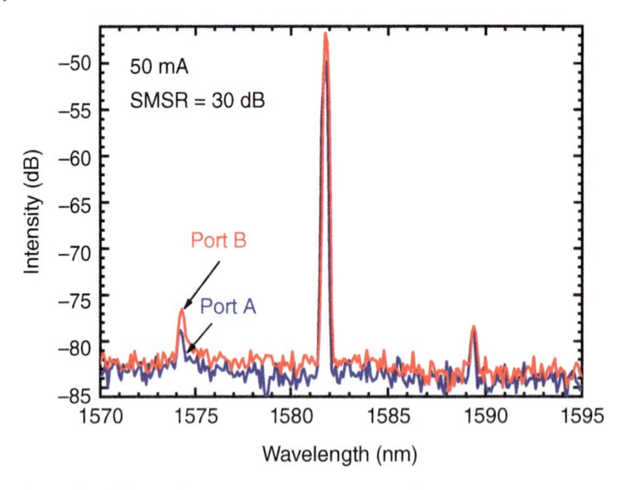

Figure 5.23 Lasing spectra of the microdisk laser at room temperature with an injection current of 50 mA. Source: Lin et al. [28]. © 2012, Optical Society of America.

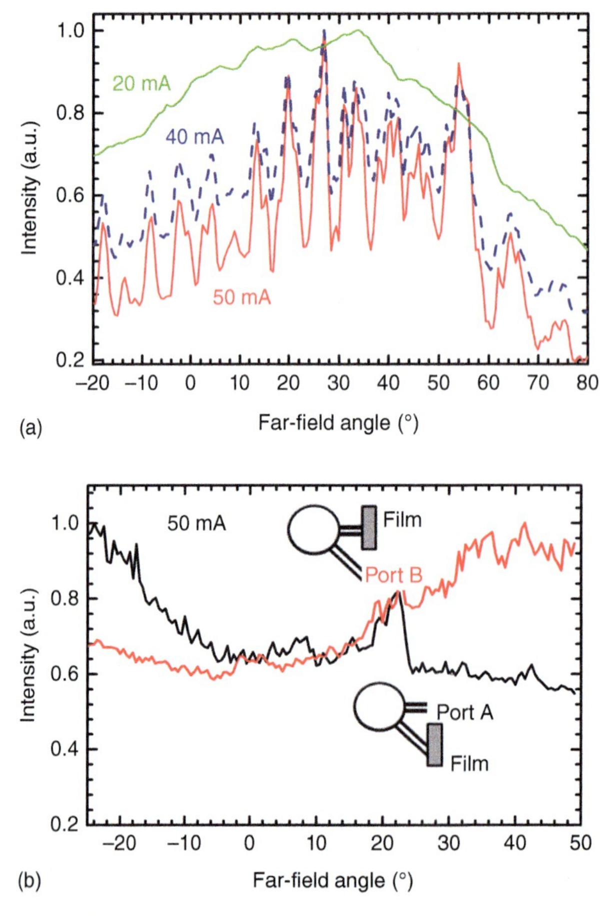

Figure 5.24 Normalized far-field patterns of (a) the microlaser with two output waveguides at the injection currents of 20, 40, and 50 mA, and (b) from ports A and B at 50 mA, respectively. Source: Lin et al. [28]. © 2012, Optical Society of America.

existence of degenerate modes. The far-field patterns were also measured for each port by blocking the output from the other port using a film. Extremely wide far-field patterns are observed for ports A and B, with the peaks appearing at the far-field angle of −25° and 40°, respectively. In fact, the WGM in the circular resonator couples to the high-order transverse modes in the 2-μm-wide output waveguide, so the far-field pattern is very wide for each port. The measured far-field pattern for the microlaser with one output waveguide usually has a FWHM larger than 100°. We expect that the coherence characteristics can be used to construct a compact optical sensor by measuring the variation of the light path difference of the two light beams.

5.7 Summary

We have shown that the waveguide directly connecting microdisk is a robust design to obtain efficient waveguide-coupled unidirectional emission microdisk lasers. Due to the introduced output waveguide, mode coupling between the WGMs with close wavelengths occurs and results in high *Q*-coupled modes, which can result in high-efficiency output coupling into the waveguide. In addition, waveguide-connected microring lasers are analyzed and demonstrated to suppress higher transverse-order WGMs and improve the lasing performance. Furthermore, locally deformed microrings are proposed to form near CCW traveling-wave modes, and unidirectional emission hybrid lasers on silicon are fabricated by wafer-bonding technique. Finally, deformed-circular microlasers with a flat side on the perimeter are proposed to realize wide-far field emission for potential application of microscale illumination, and the far-field patterns of microdisk lasers with two output waveguides are suggested to be used as a compact optical sensor.

References

1 McCall, S.L., Levi, A.F.J., Slusher, R.E. et al. (1992). Whispering-gallery mode microdisk lasers. *Appl. Phys. Lett.* 60: 289–291.
2 Gmachl, C., Capasso, F., Narimanov, E.E. et al. (1998). High-power directional emission from microlasers with chaotic resonators. *Science* 280: 1556–1564.
3 Baryshnikov, Y., Heider, P., Parz, W., and Zharnitsky, V. (2004). Whispering gallery modes inside asymmetric resonant cavities. *Phys. Rev. Lett.* 93: 133902.
4 Lee, S.Y., Rim, S., Ryu, J.W. et al. (2004). Quasi-scarred resonances in a spiral-shaped microcavity. *Phys. Rev. Lett.* 93: 164102.
5 Chern, G.D., Tureci, H.E., Stone, A.D. et al. (2003). Unidirectional lasing from InGaN multiple-quantum-well spiral-shaped micropillars. *Appl. Phys. Lett.* 83: 1710–1712.
6 Wiersig, J. and Hentschel, M. (2008). Combining directional light output and ultralow loss in deformed microdisks. *Phys. Rev. Lett.* 100: 033901.

7 Choi, S.J., Djordjev, K., Choi, S.J., and Dapkus, P.D. (2003). Microdisk lasers vertically coupled to output waveguides. *IEEE Photonics Technol. Lett.* 15: 1330–1332.

8 Campenhout, J.V., Rojo-Romeo, P., Regreny, P. et al. (2007). Electrically pumped InP-based microdisk lasers integrated with a nanophotonic silicon-on-insulator waveguide circuit. *Opt. Express* 15: 6744–6749.

9 Huang, Y.Z., Hu, Y.H., Chen, Q. et al. (2007). Room-temperature continuous-wave electrically injected InP–GaInAsP equilateral-triangle-resonator lasers. *IEEE Photonics Technol. Lett.* 19: 963–965.

10 Huang, Y.Z., Che, K.J., Yang, Y.D. et al. (2008). Directional emission InP/InGaAsP square-resonator microlasers. *Opt. Lett.* 33: 2170–3172.

11 Yang, Y.D., Wang, S.J., and Huang, Y.Z. (2009). Investigation of mode coupling in a microdisk resonator for realizing directional emission. *Opt. Express* 17 (25): 23010–23015.

12 Wang, S.J., Lin, J.D., Huang, Y.Z. et al. (2010). AlGaInAs/InP microcylinder lasers connected with an output waveguide. *IEEE Photonics Technol. Lett.* 22: 1349–1351.

13 Huang, Y.Z., Wang, S.J., Yang, Y.D. et al. (2010). Investigation on multiple-port microcylinder lasers based on coupled modes. *Semicond. Sci. Technol.* 25 (10): 105005.

14 Taflove, A. and Hagness, S.C. (1998). *Computational Electrodynamics: The Finite-Difference Time-Domain Method*, 2e. Boston, MA/London: Artech House.

15 Guo, W.H., Li, W.J., and Huang, Y.Z. (2001). Computation of resonant frequencies and quality factors of cavities by FDTD technique and Padé approximation. *IEEE Microwave Wirel. Compon. Lett.* 11: 223–225.

16 Yang, Y.D., Weng, H.Z., Liu, B.W. et al. (2015). Localized-cavity-loss-induced external mode coupling in optical microresonators. *J. Opt. Soc. Am. B: Opt. Phys.* 32: 2376–2381.

17 Wiersig, J. (2006). Formation of long-lived, scarlike modes near avoided resonance crossings in optical microcavities. *Phys. Rev. Lett.* 97: 253901.

18 Lv, X.M., Zou, L.X., Lin, J.D. et al. (2012). Unidirectional-emission single-mode AlGaInAs–InP microcylinder lasers. *IEEE Photonics Technol. Lett.* 24: 963–965.

19 Zou, L.X., Huang, Y.Z., Liu, B.W. et al. (2015). Thermal and high speed modulation characteristics for AlGaInAs/InP microdisk lasers. *Opt. Express* 23: 2879–2888.

20 Lv, X.M., Huang, Y.Z., Zou, L.X. et al. (2013). Optimization of direct modulation rate for circular microlasers by adjusting mode *Q* factor. *Laser Photonics Rev.* 7: 818–829.

21 Lv, X.M., Yang, Y.D., Zou, L.X. et al. (2014). Mode characteristics and optical bistability for AlGaInAs/InP microring lasers. *IEEE Photonics Technol. Lett.* 26: 1703–1706.

22 Ma, X.W., Lv, X.M., Huang, Y.Z. et al. (2014). Mode characteristics for unidirectional-emission microring resonator lasers. *J. Opt. Soc. Am. B: Opt. Phys.* 31: 2773–2778.

23 Lv, X.M., Huang, Y.Z., Yang, Y.D. et al. (2014). Influences of carrier diffusion and radial mode field pattern on high speed characteristics for microring lasers. *Appl. Phys. Lett.* 104: 161101.

24 Sui, S.S., Huang, Y.Z., Tang, M.Y. et al. (2016). Locally deformed-ring hybrid microlasers exhibiting stable unidirectional emission from a Si waveguide. *Opt. Lett.* 41: 3928–3931.

25 Wiersig, J., Kim, S.W., and Hentschel, M. (2008). Asymmetric scattering and nonorthogonal mode patterns in optical microspirals. *Phys. Rev. A* 78: 053509.

26 Sui, S.S., Huang, Y.Z., Tang, M.Y. et al. (2017). Hybrid deformed-ring AlGaInAs/Si microlasers with stable unidirectional emission. *IEEE J. Sel. Top. Quantum Electron.* 23: 1500308.

27 Lin, J.D., Zou, L.X., Huang, Y.Z. et al. (2012). Wide-angle emission and single mode deformed circular microlasers with a flat side. *Appl. Opt.* 51: 3930–3935.

28 Lin, J.D., Huang, Y.Z., Yang, Y.D. et al. (2012). Coherence of a single mode InAlGaAs/InP cylinderical microlaser with two output ports. *Opt. Lett.* 37: 1977–1979.

29 Lin, J.D., Huang, Y.Z., Yao, Q.F. et al. (2011). InAlGaAs/InP cylinder microlaser connected with two waveguides. *Electron. Lett.* 47: 929–930. https://doi.org/10.1049/el.2011.2098.

6 Equilateral-Triangle-Resonator Microlasers

6.1 Introduction

Equilateral-triangle resonator (ETR) was reported for fabricating semiconductor lasers cleaved from a laser wafer grown on (111) semiconductor substrate in 1964 [1], in the very beginning of semiconductor lasers. But lasing mode intervals are agreement with a light ray path vertically impinging on the side of the resonator instead of totally internal reflection light rays of whispering-gallery modes. In 1996, triangular-facet semiconductor lasers were directly grown by selective area metalorganic chemical vapor deposition on (111)B GaAs substrate [2, 3]. Mode-field patterns of the ETR were analyzed under perfectly confined approximation and triangular lasers were fabricated with the resonators cleaved from (111)B InGaAs/GaAs laser wafer [4]. However, totally internal reflection mode light rays still have evanescent electromagnetic fields in the external regions for the resonator with real refractive index step, and corresponding reflection phase shift and radiation loss affect mode wavelengths and passive mode Q factors.

In this chapter, mode characteristics for the ETR are analyzed using strict boundary condition, and analysis solutions of mode wavelengths and field distributions are compared with the finite-difference time-domain (FDTD) simulation results [5–7, 10]. Furthermore, ETR semiconductor microlasers were fabricated using a common semiconductor laser wafer by inductively coupled plasma (ICP) etching technique [11].

6.2 Mode Analysis Based on the ETR Symmetry

6.2.1 Wave Equations for TE and TM Modes

Assigning transverse-magnetic (TM) and transverse-electric (TE) modes as in three-dimensional (3D) ETR, we analyze mode characteristics with nonzero electromagnetic fields of (E_z, H_x, H_y) and (H_z, E_x, E_y) for TM and TE modes, respectively, based on the symmetry operation of two-dimensional (2D) ETR [6, 7]. Using $F_j(x, y)\exp(-i\omega t)$ to present one component of the electromagnetic fields

Microcavity Semiconductor Lasers: Principles, Design, and Applications, First Edition.
Yong-zhen Huang and Yue-de Yang.

with $j = x, y, z$, we have the following wave equation for the ETR

$$\left(\frac{\partial^2}{\partial x^2} + \frac{\partial^2}{\partial y^2}\right) F_j(x,y) + n^2(x,y)k_0^2 F_j(x,y) = 0, \tag{6.1}$$

where $n(x, y)$ is the refractive index distribution, $k_0 = 2\pi/\lambda$ is the free space wave number, and λ is the mode wavelength. For TM modes, nonzero magnetic fields $H_x(x, y)$ and $H_y(x, y)$ can be obtained from $E_z(x, y)$ by

$$\begin{cases} H_x(x,y) = -\dfrac{i}{\omega\mu}\dfrac{\partial E_z(x,y)}{\partial y} \\ H_y(x,y) = \dfrac{i}{\omega\mu}\dfrac{\partial E_z(x,y)}{\partial x} \end{cases}. \tag{6.2}$$

Similar, nonzero electric fields $E_x(x, y)$ and $E_y(x, y)$ for TE modes are given by

$$\begin{cases} E_x(x,y) = \dfrac{i}{\varepsilon_0 n^2(x,y)\omega}\dfrac{\partial H_z(x,y)}{\partial y} \\ E_y(x,y) = -\dfrac{i}{\varepsilon_0 n^2(x,y)\omega}\dfrac{\partial H_z(x,y)}{\partial x} \end{cases}. \tag{6.3}$$

6.2.2 Transverse Modes by Unfolding Light Ray in the ETR

Under an effective index approximation, we can reduce 3D ETR to a 2D case in the x–y plane as shown in Figure 6.1a. For the 2D ETR with a uniform real refractive index $n_1 > 2$ surrounded by air, a light ray propagating parallel to the ETR sides will experience total internal reflection and return to the starting point after six reflections. Unfolding the total internal reflection light rays, we can model transverse mode field distributions as in a three-layer slab waveguide in Figure 6.1b [6], with a width of $\sqrt{3}a/2$. The transverse mode-field distribution for waves propagating along the x-direction inside the ETR, i.e. Ω region, can be written as

$$F_{z,\Omega}(x,y) = \cos[\kappa_m(y - \sqrt{3}a/12) + m\pi/2]\exp(i\beta_l x). \tag{6.4}$$

We simply express $E_z(x, y)$ and $H_z(x, y)$ for TM and TE modes as $F_z(x, y)$. By submitting the wavefunction (6.4) into wave Eq. (6.1), we have the transverse and longitudinal propagation constants κ_m and β_l satisfying

$$\kappa_m{}^2 + \beta_l{}^2 = n_1^2 k_0{}^2. \tag{6.5}$$

In the external region Ω_1, decaying wavefunction can be written as $A\exp(-\gamma_t y)\exp(i\beta_l x)$ with the decay constant of

$$\gamma_t = \sqrt{(n_1^2 - 1)k_0{}^2 - \kappa_m{}^2}, \tag{6.6}$$

for the ETR with a refractive index n_1 and the external region of free space. An eigenvalue equation can be obtained for the transverse mode field from the boundary continuity of electromagnetic fields. However, for an ETR waveguide with a large refractive index step and the side length of several times of mode wavelength, we can approximately apply perfectly confined approximation to the transverse mode

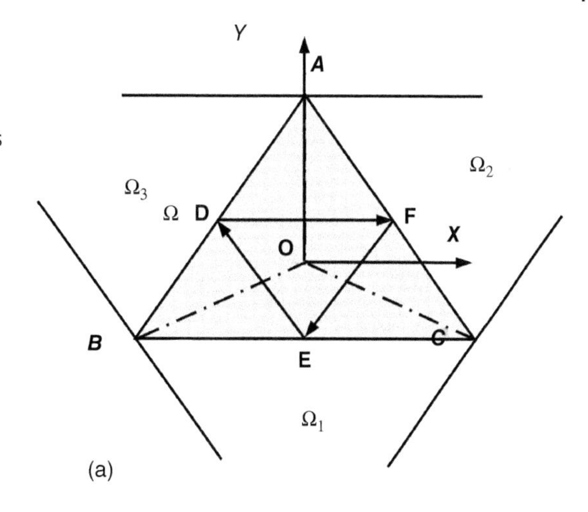

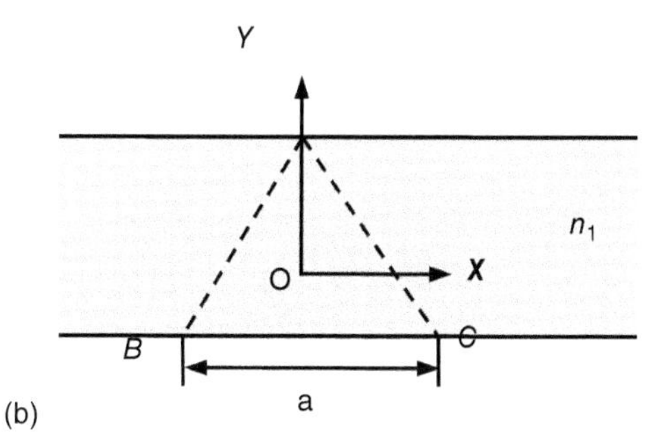

Figure 6.1 (a) Mode light rays ED, DF, and FE inside an ETR confined by totally internal reflection as in slab waveguides with the other three sides across vertices A, B, and C and parallel to the sides BC, AC, and AB, respectively. (b) The equivalent slab waveguide for transverse mode distribution mode light rays propagating in the x-direction. Source: Based on Huang et al. [6].

wavefunction (6.4) by taking $F_{z,\Omega}(x, y) = 0$ at the side BC of $y = -\sqrt{3}a/6$ and obtain the following eigenvalue equation

$$\sqrt{3}\kappa_m a = 2(m+1)\pi, \quad m = 0, 1, 2, \ldots, \tag{6.7}$$

with the transverse mode number m. In Section 6.3, the perfectly confined approximation is applied for the transverse mode wavefunction.

6.2.3 Evanescent Fields in External Regions

The eigenvalue equation for longitudinal mode propagation can be obtained from the boundary continuous conditions for electromagnetic fields at three sides of the ETR. The evanescent fields in the external regions and eigenvalue equation are first derived from the boundary conditions of the ETR according to the transmission of mode light rays [6]. Here, we present mode-field distributions according to the symmetry of the ETR and derive evanescent fields and eigenvalue equation–based boundary conditions as in [7]. The symmetry of an ETR can be described by the point group C_{3v}, which consists of the identity element E, rotational elements C_3 and $C_3^2 = C_3^{-1}$ rotating anticlockwise and clockwise angles of $2\pi/3$, and mirror reflection elements σ_x, $C_3^{-1}\sigma_x$, and $C_3\sigma_x$ relative to the lines of OA, OB, and OC as shown in

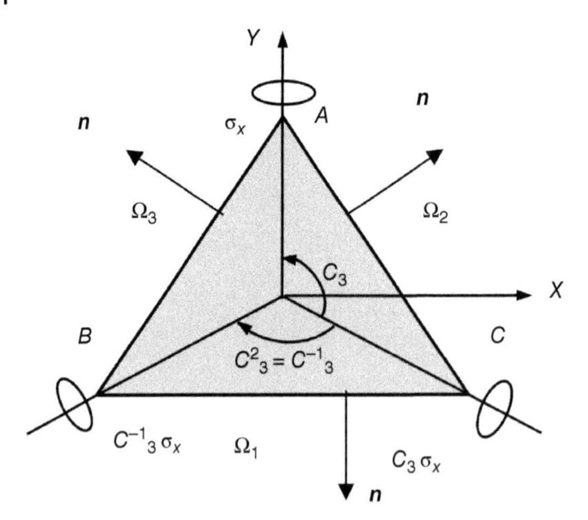

Figure 6.2 Schematic diagram of an ETR marked by three vertices A, B, and C, and the corresponding symmetry operation elements. The external regions of the ETR are divided into three regions of Ω_1, Ω_2, and Ω_3. Source: Huang et al. [7]. © 2006, IEEE.

Figure 6.2. The group operation satisfies $C_3^{-1}f_z(\mathbf{r}) = f_z(C_3\mathbf{r})$ and $C_3\mathbf{r}$ with $\varphi = 2\pi/3$ as

$$C_3\begin{pmatrix} x \\ y \end{pmatrix} = \begin{pmatrix} \cos\varphi & -\sin\varphi \\ \sin\varphi & \cos\varphi \end{pmatrix}\begin{pmatrix} x \\ y \end{pmatrix}, \tag{6.8}$$

and $\varphi = -2\pi/3$ for C_3^{-1}.

To guarantee the recovery of the field pattern under C_3^3 operation, the eigenvalues are taken as $\exp(-i2l_1\pi/3)$ and $\exp(i2l_1\pi/3)$ with an integer l_1 for the operation of C_3

$$C_3F_z^+(x,y) = \exp(-i2l_1\pi/3)F_z^+(x,y), \tag{6.9}$$

and

$$C_3F_z^-(x,y) = \exp(i2l_1\pi/3)F_z^-(x,y). \tag{6.10}$$

We can constitute the following wavefunctions inside the ETR satisfying (6.9) and (6.10)

$$F_z^+(x,y) = F_z(x,y) + C_3^{-1}F_z(x,y)\exp\left(-i\frac{2l_1\pi}{3}\right) + C_3^{-2}F_z(x,y)\exp\left(-i\frac{4l_1\pi}{3}\right), \tag{6.11}$$

$$\begin{aligned} F^-{}_z(x,y) &= \sigma_x F^+{}_z(x,y) \\ &= \sigma_x F_z(x,y) + C_3\sigma_x F_z(x,y)\exp\left(-i\frac{2l_1\pi}{3}\right) + C_3{}^2\sigma_x F_z(x,y)\exp\left(i\frac{2l_1\pi}{3}\right), \end{aligned} \tag{6.12}$$

where l is the longitudinal mode number, and $\sigma_xF_z(x,y) = F_z(-x,y)$ from (6.4). The wavefunctions of (6.11) and (6.12) are clockwise and anticlockwise propagation traveling waves, respectively. Accounting for evanescent fields in the external regions, we have

$$F_z(x,y) = F_{z,\Omega}(x,y)\delta(\mathrm{ETR}) + F_{z,\Omega_1}(x,y)\delta(\Omega_1), \tag{6.13}$$

where $\delta(\mathrm{ETR}) = 1$ and 0 for the position (x, y) locates the internal and external regions of the ETR, and $\delta(\Omega_1) = 1$ and 0 for (x, y) at the external region Ω_1 and the other regions, respectively. If $F_z(x, y)$ satisfies wave Eq. (6.1), we can prove that the other terms in (6.11) and (6.12) also satisfy (6.1). Submitting perfectly confined transverse mode function (6.4) into (6.11) and (6.12), we can obtain mode-field patterns inside the ETR for the traveling waves. And the mode-field distribution (6.11) along the side BC of $y = -\sqrt{3}a/6$ can be expressed as:

$$
\begin{aligned}
&F_z^+(x, -\sqrt{3}a/6) \\
&= f_z\left(-\frac{x}{2} + \frac{a}{4}, \frac{\sqrt{3}x}{2} + \frac{\sqrt{3}a}{12}\right) \exp\left(-i\frac{2l_1\pi}{3}\right) \\
&\quad + f_z\left(-\frac{x}{2} - \frac{a}{4}, -\frac{\sqrt{3}x}{2} + \frac{\sqrt{3}a}{12}\right) \exp\left(-i\frac{4l_1\pi}{3}\right) \\
&= \exp\left(-i\frac{\beta_l x}{2}\right)\left\{\cos\left(\kappa_m\frac{\sqrt{3}}{2}x + \frac{m\pi}{2}\right)\exp\left[-i\left(\frac{2l_1\pi}{3} - \frac{\beta_l a}{4}\right)\right]\right. \\
&\quad \left. + \cos\left(-\kappa_m\frac{\sqrt{3}}{2}x + \frac{m\pi}{2}\right)\exp\left[i\left(\frac{2l_1\pi}{3} - \frac{\beta_l a}{4}\right)\right]\right\}
\end{aligned}
\tag{6.14}
$$

The corresponding evanescent field in the external region Ω_1 can be constituted based on mode-field distribution (6.14). Assuming mode field in the external region is an exponentially decay function for confined modes, we can express the evanescent fields in the external region Ω_1 by multiplying $\exp[\gamma(y + \sqrt{3}a/6)]$ and (6.14), according to the boundary continuity condition of E_z and H_z, as:

$$
\begin{aligned}
&F_{z,\Omega_1}^+(x, y) \\
&= \exp\left(-i\frac{\beta_l x}{2} + \gamma\left(y + \frac{\sqrt{3}a}{6}\right)\right)\left\{\cos\left(\kappa_m\frac{\sqrt{3}}{2}x + \frac{m\pi}{2}\right)\exp\left[-i\left(\frac{2l_1\pi}{3} - \frac{\beta_l a}{4}\right)\right]\right. \\
&\quad \left. + \cos\left(-\kappa_m\frac{\sqrt{3}}{2}x + \frac{m\pi}{2}\right)\exp\left[i\left(\frac{2l_1\pi}{3} - \frac{\beta_l a}{4}\right)\right]\right\}.
\end{aligned}
\tag{6.15}
$$

Submitting (6.15) into wave Eq. (6.1) at free space with $n(x, y) = 1$, we have the following relation for the decay constant:

$$
\gamma \approx \sqrt{\beta_l^2/4 - k_0^2},
\tag{6.16}
$$

under the approximation $\beta_l >> \kappa_m$. In fact, the boundary conditions are approximately satisfied by ignoring the small terms of κ_m/β_l.

6.2.4 Eigenvalue Equation

For TM modes, $H_x(x, y)$ inside the ETR and in the external region Ω_1 can be obtained from $E_z(x, y)$ using (6.2), and $E_x(x, y)$ inside the ETR and in the external region Ω_1 can be obtained from $H_z(x, y)$ using (6.3) for TE modes. Here, TE and TM modes are defined with electric and magnetic fields polarized in triangle plane, respectively. Then applying the boundary continuity conditions to $H_x(x, y)$ and $E_x(x, y)$ at the side BC, it can be found that $(-1)^m \exp(-i\phi) = \exp[-i(2l_1\pi/3 - \beta_l a/2)]$; thus, we can obtain an eigenvalue equation as:

$$\beta_l \frac{3a}{2} + 3\phi = 2(l_1 + 3l_2)\pi + 3\delta\pi, \tag{6.17}$$

where δ is 0 and 1 for even and odd transverse modes, l_2 is an integer, and ϕ is the phase shift of the total internal reflection mode light at the sides

$$\phi = -2\tan^{-1}\left(\frac{2\xi\gamma}{\sqrt{3}\beta_l}\right), \tag{6.18}$$

with $\xi = n_1^2$ (squared refractive index ratio of the ETR to the external region) for the TE modes and $\xi = 1$ for the TM modes. The right term in Eq. (6.17) is integral and half-integral multiples of 2π for even and odd transverse modes, respectively. The longitudinal mode number is defined as

$$l = 2(l_1 + 3l_2) + 3\delta, \tag{6.19}$$

the longitudinal mode number is even and odd numbers for the even and odd transverse modes, respectively, but is different from that used in [6, 7]. Then, submitting the propagation constant Eqs. (6.7) and (6.17) into (6.5), we can express the resonant mode wavelength as

$$\lambda_{m,l} \approx \frac{3n_1 a}{\sqrt{(l - 3\phi/\pi)^2 + 3(m+1)^2}}. \tag{6.20}$$

6.3 Mode-Field Distributions

In this section, we present mode-field distributions of the analysis solutions as in [7] and compare them with simulation results of EDTD technique.

6.3.1 Mode Degeneracy and Classify

The wavefunctions (6.11) and (6.12) are complex transmission waves propagating oppositely. We can constitute following real standing wave functions from them as new basis functions

$$F_z^e(x,y) = \frac{1}{2}[F^+{}_z(x,y) + F^-{}_z(x,y)], \tag{6.21}$$

$$F_z^o(x,y) = \frac{i}{2}[F^+{}_z(x,y) - F^-{}_z(x,y)]. \tag{6.22}$$

The functions (6.21) and (6.22) are even and odd symmetry functions relative to mirror reflection element σ_x with superscripts "e" and "o" indicating even and odd symmetries, respectively. Furthermore, we can prove the following relation under rotation operation C_3 using the relations (6.9) and (6.10):

$$\begin{aligned} &C_3[F^e{}_{z,\Omega}(x,y), F^o{}_{z,\Omega}(x,y)] \\ &= [F^e{}_{z,\Omega}(x,y), F^o{}_{z,\Omega}(x,y)] \begin{bmatrix} \cos\left(\frac{2l_1\pi}{3}\right) & \sin\left(\frac{2l_1\pi}{3}\right) \\ -\sin\left(\frac{2l_1\pi}{3}\right) & \cos\left(\frac{2l_1\pi}{3}\right) \end{bmatrix}. \end{aligned} \tag{6.23}$$

Table 6.1 Character table of the point group C_{3v}.

	E	$2C_3$	$3\sigma_x$
A_1	1	1	1
A_2	1	1	−1
E	2	−1	0

Examining the transformation of the wavefunctions (6.21) and (6.22) under different symmetry operation and comparing them with the character table of the point group C_{3v} in Table 6.1, we can confirm that mode fields $F_z^e(x,y)$ and $F_z^o(x,y)$ form doubly degenerate irreducible representations E as the longitudinal mode number $l = 3p + 1$ and $3p + 2$ with p is an integer number. As $l = 3p$, $F_z^e(x,y)$ and $F_z^o(x,y)$ are symmetry under rotation operation C_3 and form A_1 and A_2 representations of point group C_{3v}.

6.3.2 Comparisons of Analytical Solutions with Simulated Results

Here, we simulate mode-field patterns using FDTD technique under symmetry and antisymmetry boundary conditions and compare them with analytical solutions. In the simulation, an exciting source with a cosine impulse modulated by a Gaussian function $F(t) = \exp(-(t - t_0)^2/t_w{}^2)\cos(2\pi f_0 t)$ is applied, with pulse half-width $t_w = 5000\Delta t$, pulse center $t_0 = 2t_w$, and Δt is time step of the Courant limit. The narrow band pulse centered at a resonant frequency f_0 is added to z-directional electromagnetic field E_z or H_z for simulating mode-field pattern of single TM or TE mode. To separate degenerate eigenstate with the same resonant frequency, we apply symmetric and antisymmetric conditions to H_z and E_z relative to the y-axis for TE and TM modes, respectively. For other boundaries, the perfect matched layer absorbing boundary condition is used to conclude the computing window.

For a 2D ETR with a refractive index $n_1 = 3.2$ and a side length $a = 3$ μm, the analytical solutions and FDTD simulation results of z-directional electric field are plotted in Figure 6.3 for the fundamental transverse mode $TM_{0,18}$ and $TM_{0,20}$, where the left and the right sides are even and odd modes $E_z^e(x,y)$ and $E_z^o(x,y)$. Furthermore, we plot the analytical and simulated mode-field patterns of z-directional magnetic fields $H_z^e(x,y)$ and $H_z^o(x,y)$ in Figure 6.4 for the even and odd modes of $TE_{0,18}$. The results indicate that the simulated mode-field patterns are in agreement very well with the analytical solutions. In addition, the results show that the mode-field patterns of $TM_{0,18}$ and $TE_{0,18}$ are symmetrical under C_3 operation.

6.3.3 Size Limit for ETR

For a 2D ETR with refractive index $n_1 = 3.2$ surrounded by air, we calculate the mode frequency and quality factor using FDTD technique [10] and Padé approximation [11]. The Gaussian function–modulated cosine impulse is added to $E_z(x, y)$ and $H_z(x, y)$ at points inside the ETR cavity with low symmetry to excite all resonant modes in a selected frequency range for TE and TM modes,

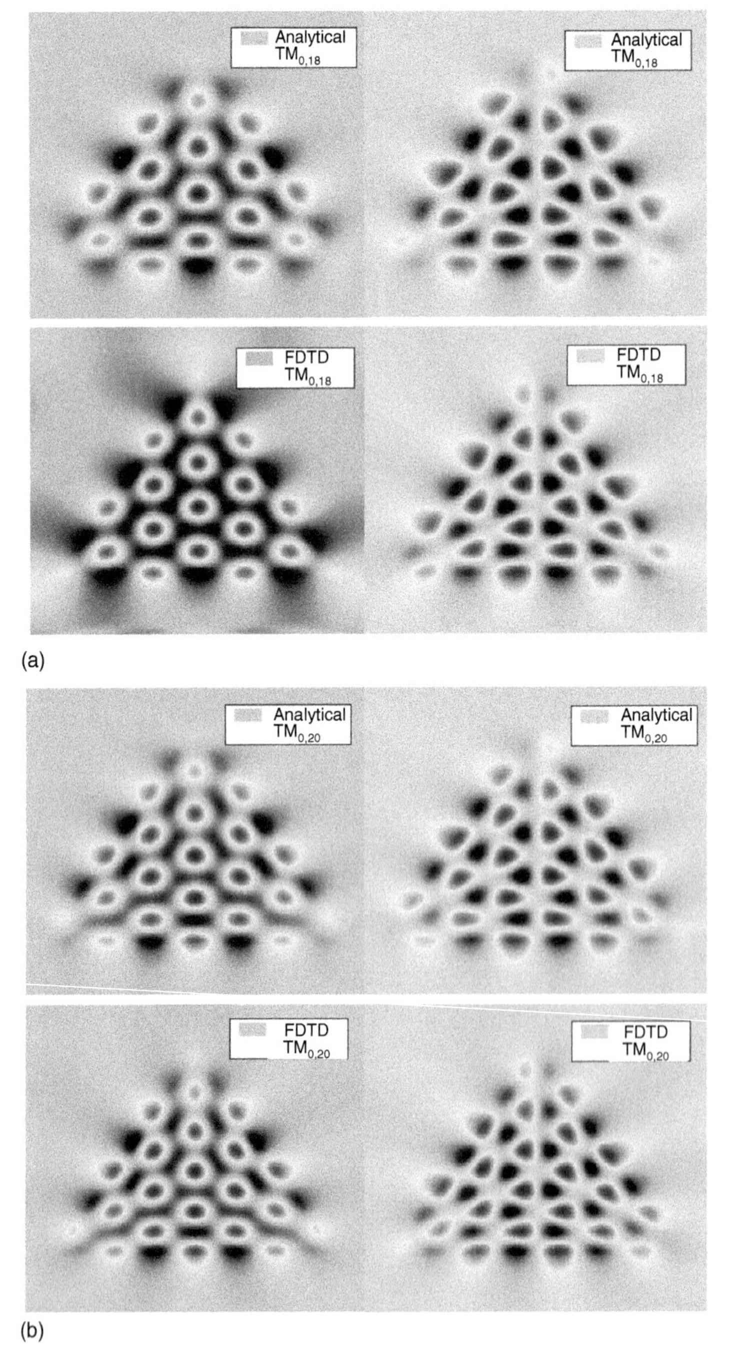

Figure 6.3 Analytical and FDTD simulated mode-field distributions of $E_z^e(x, y)$ and $E_z^o(x, y)$ for (a) $TM_{0,18}$ and (b) $TM_{0,20}$ with even and odd modes at the left and right sides, respectively, in an ETR with a side length of 3 μm and a refractive index of 3.2 surrounded by air.

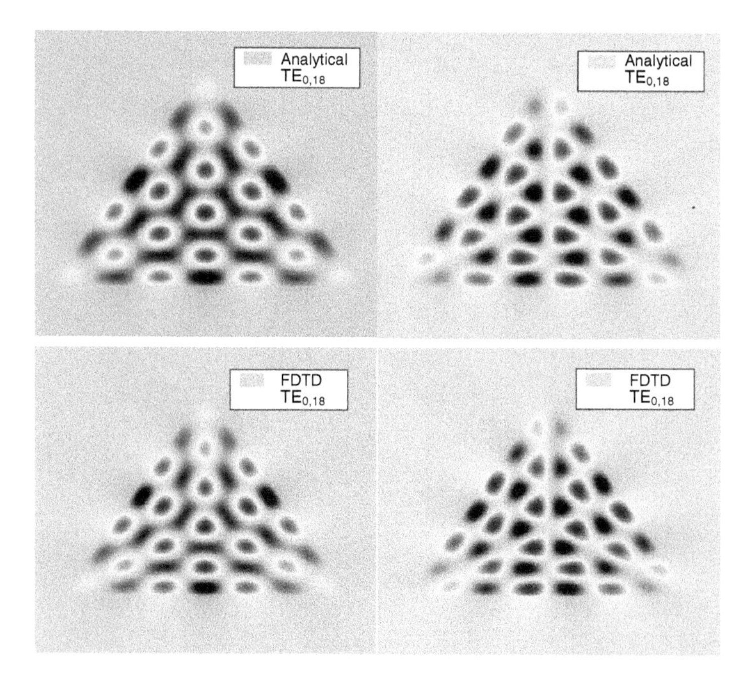

Figure 6.4 Analytical and simulated mode-field distributions $H_z^e(x,y)$ and $H_z^o(x,y)$ for $TE_{0,18}$ with even and odd mode distributions at the left and right sides, respectively, in an ETR with a side length of 3 μm and a refractive index of 3.2 surrounded by air.

respectively. Figure 6.5a,b show the quality factors vs. longitudinal mode number for the fundamental transverse TE and TM modes with wavelengths around 1550 nm, respectively, in the ETR with the side length a = 3, 5, and 6 μm. The numerical results show that the symmetric and antisymmetric modes have the same quality factor as $l = 3p+1$ and $3p+2$, so they are perfect degenerate modes. And the symmetric and antisymmetric modes have different quality factor as $l = 3p$, so they are accidentally degenerate modes. The variation of Q-factors agrees very well with the above mode qualification. In addition, the numerical results show that the quality factors usually decrease with the decrease of longitudinal mode number for TM modes. However, the quality factors of TE modes are more complicated and reach some maximum values as shown in Figure 6.5a.

6.4 Far-Field Emission and Waveguide-Output Coupling

6.4.1 Mode Q-Factor Calculated by Far-Field Emission

The mode Q-factors can be calculated by time-consuming FDTD numerical simulation. In this section, we calculate mode Q-factors for the ETR by far-field emission based on the analytical field distribution, and compare the results with those obtained by FDTD simulation.

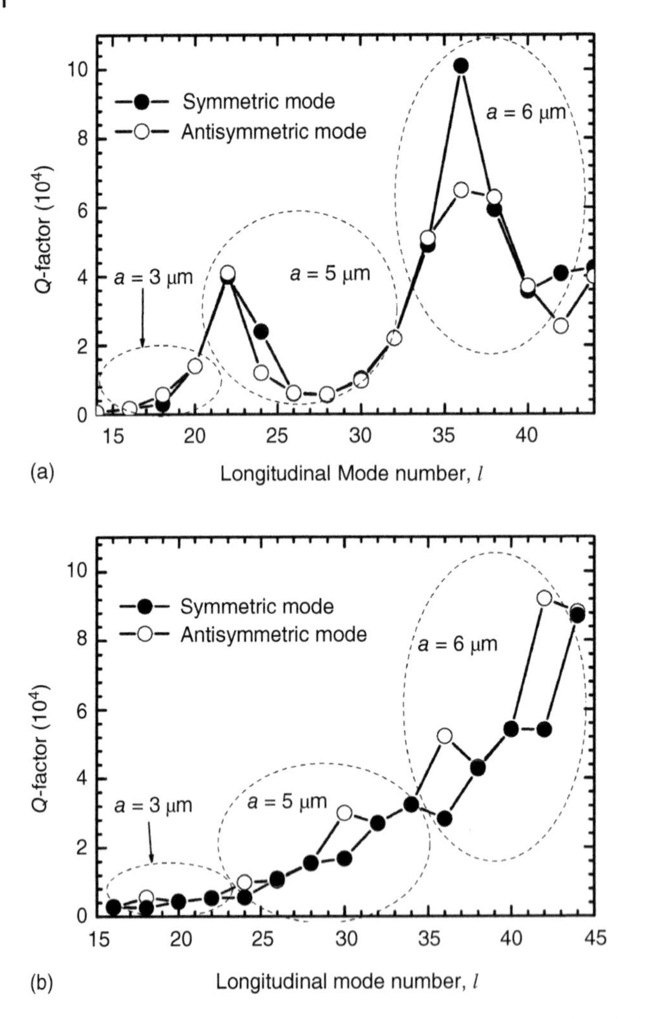

Figure 6.5 Mode Q-factors vs. longitudinal mode number for the fundamental transverse (a) TE and (b) TM modes, with the mode wavelength around 1.55 μm at the ETR with the side length of 3, 5, and 6 μm. Source: Huang et al. [7]. © 2006, IEEE.

For an ETR with unit vectors $\boldsymbol{n}$ normal to the sides AB, BC, and CA as shown in Figure 6.2, we can deduce the far field as [10]:

$$F_{fz}(\boldsymbol{r}') = \frac{\exp(-ikr')}{\sqrt{r'}} K(\varphi), \tag{6.24}$$

$$K(\varphi) = \frac{\exp(i3\pi/4)}{\sqrt{8\pi k}} \oint_{C_s} \left[\frac{\partial F_z(\boldsymbol{r})}{\partial \boldsymbol{n}} - ikF_z(\boldsymbol{r})\boldsymbol{n}\cdot\hat{\boldsymbol{r}}' \right] \exp(ik\boldsymbol{r}\cdot\hat{\boldsymbol{r}}')\, dC, \tag{6.25}$$

where F_{fz} represents far-field electric field E_z for TM and magnetic field H_z for TE modes, respectively, and $\hat{\boldsymbol{r}}' = \boldsymbol{i}\cos(\varphi) + \boldsymbol{j}\sin(\varphi)$ is the unit vector of the position vector $\boldsymbol{r}$' in the external region for describing the far field, $\boldsymbol{i}$ and $\boldsymbol{j}$ are the unit vectors in the x- and y-directions, $\boldsymbol{r}$ is the position vector for expressing the near-field distribution, $\partial/\partial \boldsymbol{n} = \boldsymbol{n}\cdot(\boldsymbol{i}\partial/\partial x + \boldsymbol{j}\partial/\partial y)$, and the integral is along the perimeter of the ETR. In addition to the near field $F_z(\boldsymbol{r})$, the normal derivatives at the ETR sides can also be obtained from the analytical solution of the electric field E_z and magnetic field H_z for the TM and TE modes, respectively. The power emitting from the ETR averaged in a cycle can be calculated by

$$P = \int_0^{2\pi} \frac{\eta}{2} |K(\varphi)|^2 d\varphi. \tag{6.26}$$

where $\eta = \sqrt{\varepsilon_0/\mu_0}$ and $\sqrt{\mu_0/\varepsilon_0}$ for the TM and TE modes, respectively, and $\eta|K(\phi)|^2/2$ is the power angular spectrum. We can also calculate the energy W stored in the ETR averaged in a cycle using the analytical field distribution by

$$W = \oiint \frac{\varsigma}{2} \left[\frac{n_1^2 |F_z|^2}{2} + \frac{|\partial F_z/\partial x|^2 + |\partial F_z/\partial y|^2}{2k^2} \right] dx\, dy \tag{6.27}$$

where $\zeta = \mu_0/n_1^2$ and ε_0 for TE and TM modes, respectively, and n_1 is the distribution of refractive index. Based on the emitted power and the stored energy, we can calculate mode Q-factor by

$$Q = \omega \frac{W}{P}, \tag{6.28}$$

where ω is the mode angular frequency.

The mode Q-factors obtained from (6.28) are plotted in Figure 6.6 (a) TE and (b) TM modes, as solid and open squares for symmetric and antisymmetric modes, respectively. The degenerate modes of the representations E with $l = 3p + 1$ and $3p + 2$ have the same mode Q-factor, and the accidentally degenerate modes with $l = 3p$, which form the A_1 and A_2 representations, have different mode Q-factors. The mode Q-factors calculated by FDTD simulation are also plotted in Figure 6.6 as solid and open circles for symmetric and antisymmetric modes, respectively. The results show that the far-field emission based on the analysis field distribution is rather good in estimating mode Q-factors. However, the mode Q factors calculated by FDTD simulation have an evident peak around $l = 22$, which is four times the value obtained from the far-field emission based on the analysis solutions. The evanescent field distributions (6.15) and (6.16) in the external regions of the ETRs are obtained under approximate conditions, which limit the accuracy of the analytical solution.

6.4.2 Output Coupling by Connecting a Waveguide

Due to the break of symmetries, the field distributions of whispering gallery like modes in the ETR have multiple angular components and are not homogeneous along the sides of the resonators. For the ETR with a waveguide connected to the place with a weak mode-field distribution, we expect that the confined modes can still have high Q-factors suitable for realizing directional emission microlasers. By connecting an output waveguide to one of vertices of the ETR, we simulate the mode Q factor and output coupling efficiency vs. the width of the output waveguide by FDTD technique. Firstly, we use the pulse sources covering a wide frequency range to excite all the considered TE modes, and then apply the Padé

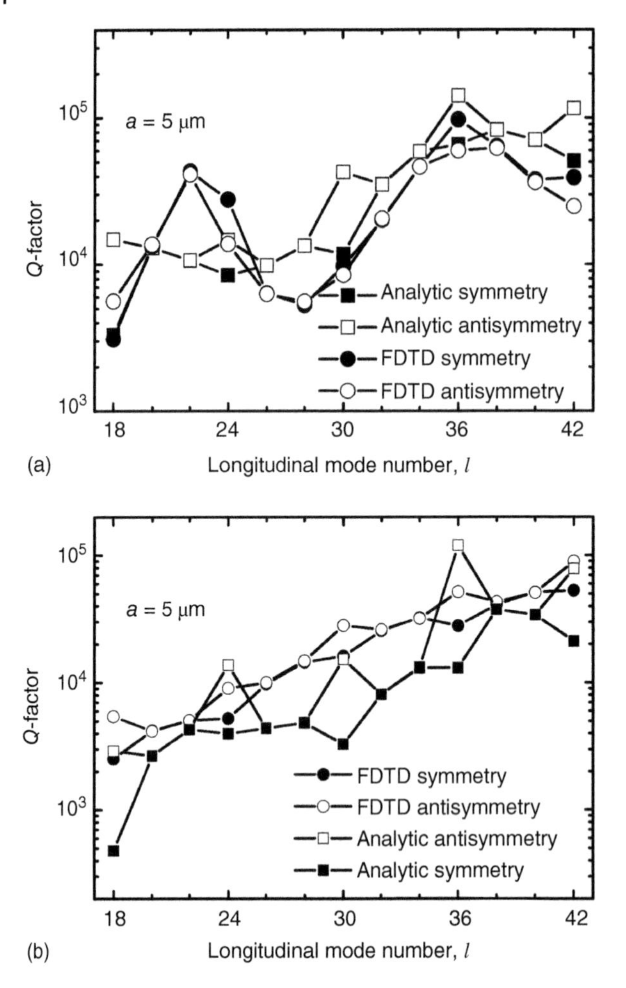

Figure 6.6 Mode Q-factors vs. longitudinal mode number for (a) TE modes and (b) TM modes in an ETR with side length 5 μm, calculated by FDTD simulation and far-field emission based on analytical near field pattern. Source: Huang et al. [7]. © 2006, IEEE.

approximation method to calculate the intensity spectra from the FDTD output. After obtaining the mode frequencies and Q-factors, we use an exciting source to cover a narrow-frequency range centered at an obtained mode frequency, to excite the field distribution of the corresponding mode. Based on the obtained mode field distribution, the output efficiency is estimated as the ratio of the energy flux confined in the output waveguide to the total energy flux around the whole resonator, with the energy flux calculated by integrating the Poynting vector over one period.

Symmetric and antisymmetric exciting sources are used to simulate the symmetric and antisymmetric modes relative to the midline of the output waveguide and the ETR. The mode Q-factors and output efficiencies of $TE_{0,60}$ with a mode wavelength of 1.537 μm are plotted in Figure 6.7 as functions of the width of the output waveguide, for an ETR with a side length of 10 μm and refractive index of 3.2 surrounded by air. The open squares and circles are the output efficiencies of the symmetric and antisymmetric modes of $TE_{0,60}$, respectively, and the solid squares and circles are the

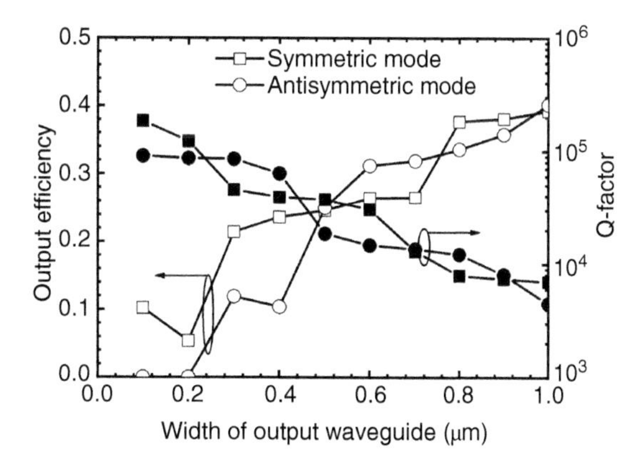

Figure 6.7 Mode Q factor and output efficiency vs. the width of output waveguide for $TE_{0,60}$ mode in an ETR with a side length of 10 μm. The solid and open symbols are mode Q factors and output efficiencies, respectively.

corresponding Q-factors. With the increase of the width of the output waveguide, the mode Q-factors decrease and the output efficiencies increase. At the output waveguide width of 1 μm, the Q-factors are 7.0×10^3 and 4.5×10^3 for the symmetric and antisymmetric $TE_{0,60}$ modes, respectively, and the output efficiencies are about 0.4. The mode Q factors are in the same magnitude as a Fabry–Perot cavity with a cavity length of 300 μm and a reflectivity of 0.3 with cleaved mirrors. Therefore, directional emission ETR microlasers with an output waveguide connected to one of the vertices of the ETR can be realized with a high-efficiency output power.

6.5 Mode Analysis Using Reflected Phase Shift of Plane Wave

6.5.1 Mode Analysis Using Mode Light Ray Approximation

In this section, we analyze mode solutions for the ETR based on the reflection phase shift of plane waves on the boundaries of the ETR as in [8]. For the ETR with refractive index of $n_1 > 2$ surrounded by air, a plane wave parallel to the sides of the ETR will experience total internal reflection on the boundaries of the ETR. Through the image operation, the origin ETR Ω_1, and the image regions $\Omega_j (j = 2, 3, 4, 5, 6)$ form a waveguide with width of $\sqrt{3}a/2$ as shown in Figure 6.8, where a is the side length of the ETR, and k_1 and k_2 are unfolded plane waves.

The total reflection of the plane wave on the boundary between the different regions is equivalent to transmitting into the next image region with a phase shift of $\phi(\theta)$, where θ is the incident angle and satisfies $\theta > \sin^{-1}(1/n_1)$ for the total internal reflection. The boundary conditions of the Maxwell equations will yield a unity reflection coefficient with the phase shift $\phi(\theta)$ as:

$$\phi(\theta) = -2\tan^{-1}\left(\frac{\xi\sqrt{n_1^2\sin^2\theta - 1}}{n_1\cos\theta}\right) \tag{6.29}$$

with $\xi = n_1^2$ and 1 for the TE and TM modes. In Figure 6.8, a plane wave is supposed to emerge from side AB of the origin region Ω_1, and then transmits through the

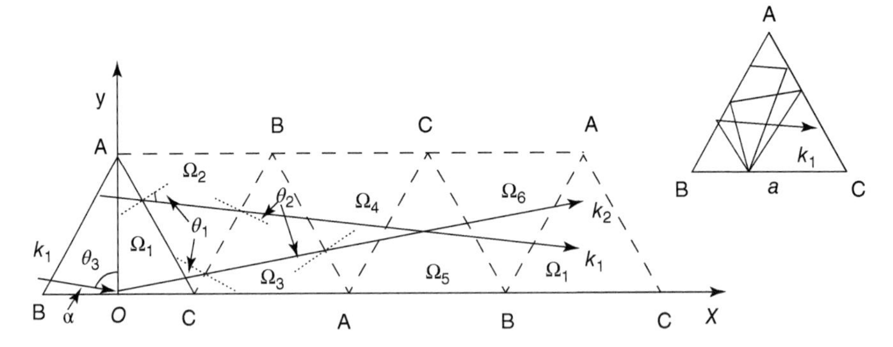

Figure 6.8 The unfolded plane wave and the corresponding cavity of one-period mode light rays transmit inside the ETR as shown in the upper right side. Source: Yang et al. [8]. © 2009, IEEE.

image regions in the sequence Ω_2, Ω_3, Ω_4, etc. Finally, the plane wave emerges from side AB of region Ω_1' with the same angle in Ω_1; thus, region Ω_1' is equivalent to the origin region Ω_1, and the ETR can be modeled as a cavity with the width of $\sqrt{3}a/2$ and the one-period length of $3a$.

Assuming F_z is the electric field E_z for the TM mode or the magnetic field H_z for the TE mode, which satisfies wave Eq. (6.1), we can express the mode field inside the waveguide propagating along the x direction by two plane waves with field distributions $\exp[in_1k_0(x\cos\alpha + y\sin\alpha)]$ and $\exp[in_1k_0(x\cos\alpha - y\sin\alpha)]$, where α is angle between the plane waves k_1 and k_2 and x-axis, and the time-dependence factor $\exp(-i\omega t)$ is omitted. The angles between the plane wave k_1 and sides BC, CA, and AB of origin region Ω_1 are α, $\pi/3-\alpha$, and $\pi/3+\alpha$, respectively. Due to the symmetry of the ETR, we also can choose the side CA or AB as the x-axis; thus, angles between plane wave and the x-axis equal to α, $\pi/3-\alpha$, and $\pi/3+\alpha$ are corresponding to the same mode. The incident angles of the plane waves on the three sides of the ETR θ_1, θ_2, and θ_3 are expressed by α as:

$$\theta_1 = \pi/6 + \alpha, \quad \theta_2 = \pi/6 - \alpha, \quad \theta_3 = \pi/2 - \alpha. \tag{6.30}$$

The angle α is set in the range from 0 to $\pi/6$ to avoid repeated calculation in the following discussions. All of the incident angles should be larger than the critical angle for total internal reflection; thus, α should satisfy $0 \le \alpha < \pi/6 - \sin^{-1}1/n_1$. Considering the phase shifts on the boundaries inside the equivalent cavity, we can express the field distributions for the plane waves k_1 and k_2 as

$$F_{z1}(x,y) = \exp[in_1k_0(x\cos\alpha - y\sin\alpha) + i\phi_{1j}], \tag{6.31}$$

$$F_{z2}(x,y) = \exp[in_1k_0(x\cos\alpha + y\sin\alpha) + i\phi_{2j}], \tag{6.32}$$

where ϕ_{1j} and ϕ_{2j} are the phase shifts of the plane waves k_1 and k_2 in Ω_j, respectively, and satisfy

$$\phi_{12} = \phi_{11} + \phi(\theta_1), \quad \phi_{13} = \phi_{12} + \phi(\theta_2), \quad \phi_{14} = \phi_{13} + \phi(\theta_1),$$
$$\phi_{15} = \phi_{14} + \phi(\theta_2), \quad \phi_{16} = \phi_{15} + \phi(\theta_1), \quad \phi_{11'} = \phi_{16} + \phi(\theta_2),$$

$$\phi_{22} = \phi_{21} + \phi(\theta_2),\ \ \phi_{23} = \phi_{22} + \phi(\theta_1),\ \ \phi_{24} = \phi_{23} + \phi(\theta_2),$$
$$\phi_{25} = \phi_{24} + \phi(\theta_1),\ \ \phi_{26} = \phi_{25} + \phi(\theta_2),\ \ \phi_{21'} = \phi_{26} + \phi(\theta_1). \tag{6.33}$$

The phase shift Eq. (6.33) guarantee that the total internal reflections of the plane waves on the boundaries inside the cavity satisfy Maxwell's equations. In addition, according to the boundary conditions at the $y = 0$ and $y = \sqrt{3}a/2$ planes, the field distributions of the plane waves k_1 and k_2 satisfy

$$F_{z1}(x,y) = F_{z2}(x,y)\exp[i\phi(\theta_3)] \quad y = 0, \tag{6.34}$$

$$F_{z2}(x,y) = F_{z1}(x,y)\exp[i\phi(\theta_3)] \quad y = \sqrt{3}a/2. \tag{6.35}$$

Based on Eqs. (6.31)–(6.35), we obtain

$$\phi_{21} = \phi_{11} + \phi(\theta_3) + 2m_1\pi, \tag{6.36}$$

$$\phi_{11} + \phi(\theta_1) - n_1 k_0 \frac{\sqrt{3}}{2} a \sin\alpha = \phi_{21} + \phi(\theta_2) + n_1 k_0 \frac{\sqrt{3}}{2} a \sin\alpha + \phi(\theta_3) + 2m_2\pi, \tag{6.37}$$

where m_1 and m_2 are integer numbers. From Eqs. (6.36) and (6.37), we obtain

$$n_1 k_0 \sqrt{3}a \sin\alpha + 2\phi(\theta_3) - \phi(\theta_1) + \phi(\theta_2) = -2(m_1 + m_2)\pi. \tag{6.38}$$

Let $m = -m_1 - m_2$, then we have

$$n_1 k_0 \sqrt{3}a \sin\alpha + 2\phi(\theta_3) - \phi(\theta_1) + \phi(\theta_2) = 2m\pi, \tag{6.39}$$

where $m = 0, 1, 2, \ldots$, is the transverse mode number. The definition is similar to that in (6.7) without perfectly confined approximation. The fields Ω_1' must be in phase with the fields in Ω_1 to produce a consistent solution, thus

$$n_1 k_0 3a \cos\alpha + \phi_{11'} - \phi_{11} = 2l\pi, \tag{6.40}$$

$$n_1 k_0 3a \cos\alpha + \phi_{21'} - \phi_{21} = 2l\pi. \tag{6.41}$$

From Eqs. (6.33), (6.40) and (6.41), we obtain

$$\phi_{11'} - \phi_{11} = \phi_{21'} - \phi_{21} = 3\phi(\theta_1) + 3\phi(\theta_2), \tag{6.42}$$

and then

$$n_1 k_0 3a \cos\alpha + 3\phi(\theta_1) + 3\phi(\theta_2) = 2l\pi, \tag{6.43}$$

where l is the longitudinal mode number. From (6.29, 6.30, 6.39, 6.43), we can obtain the mode wavelength λ and the angle α for different modes. The above technique only suits the geometry that can be mapped into a slab waveguide, such as isosceles right triangle and rectangle besides the ETR. The real field distribution in the ETR $F_z(x, y)$ at $(x, y) \in \Omega_1$ can be obtained as a sum of the fields in the six image points (x_j, y_j),

$$F_z(x,y) = \sum_{j=1}^{6} F_{z1}(x_j, y_j) + \sum_{j=1}^{6} F_{z2}(x_j, y_j), \tag{6.44}$$

where

$$x_1 = x, \; y_1 = y,$$
$$x_2 = 3a/4 - x/2 - \sqrt{3}y/2, \; y_2 = \sqrt{3}a/4 - \sqrt{3}x/2 + y/2,$$
$$x_3 = 3a/4 - x/2 + \sqrt{3}y/2, \; y_3 = \sqrt{3}a/4 - \sqrt{3}x/2 - y/2,$$
$$x_4 = 3a/2 + x_1, \; y_4 = \sqrt{3}a/2 - y_1,$$
$$x_5 = 3a/2 + x_2, \; y_5 = \sqrt{3}a/2 - y_2,$$
$$x_6 = 3a/2 + x_3, \; y_6 = \sqrt{3}a/2 - y_3. \tag{6.45}$$

The field distributions of Eq. (6.44) can be combined into even and odd modes relative to the y-axis as:

$$F_z^e(x,y) = F_z(x,y) + F_z(-x,y), \tag{6.46}$$

$$F_z^o(x,y) = F_z(x,y) - F_z(-x,y). \tag{6.47}$$

The superscripts "e" and "o" indicate even and odd symmetry modes, respectively. In addition, if l_2 and m have different parities, we have the following equation based on (6.31)–(6.33), (6.39) and (6.43):

$$F_{z1}(x_j, y_j) + F_{z2}(x_j, y_j) + F_{z1}(x_{j+3}, y_{j+3}) + F_{z2}(x_{j+3}, y_{j+3}) = 0, \tag{6.48}$$

where $j = 1, 2, 3$. Thus, (6.44) leads to a trivial solution $F_z(x, y) = 0$ for all $(x, y) \in \Omega_1$ when the longitudinal mode number l and the transverse mode number m have different parities. So the mode numbers should satisfy the following relation for confined modes

$$l + m = 2, 4, 6, \ldots \tag{6.49}$$

The longitudinal mode number l is defined as in [6, 8] with one period of $3a$, and the definition is the same as in Sections 6.2 and 6.3. The difference is that the phase shifts for the light rays with all possible incident angles are considered, and no approximation is used in the estimation of internal field distributions. Therefore, this method can give more accurate mode wavelength and mode Q value, especially for the high-order transverse modes and the modes in small-size ETRs.

6.5.2 Comparison of Mode Q Factors

The mode Q-factors obtained from (6.28) by far-field emission using analytical field distributions (6.46) and (6.47) are plotted in Figure 6.9 as solid and open squares, respectively, for even and odd modes of the fundamental and the first-order transverse modes. The degenerate mode behaviors are predicted in Section 6.3. The degenerate modes with longitudinal mode numbers $l = 3p+1$ and $3p+2$ have the same Q-factors, and the accidentally degenerate modes with $l = 3p$ have different Q-factors, where p is an integer. The mode Q-factors calculated by the FDTD simulations are also plotted in Figure 6.9 as solid and open circles for even and odd modes, respectively. The mode Q factors calculated from analysis solution agree very well with the FDTD results, especially for TM modes. The differences of

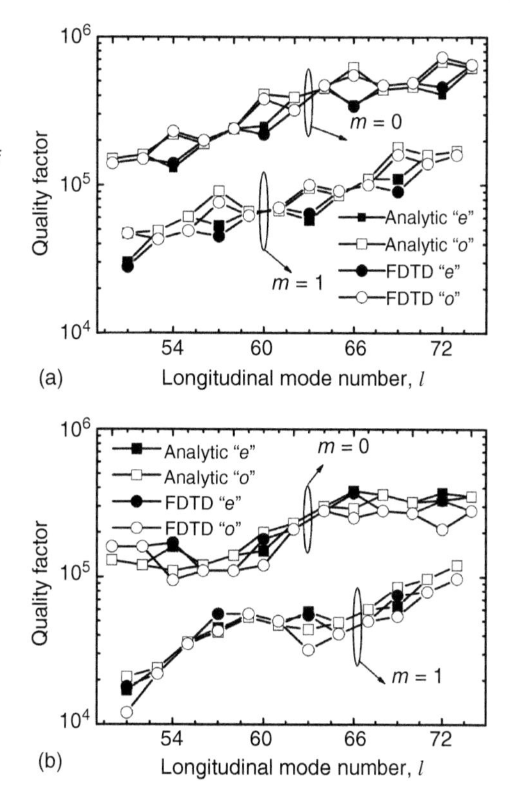

Figure 6.9 Mode Q-factors vs. longitudinal mode number l_2 for (a) TM modes and (b) TE modes in an ETR with side length of 10 μm and a refractive index of 3.2 confined by air. Solid and open squares are Q-factors of the even and odd modes, respectively, obtained by far-field emission based on analytical near-field pattern. Solid and open circles are Q-factors of the even and odd modes, respectively, obtained by the FDTD simulation. Source: Yang et al. [8]. © 2009, IEEE.

Q-factors obtained by two methods are less than 10%. For the TM modes with $l = 3p$, both the analytical model and FDTD simulation show that the Q-factors of odd modes are larger than those of even modes, but the phenomenon is not observed for TE modes. The numerical results show that the Q-factors usually increase with the increase of the longitudinal mode numbers. The Q-factors of TE modes oscillate with the longitudinal mode numbers similarly as in Figure 6.6, but with much weaker peak values for large longitudinal mode numbers.

In addition, the mode wavelengths of $TM_{0,30}$ and $TM_{2,30}$ modes obtained from (6.43) are 1.556 and 1.542 μm, which are only 0.26% smaller than the FDTD simulation results. The light ray method was applied for analyzing mode characteristics of the ETR and the radiation loss was overestimated by adding emission power from all sides of the ETR [12]. But the radiation fields in the external vertex regions are composed of radiation fields from two adjacent sides with potential of coherence canceling of the radiation fields, especially for odd symmetry modes. Furthermore, the method can be applied for ETR confined by a multilayer structure, with the reflection factor calculated from the boundary conditions of Maxwell's equations.

6.5.3 Effect of Metal Layer on Mode Confinement

The ETR lasers were fabricated with the ETR side walls surrounded by an insulator SiO_2 layer and Ti/PtAu p-electrode [9]. Thus, a complicated boundary condition

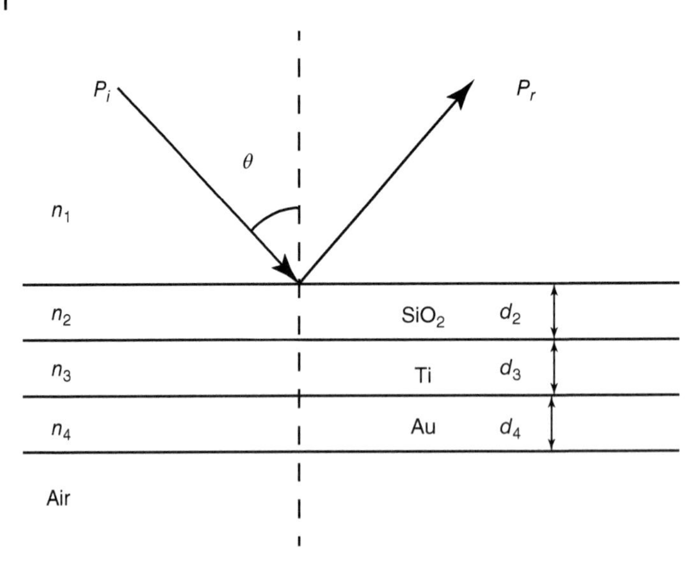

Figure 6.10 Schematic diagram of the reflection of plane wave on a multilayer boundary for the ETR with side walls confined by an insulator layer and p-electrode. Source: Yang et al. [8]. © 2009, IEEE.

should be considered for modeling mode characteristics. We consider an ETR with an insulator layer, a titanium layer, and a gold layer between the semiconductor material and air. The insulator used in the calculation is silica with a refractive index of 1.45, and the refractive indices of titanium and gold are taken to be $3.7 + 4.5i$ and $0.18 + 10.2i$, respectively. The schematic diagram of the multilayer structure is shown in Figure 6.10, the reflection factor $r(\theta) = P_r/P_i$ can be calculated from the boundary conditions of Maxwell's equations, where P_r and P_i are the reflected and incident field amplitudes.

Taking the thicknesses of the insulator and gold layers to be 0.4 and 0.2 μm, we plot the power reflectivity $R(\theta) = |r(\theta)|^2$ for the multilayer structure with different thicknesses of titanium layers in Figure 6.11 for (a) TM and (b) TE plane waves. With the increase of the thickness of the titanium layer, the power reflectivity $R(\theta)$ of the plane wave with incident angle less than 30° decreases rapidly, which results in a decrease of the Q factors for the modes in the ETR. When $d_3 = 0.02$ μm, the power reflectivity of TM plane waves at 30° is greater than 95%, but that of TE plane waves is less than 60%. The TE modes with the incident angles near 28°, which is close to the Brewster angle of the semiconductor–insulator interface, will be suppressed by the absorption of titanium layer. When the thickness of titanium layer is 0, we find that the reflectivity $R(\theta)$ is close to 1 for all incident angles in the range between 0° and 90°.

6.6 Mode Characteristics of ETR Microlasers

6.6.1 Device Fabrication

ETR microlasers with an output waveguide connected to one of the vertices were fabricated using InP/GaInAsP edge-emitting laser wafer grown by metal–organic

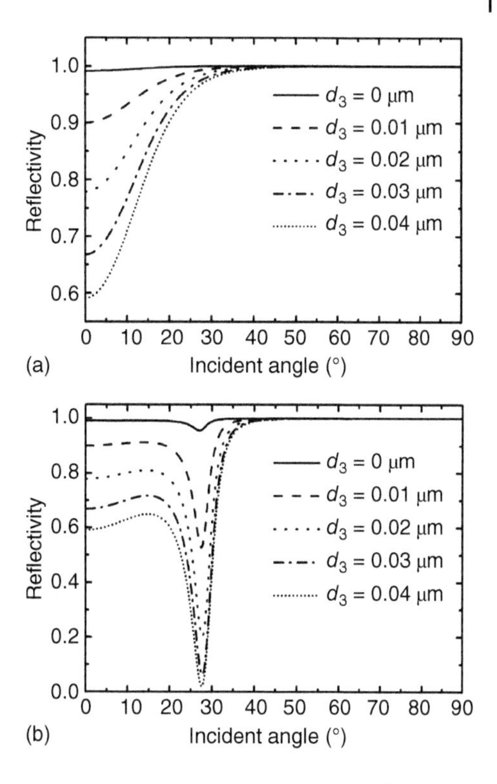

Figure 6.11 The power reflectivity of (a) TM and (b) TE plane waves on the multilayer boundary with different d_3 vs. the incident angle. Source: Yang et al. [8]. © 2009, IEEE.

chemical vapor deposition on (001) InP substrate, with the thicknesses of the quantum wells and the barrier layers being 6 and 10 nm, respectively. The active region is sandwiched by up and down 110-nm-thick GaInAsP cladding layers, and the upper layers are a 1.5-μm-thick p-InP and a p^+-InGaAs ohmic contact layer. Firstly, an 800-nm SiO_2 was deposited by plasma-enhanced chemical vapor deposition (PECVD) on the as-grown InP/GaInAsP laser wafer as a hard mask for dry etching. Next, the ETR patterns are transferred onto the SiO_2 layer using standard photolithography and ICP etching techniques. The patterned SiO_2 hard masks define the ETR geometry for subsequent ICP process to etch InP and GaInAsP with BCl_3–Cl_2 gas mixtures. The ICP etching depth is about 4 μm, which is greater than the vertical penetration depth of the optical field of the confined modes. After the ICP etching, a chemical etching process using the solution of $Br_2 + HBr + H_2O$ is applied to remove the reactants on the side walls formed in the ICP etching process and improve the smooth of the side walls. Then, the residual SiO_2 hard masks are removed using diluted HF solution. Finally, a 300-nm SiO_2 insulating layer is deposited and a contact window is opened on top of each ETR pattern using wet etching process. A top Ti–Pt–Au p-contact is formed using standard metal deposition and the wafer is lapped down to a thickness of about 100 μm, and an Au/Ge/Ni/Au metallization is used for the n-type contacts. The scanning electron microscope (SEM) pictures of an ETR formed just after the ICP etching process and one of the vertices of an ETR after the wet chemically etching process are presented in Figure 6.12a,b, respectively. The vertical side walls of the ETR in Figure 6.12a are a little rough after the ICP etching process. In addition, some vertical lines exist

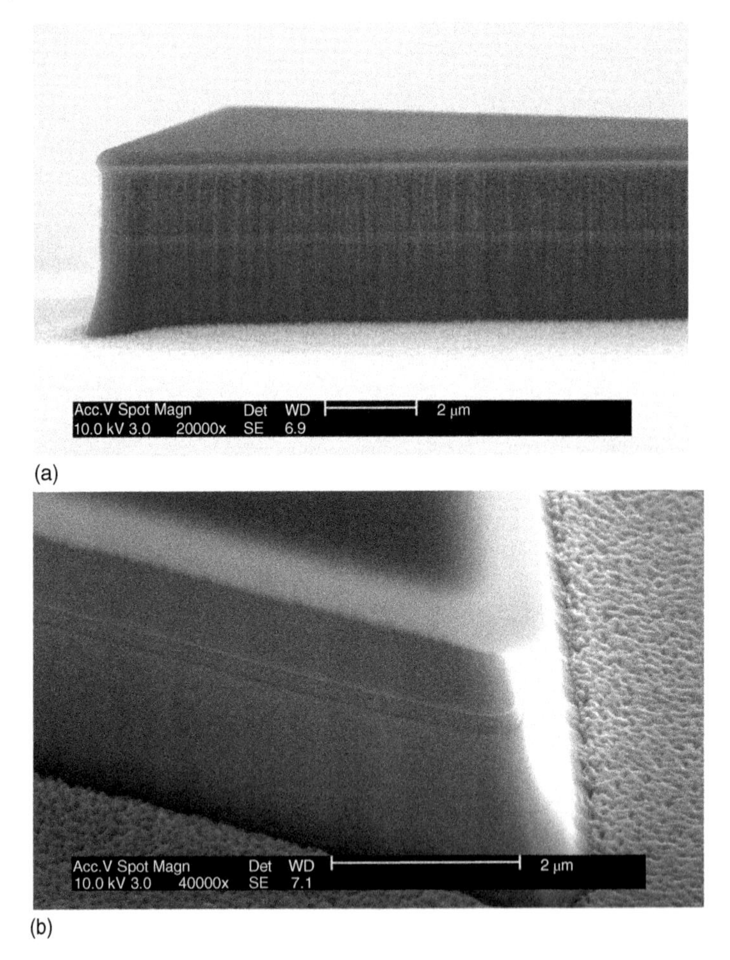

Figure 6.12 Scanning electron microscope images of (a) an ETR formed just after the ICP etching process and (b) close-up view of one of the vertices of an ETR after the ICP and the wet chemically etching processes. Source: Huang et al. [9]. © 2007, IEEE.

due to the imperfect mask pattern. After the wet chemically etching process, the vertical side walls are very smooth.

6.6.2 Lasing Characteristics

The fabricated InP/GaInAsP ETR lasers are cleaved across the output coupling waveguide and mounted on heat sink for measuring output power and lasing spectra. The laser output power vs. injection current for an ETR laser with the side length of 20 μm are plotted in Figure 6.13a under continuous-wave (CW) and pulsed injection currents, with a duty circle of 0.2% for the pulse cases. The output powers of the pulsed operation are peak powers as measured power divided by the duty circle. The maximum CW output power is 0.17 and 0.067 mW at 290 and 295 K, with the threshold current of 34 and 43 mA, respectively. The CW output power is

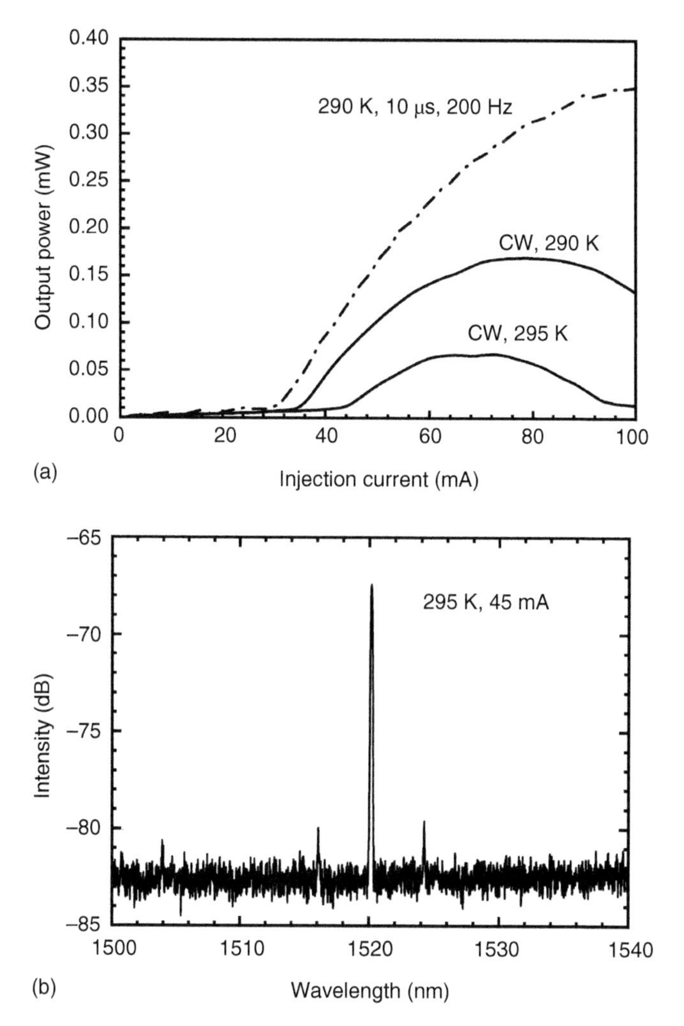

Figure 6.13 (a) Output power vs. continuous and pulsed injection current at 290 K, and continuous current at 295 K, (b) lasing spectra at $T = 295$ K and $I = 45$ mA, for an InP/GaInAsP ETR laser with a side length of 20 μm. Source: Huang et al. [9]. © 2007, IEEE.

limited by the temperature rising with the increase of the injection current. Due to the heating, the laser returns spontaneous emission state again as CW injection current is larger than 96 mA at 295 K. The threshold current of 34 mA corresponds to threshold current density of 19.6 kA cm^{-2}. The high threshold current can be partly explained by a low mode Q-factor as a 2-μm-wide output waveguide connected to one of the vertices of the ETR. The lasing spectrum at temperature of 295 K and CW injection current of 45 mA is presented in Figure 6.13b. The lasing mode wavelength is 1520.16 nm with two minor modes at 1516.10 and 1524.24 nm.

Finally, we compare the lasing spectra with the analyzed mode wavelength for an ETR microlaser with a side length of 30 μm. The room-temperature lasing spectra of the ETR laser are presented in Figure 6.14a with evident peaks of different

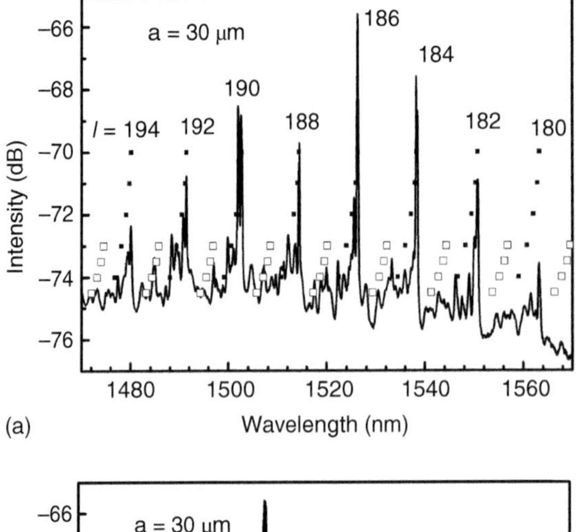

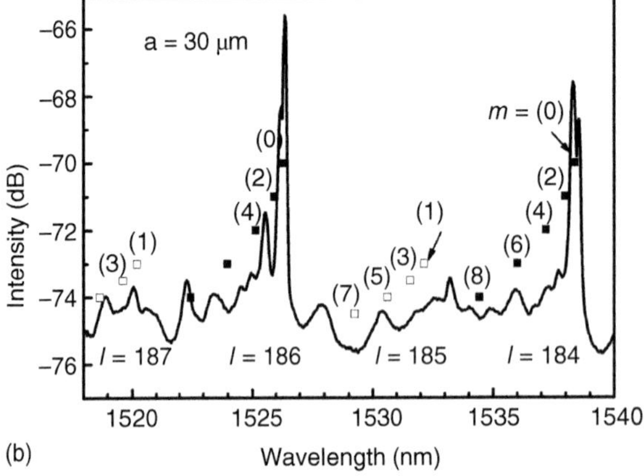

Figure 6.14 (a) Lasing spectrum of an ETR microlaser with a side length of 30 μm at room temperature and longitudinal mode numbers of even transverse modes, and (b) detailed spectrum compared with theoretical mode wavelength marked by mode numbers (m, l), and the corresponding longitudinal mode number l_2. The solid and open squares are analytical mode wavelengths for the even and odd transverse modes, respectively.

longitudinal and transverse mode numbers. Due to the p-electrode confinement, we can find resonant peaks up to the eighth-order transverse mode. The mode refractive index is approximately obtained from the group index by $n_g = n_1 + Edn_1/dE$, which is derived from longitudinal mode interval of a Fabry–Perot cavity laser fabricated from the same laser wafer. The mode refractive index is approximately expressed as $n_1 = 2.0642 + 1.396E$, where E is the photon energy in the unit of eV. The mode wavelengths calculated by (6.20) are marked in Figure 6.14a by solid and open squares for the even and odd transverse modes, respectively, with longitudinal mode numbers from $l = 180$ to 194. In Figure 6.14b, the detailed lasing spectra are compared with analytical mode wavelengths with mode numbers (m, l) for the even order (zeroth, second, fourth, sixth, eighth) and odd order (first, third, fifth, seventh) transverse modes. The side walls of the fabricated ETR lasers surrounded by SiO_2 and the golden metal of the p-electrode as shown in Figure 6.15 can enhance the mode confinement, especially for the higher-order transverse modes. Different transverse modes may have different output coupling efficiencies, especially the even and odd transverse modes, and degenerate modes can also influence the lasing spectra. Accounting the practical complexity, we conclude that the lasing spectra agree very well with the analytical results.

Figure 6.15 Schematic diagram of an ETR microlaser with sidewalls surrounded by SiO_2 and p-electrode.

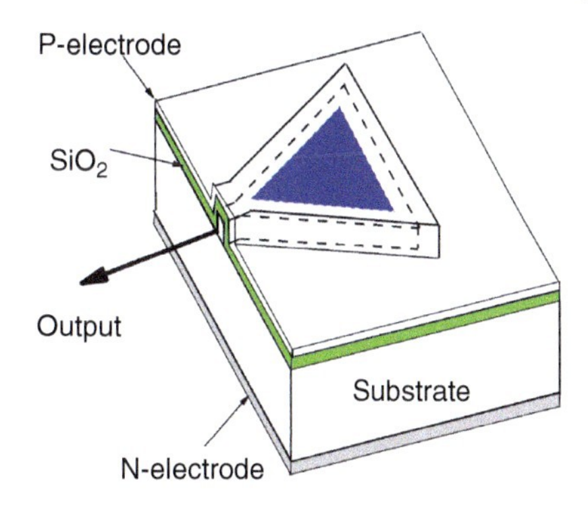

6.7 Summary

The mode characteristics have been analyzed for an ETR with eigenstates according to the irreducible representations of the point group C_{3v}. The analytical solutions are obtained with boundary conditions satisfied under an approximation $\beta_l >> \kappa_m$. Doubly degenerate eigenstates of the A_1 and A_2 irreducible representations of C_{3v} are classified as the longitudinal mode number l is a multiple of 3, and the others are the E irreducible representation for the other values of l. The ETR with standing mode-field patterns is suitable for realizing directional emission microlasers by connecting an output waveguide to weak mode field region, such as a vertex of the ETR.

Furthermore, the mode characteristics are analyzed based on the reflection phase shifts of plane waves on the boundaries. The analytical mode wavelengths and mode Q-factors calculated by far-field emission based on analyzed field patterns agree very well with those calculated by FDTD simulation.

Continuous-wave electrically injected InGaAsP/InP ETR microlasers are realized at room temperature with a side length down to 10 μm. Clear lasing spectra with evident peaks up to eighth transverse mode are observed for an ETR microlaser with a side length of 30 μm, and the lasing peaks are assigned with corresponding mode numbers by comparing with the analytical solution of mode wavelength.

References

1 Marinace, J.C., Michel, A.E., and Nathan, M.I. (1964). Triangular injection lasers. *Proc. IEEE* 52: 722–723.

2 Ando, S., Kobayashi, N., and Ando, H. (1996). Triangular-facet laser with optical waveguides grown by selective area metalorganic chemical vapor deposition. *Jpn. J. Appl. Phys.* 35 (Pt. 2, 4A): L411–L413.

3 Ando, S., Kobayashi, N., and Ando, H. (1997). Triangular-facet lasers coupled by a rectangular optical waveguide. *Jpn. J. Appl. Phys.* 36 (Pt. 2, 2A): L76–L78.

4 Chang, H.C., Kioseoglou, G., Lee, E.H. et al. (2000). Lasing modes in equilateral-triangle laser cavities. *Phys. Rev. A* 62: 013816.

5 Huang, Y.Z., Guo, W.H., and Wang, Q.M. (2000). Influence of output waveguide on mode quality factor in semiconductor microlasers with an equilateral triangle resonator. *Appl. Phys. Lett.* 77: 3511–3513.

6 Huang, Y.Z., Guo, W.H., and Wang, Q.M. (2001). Analysis and numerical simulation of eigenmode characteristics for semiconductor lasers with an equilateral triangle micro-resonator. *IEEE J. Quantum Electron.* 37 (1): 100–107.
7 Huang, Y.Z., Chen, Q., Guo, W.H. et al. (2006). Mode characteristics for equilateral triangle optical resonator. *IEEE J. Sel. Top. Quantum Electron.* 12 (1): 59–65.
8 Yang, Y.D., Huang, Y.Z., and Wang, S.J. (2009). Mode analysis for equilateral-triangle-resonator microlasers with metal confinement layers. *IEEE J. Quantum Electron.* 45 (12): 1529–1536.
9 Huang, Y.Z., Hu, Y.H., Chen, Q. et al. (2007). Room-temperature continuous-wave electrically injected InP–GaInAsP equilateral-triangle-resonator lasers. *IEEE Photonics Technol. Lett.* 19: 963–965.
10 Taflove, A. and Hagness, S.C. (2005). *Computational Electrodynamics: The Finite-Difference Time-Domain Method*. Boston, MA: Artech House.
11 Guo, W.H., Li, W.J., and Huang, Y.Z. (2001). Computation of resonant frequencies and quality factors of cavities by FDTD technique and Padé approximation. *IEEE Microwave Wirel. Compon. Lett.* 11: 223–225.
12 Wysin, G.M. (2006). Electromagnetic modes in dielectric equilateral triangle resonators. *J. Opt. Soc. Am. B: Opt. Phys.* 23: 1536–1549.

7

Square Microcavity Lasers

7.1 Introduction

Regular polygonal microcavities supporting high-quality (Q) factor whispering-galley-like (WG-like) modes have been widely studied for realizing low-power-consumption compact-size photonic devices [1–6]. The field distributions of the WG-like modes exhibit sinusoidal-modulated envelope along the sidewalls; thus, an output waveguide directly connected to the position with weak mode field can achieve efficient unidirectional emission without a distinct degradation of mode Q factors [7]. For the square microcavities with integrable internal dynamic, a quasi-analytical solution can be obtained for the WG-like modes, which makes the performance of the square microcavity devices more predictable [8–10]. Compared with the WG modes in circular microdisks, the fields of the WG-like modes in the square microcavities distribute more uniformly over the whole cavity region, which promises high injection efficiency and avoids burning-induced carrier diffusion in microdisk lasers [11]. In addition, the high-Q WG-like modes in the square microcavities are nondegenerate modes with the free spectra range (FSR) twice that of the WG modes in the microdisks with similar sizes, which makes the square microlasers more suitable for stable single-mode operation [12].

Square microcavities have been studied intensively for potential applications in photonic integrated circuits and optical interconnects [13–23]. Square microcavity lasers with different materials were experimentally demonstrated with side lengths from hundreds of micron to submicron [14, 20, 23]. Waveguide-coupled unidirectional emission square microcavity lasers were proposed and demonstrated for multiple functional applications [24, 25]. In this chapter, the mode characteristics of square optical microcavities and the lasing properties of square microcavity lasers are summarized. In Section 7.2, an analytical method is introduced to describe the confined modes in square microcavities. In Section 7.3, the symmetry properties and the mode-coupling theory are employed to analyze the degeneracies of the modes and the generation of high-Q WG-like modes. In Section 7.4, the mode characteristics of the WG-like modes are presented. In Section 7.5, an output waveguide directly connected to the square microcavities is applied to realize unidirectional emission. In Section 7.6, the waveguide-coupled square microcavity semiconductor lasers are experimentally demonstrated. In Section 7.7, dual-transverse-mode lasing

Microcavity Semiconductor Lasers: Principles, Design, and Applications, First Edition.
Yong-zhen Huang and Yue-de Yang.

with tunable wavelength intervals is achieved by designing the cavity geometry and current injection region. In Section 7.8, the applications of dual-mode square microcavity lasers in the generation of microwave and optical frequency comb (OFC) are presented. In Section 7.9, mode control for tunable single-mode lasing is achieved for the square microcavity lasers. In Section 7.10, deformed square microcavity laser with the flat sides replaced by circular arcs are presented. A summary is finally given in Section 7.11.

7.2 Analytical Solution of Confined Modes

A two-dimensional (2D) square microcavity with a side length of a as shown in Figure 7.1 is considered, where Ω_1 is the inner region of the square with a refractive index of n_1 and Ω_2 is the external region with a refractive index of n_2. Because the high Q confined modes should satisfy the total internal reflection condition, $n_1/n_2 > \sqrt{2}$ is then required for the four bounced WG-like modes. The center of the square is placed at the coordinate origin, where the x and y axes are the symmetry axes of the square, and z direction is perpendicular to the x–y plane. The Marcatili scheme was used in the analysis of square and rectangular microcavities, where the fields of confined modes were assumed exponentially decay as moving away from the sides of the microcavity in the external region [9, 10]. The mode fields in the square microcavities are decomposed into two independent components in x and y directions under the scheme. Thus, the outer-plane magnetic field $H_z(x,y)$ for transverse-electric (TE) mode and the electric field $E_z(x,y)$ for transverse-magnetic (TM) mode in the 2D square microcavity can be expressed as

$$F_z^{p,q}(x,y) = F_z^p(x)F_z^q(y) \tag{7.1}$$

$$F_z^p(x) = \begin{cases} \cos(\kappa_x x - \varphi_x) & |x| \le a/2 \\ \cos(\kappa_x a/2 - \varphi_x)\exp[-\gamma_x(x - a/2)] & x > a/2 \\ \cos(-\kappa_x a/2 - \varphi_x)\exp[\gamma_x(x + a/2)] & x < -a/2 \end{cases} \tag{7.2}$$

$$F_z^q(y) = \begin{cases} \cos(\kappa_y y - \varphi_y) & |y| \le a/2 \\ \cos(\kappa_y a/2 - \varphi_y)\exp[-\gamma_y(y - a/2)] & y > a/2 \\ \cos(-\kappa_y a/2 - \varphi_y)\exp[\gamma_y(y + a/2)] & y < -a/2 \end{cases} \tag{7.3}$$

where p and q are the mode numbers that denote the numbers of wave nodes in the x and y directions, F_z represents H_z and E_z for the TE and TM modes, respectively. The wavefunctions are cosine and sine functions for even and odd mode numbers, but expressed in the same form by inducing the phase angles φ_x and φ_y, which are 0 or $\pi/2$ for even or odd mode numbers p and q. By submitting the above wavefunctions into wave Eq. (6.1), the relations for the propagation and decay constants κ and γ can be obtained by:

$$\kappa_x^2 + \kappa_y^2 = n_1^2 k_0^2 \tag{7.4}$$

$$\kappa_x^2 - \gamma_y^2 = \kappa_y^2 - \gamma_x^2 = n_2^2 k_0^2 \tag{7.5}$$

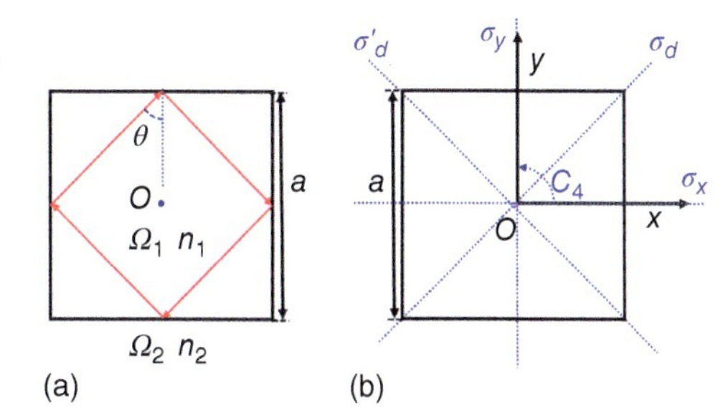

Figure 7.1 (a) The light ray of WG-like modes in square microcavities and (b) the operators of the point group C_{4v} as well as the coordinates for a square microcavity.

From (7.4, 7.5), the following relation can be obtained

$$\kappa_x^2 + \gamma_x^2 = \kappa_y^2 + \gamma_y^2 = (n_1^2 - n_2^2)k_0^2 \tag{7.6}$$

where $k_0 = 2\pi/\lambda$ is the wavenumber in vacuum. For the confined modes, their incident angles at the square sides should be greater than the critical angle of the total internal reflection at the sidewalls, which is equivalent to $\gamma_v > 0$ as

$$\kappa_v^2 < (n_1^2 - n_2^2)k_0^2 \tag{7.7}$$

where $v = x$ and y. By submitting (7.1–7.3) into wave Eq. (6.2) or (6.3), nonzero magnetic fields $H_x(x,y)$ and $H_y(x,y)$ for TM modes or nonzero electric fields $E_x(x,y)$ and $E_y(x,y)$ for TE modes can be obtained. Then, based on the continuous conditions of the tangential electromagnetic fields at the sides of the square, the mode eigenvalue equations can be obtained as

$$\begin{cases} \kappa_v \tan\frac{\kappa_v a}{2} = \gamma_v & \text{evenmode} \\ \kappa_v \tan\left(\frac{\kappa_v a}{2} - \frac{\pi}{2}\right) = \gamma_v & \text{oddmode} \end{cases} \tag{7.8}$$

for the TM modes, and the equations for the TE modes as

$$\begin{cases} n_2^2\kappa_v \tan\frac{\kappa_v a}{2} = n_1^2\gamma_v & \text{evenmode} \\ n_2^2\kappa_v \tan\left(\frac{\kappa_v a}{2} - \frac{\pi}{2}\right) = n_1^2\gamma_v & \text{oddmode} \end{cases} \tag{7.9}$$

The above equations are similar to the eigenvalue equations of three-layer symmetric slab waveguide except the definition of TE and TM modes. For the square microcavity, TE and TM modes are defined as in the original three-dimensional (3D) microcavity, and the transverse plane is chosen as the square plane. In fact, the 3D square cavity is reduced to a 2D cavity under effective index approximation, and the effective refractive indices of the TE and TM modes correspond to the mode refractive indices of the TE and TM modes in the vertically distributed slab waveguide, respectively, with the same definition. However, the eigenvalue equations of TE and TM modes in (7.8, 7.9) correspond to those of TM and TE modes of the slab waveguide as discussed in Section 2.9, respectively, due to the change of transverse plane.

Under total confinement approximations, i.e. $F_z = 0$ at the sidewalls of square, the eigenvalue equations can be simplified as $\kappa_x a = (p+1)\pi$ and $\kappa_y a = (q+1)\pi$, and the mode wavelength as

$$\lambda = 2n_1 a/\sqrt{(p+1)^2 + (q+1)^2} \tag{7.10}$$

Using the total confinement approximation solutions as initial values, the eigenvalue equations can be solved for calculating the propagation and decay constants κ_v and γ_v, and mode wavelength λ. The mode-field distributions can be obtained by substituting the results into (7.1–7.3).

In this chapter, the above-confined modes are marked as $F_z^{p,q}$ with the mode numbers p and q, i.e. $H_z^{p,q}$ and $E_z^{p,q}$ for the TE and TM modes, respectively. The obtained quasi-analytical field distributions satisfy Maxwell's equations inside the square microcavity and the continuous conditions at the sidewalls except four special corner points, so the quasi-analytical solutions provide sufficiently precise distributions in the inner region of square microcavities. The two mode numbers p and q are close for the confined modes with the incident angles around $\pi/4$ at the sidewalls of the square. The longitude mode number l is defined as $l = p + q$, and the transverse modes are defined as the modes with same l but different p and q. If one keeps the wavelength as a constant, it can be found that the transverse-mode interval is approximately in inverse proportion to the cavity area a^2, while the longitude mode interval is approximately in inverse proportion to the cavity side length a from (7.10).

7.3 Symmetry Analysis and Mode Coupling

In this section, the group theory is used to analyze the mode symmetry for a square resonator as in [2, 26]. The symmetry of a square microcavity can be described by the point group C_{4v} as shown in Figure 7.1b. The point group C_{4v} includes a rotational subgroup $C_4 = \{E, C_4^1, C_4^2, C_4^3\}$, mirror elements relative to x and y axes $\{\sigma_x, \sigma_y\}$, and mirror elements relative to diagonal $\{\sigma_d, \sigma'_d\}$. The character of the point group C_{4v} is listed in Table 7.1. The point group C_{4v} has four one-dimensional (1D) and one 2D irreducible representations.

According to the variations of the wavefunctions under the group element, we can classify the mode-field distributions into the irreducible representations of the point group:

(a) If p and q have different parities, the doubly degenerate modes $F_z^{p,q}$ and $F_z^{q,p}$ form the 2D representation E_1.
(b) If mode numbers $p = q$, $F_z^{p,q}$ form the representation of the point group A_1 and B_2 for even and odd mode number, respectively.
(c) If p and q have the same parity but p is unequal to q, $F_z^{p,q}$ and $F_z^{q,p}$ are two degenerate modes. But the doubly degenerate modes do not form any representation in Table 7.1, which means that $F_z^{p,q}$ and$F_z^{q,p}$ are no longer the eigenstates of the square microcavity, and the degeneracy is accidental.

However, $F_z^{p,q}$ and $F_z^{q,p}$ in the case (c) can be reduced into two 1D irreducible representations by combining into two new modes with field distributions as

$$\begin{aligned} F_z^{e,(p,q)} &= F_z^{p,q} + F_z^{q,p} \\ F_z^{o,(p,q)} &= F_z^{p,q} - F_z^{q,p} \end{aligned} \tag{7.11}$$

which have definite parities relative to the square diagonal mirror planes. The superscripts "e" and "o" indicate even or odd parities relative to the diagonals of the

Table 7.1 Character table of the point group C_{4v}.

	E	$C_4^1(C_4^3)$	C_4^2	$\sigma_x(\sigma_y)$	$\sigma_d(\sigma'_d)$
A_1	1	1	1	1	1
A_2	1	1	1	−1	−1
B_1	1	−1	1	1	−1
B_2	1	−1	1	−1	1
E_1	2	0	−2	0	0

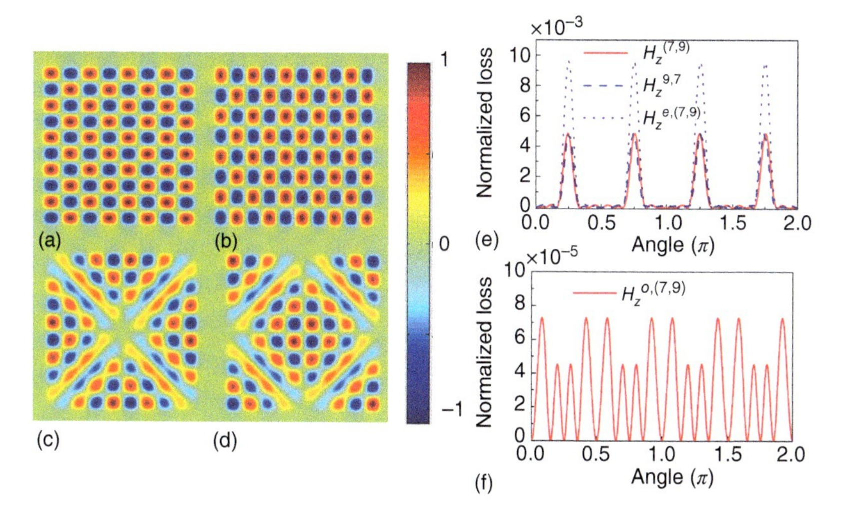

Figure 7.2 The analytical field distributions of modes (a) $H_z^{7,9}$, (b) $H_z^{9,7}$, (c) $H_z^{o,(7,9)}$ and (d) $H_z^{e,(7,9)}$, and normalized far-field emission patterns based quasi-analytical fields of (e) $H_z^{7,9}$, $H_z^{9,7}$, and $H_z^{e,(7,9)}$, and (f) $H_z^{o,(7,9)}$ for a square microcavity with a side length of 3 μm and a refractive index of 3.2. Source: Yang and Huang [26]. © 2016, IOP Publishing Ltd.

square, respectively. If both p and q are even, $F_z^{e,(p,q)}$ forms the A_1 representation and $F_z^{o,(p,q)}$ forms the B_1 representation. If both p and q are odd, $F_z^{e,(p,q)}$ forms the B_2 representation and $F_z^{o,(p,q)}$ forms the A_2 representation.

For a square microcavity with a side length of 3 μm and a refractive index of 3.2 surrounded by air, the analytical mode-field distributions are plotted in Figure 7.2a–d for TE modes $H_z^{7,9}$, $H_z^{9,7}$, $H_z^{o,(7,9)}$, and $H_z^{e,(7,9)}$, respectively. It can be expected that $H_z^{o,(7,9)}$ will have the lowest loss among the four modes, because of its weak field distribution and low scattering loss at the corners. In addition, the antisymmetric mode field patterns relative to the diagonals of the square can also result in coherence cancellation of far-field emission along the diagonal directions. The high-Q TM and TE confined modes have zero electric field (E_z) and magnetic field (H_z) along the square diagonals, respectively, which means that the square diagonals should be nodes of the standing-wave distributions.

The quasi-analytical field distributions are approximation solutions that neglect the loss from four corners of the square resonator, but the practical modes are quasi-modes with finite lifetimes. The loss from the corners can be simply estimated as boundary-wave leakage. Here, the far-field emission method is used to calculate the mode Q factors based on the analytical near-field distribution. Although the field distribution in the external region obtained by the analytical model is not correct, the internal field is expected to be sufficiently precise since it satisfies Maxwell's equations inside the microcavity and at the boundaries except for four vertices. Thus, the quasi-analytical field distributions in the square microcavity are obtained from Eqs. (7.1–7.3), and the far-field emission spectra are deduced based on the near-field F_z at the sidewalls of the square.

For the square microcavity with a side length of 3 μm and refractive index of 3.2 surrounded by air, the power angular loss spectra are calculated using the analyzed field distributions as shown in Figure 7.2a–d by Eqs. (6.24)–(6.27), and plotted in Figure 7.2e,f, which are normalized by $\omega W/2\pi$ with stored mode power W to express the angular power loss in one period. As shown in Figure 7.2e, the angular loss spectra of modes $H_z^{7,9}$ and $H_z^{9,7}$ are similar, and those of the $H_z^{e,(7,9)}$ correspond to interference increase of the angular radiation loss. In Figure 7.2f, the power angular loss spectra of $H_z^{o,(7,9)}$ are two orders of magnitude smaller than those in Figure 7.2e, because of interference decrease of the radiation loss in the diagonal directions.

The mode Q-factors obtained by the far-field emission from Eq. (6.28) are 1.12×10^3, 1.12×10^3, 5.72×10^2, and 3.26×10^4 for the modes $H_z^{7,9}$, $H_z^{9,7}$, $H_z^{e,(7,9)}$, and $H_z^{o,(7,9)}$, respectively. Mode Q-factors of 3.33×10^4 and 2.23×10^2 are obtained for $H_z^{o,(7,9)}$and $H_z^{e,(7,9)}$ in the square microcavity by the finite-element method (FEM) simulation. The results show that the quasi-analytical solutions give very accurate Q factor for the high-Q mode. Because the field distribution is obtained with real value of wavenumber k_0 by neglecting the loss, the complex value of k_0 in a practical case will limit the accuracy. However, for the high-Q mode, the imaginary part of k_0 is relatively small and then the analytical near-field distribution inside square can provide sufficiently precise far-field emission.

To further understand the physical mechanism of the combination of accidentally degenerate modes $F_z^{p,q}$ and $F_z^{q,p}$ as p and q have the same parity, the mode-coupling theory is used to explain such a phenomenon [10, 26, 27]. The modes $F_z^{p,q}$ and $F_z^{q,p}$ have the same symmetries relative to the x and y axes when both p and q are even or odd. Because of the influence of the corner region, the orthogonality of the modes is broken in the dielectric square microcavity; thus, modes may couple through the continuum resulting in the so-called external coupling. The mode coupling would occur in the square resonator when the coupling is not forbidden by the mode symmetry and leads to two combined modes.

The TE modes $H_z^{7,9}$ and $H_z^{9,7}$ in a rectangular microcavity with a length of $a = 3$ μm, a width of b, and a refractive index of 3.2 surrounded by air, are numerically simulated by the FEM. In the simulation, the width b is varied from 2.8 to 3.2 μm to tune the frequency difference between $H_z^{7,9}$ and $H_z^{9,7}$. The obtained mode frequencies and Q-factors vs. b are plotted in Figure 7.3, respectively. The mode Q-factors are 5.79×10^2 and 5.85×10^2 for the uncoupled modes $H_z^{7,9}$ and $H_z^{9,7}$ at $b = 2.8$ μm.

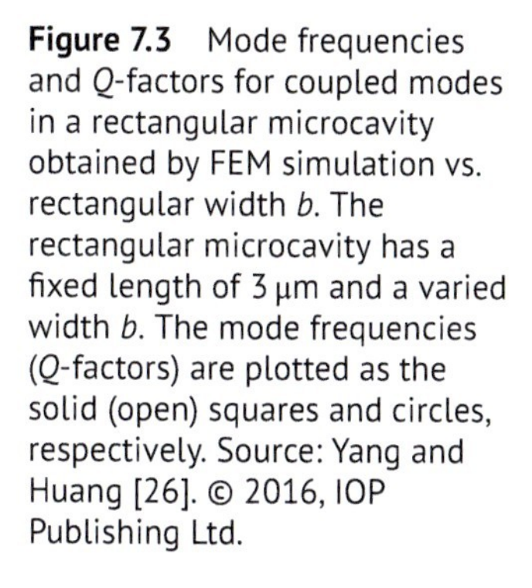

Figure 7.3 Mode frequencies and Q-factors for coupled modes in a rectangular microcavity obtained by FEM simulation vs. rectangular width b. The rectangular microcavity has a fixed length of 3 μm and a varied width b. The mode frequencies (Q-factors) are plotted as the solid (open) squares and circles, respectively. Source: Yang and Huang [26]. © 2016, IOP Publishing Ltd.

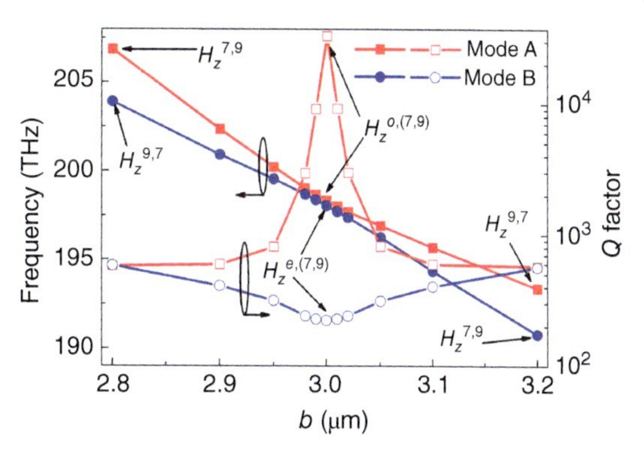

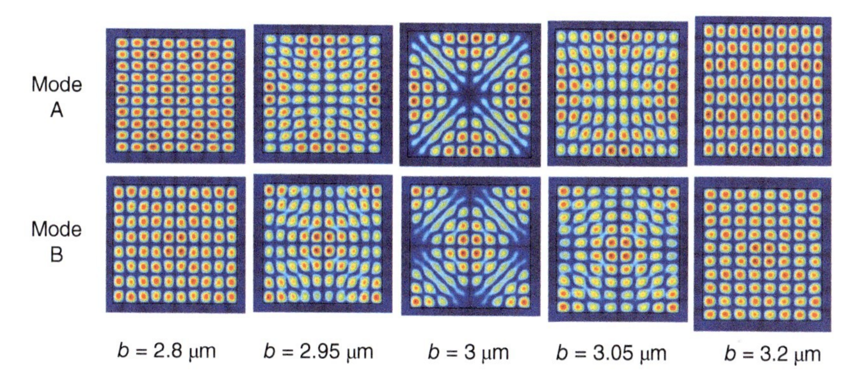

Figure 7.4 Magnetic field amplitude ($|H_z|$) patterns obtained by FEM for the rectangular microcavities with $a = 3$ μm and different b.

A high Q-factor of 3.33×10^4 is obtained for coupled mode $H_z^{o,(7,9)}$ at $b = a = 3$ μm, and the other coupled mode $H_z^{e,(7,9)}$ has a low Q-factor of 2.23×10^2. At $b = 3.2$ μm, the Q-factors drop to 5.68×10^2 and 5.77×10^2 again for the uncoupled modes $H_z^{7,9}$ and $H_z^{9,7}$ again. The frequency difference between the coupled modes is 0.297 THz at $b = a = 3$ μm.

The magnetic field amplitude ($|H_z|$) distributions of modes A and B in the rectangular microcavity with $a = 3$ μm and different b are obtained by the FEM and plotted in Figure 7.4. The transformations of the field patterns for modes A and B are consistent with the variations of the mode Q factors and frequencies in Figure 7.3. Modes A and B are the uncoupled modes $H_z^{7,9}$ and $H_z^{9,7}$ at $b = 2.8$ μm, and the coupled modes $H_z^{o,(7,9)}$ and $H_z^{e,(7,9)}$ at $b = 3$ μm, then finally become the uncoupled modes $H_z^{9,7}$ and $H_z^{7,9}$ at $b = 2.8$ μm, respectively.

The numerical simulation results demonstrate a mode-coupling phenomenon between the doubly degenerate modes with the same symmetry, which confirm that the high-Q modes in square microcavity are induced by the external mode coupling. An interesting feature for the external coupling is that one of the coupled modes may have a greatly enhanced Q-factor, while coupled mode frequencies

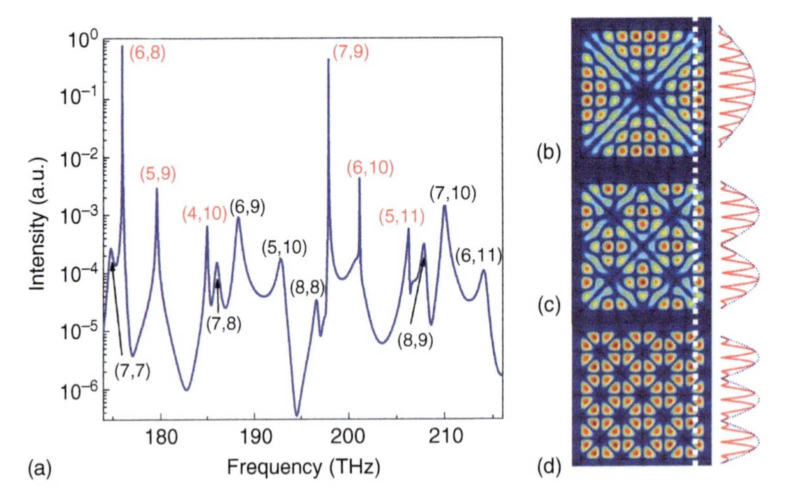

Figure 7.5 (a) Intensity spectrum calculated by the FDTD technique and Padé approximation method for the TE mode in a square microcavity with $a = 3\,\mu\text{m}$ and $n_1 = 3.2$. The mode numbers are marked by (p, q) for $H_z^{p,q}$. The magnetic field amplitude ($|H_z|$) patterns and distributions along the dotted line for modes (b) $H_z^{o,(7,9)}$, (c) $H_z^{o,(6,10)}$, and (d) $H_z^{o,(5,11)}$. Source: Yang and Huang [26]. © 2016, IOP Publishing Ltd.

exhibit anticrossing or crossing behavior in the case of strong or weak coupling, respectively.

The mode characteristics of TE modes in the square microcavity with a side length of $3\,\mu\text{m}$ and a refractive index of 3.2 are further investigated using the finite-difference time-domain (FDTD) method. The intensity spectrum obtained by the FDTD method and Padé approximation is plotted in Figure 7.5a, and all the peaks are marked by the mode numbers (p, q) for the TE modes $H_z^{p,q}$. Because the Q-factors of $H_z^{e,(p,q)}$ are much lower than those of $H_z^{o,(p,q)}$ and the frequencies are close, only one narrow peak for $H_z^{o,(p,q)}$ can be observed in the spectrum as p and q have the same parity. The simulated magnetic field amplitude ($|H_z|$) patterns and distributions along the dotted line for WG-like modes $H_z^{o,(7,9)}$, $H_z^{o,(6,10)}$ and $H_z^{o,(5,11)}$ are shown in Figure 7.5b–d, respectively. All the coupled high-Q modes have a weak distribution at the vertices and exhibit a sinusoidal envelope along the side of square as shown in Figure 7.5b–d, which can be easily understood by superposition of different cosine functions with a difference of $|p-q|/2$ in period number according to Eq. (7.11). Besides the nodes at the vertices, the node number of envelope at the sidewall, which is defined as the transverse-modes number m for the high-Q WG-like modes $H_z^{o,(p,q)}$, is equal to $|p-q|/2-1$.

7.4 Mode Analysis for High Q Modes

As shown in Section 7.3, high Q WG-like modes in a square microcavity are antisymmetric relative to the diagonals of the square. Based on the mode patterns, the transverse-mode number m and longitudinal-mode number l are related to the mode numbers p and q by

$$m = \frac{|p-q|}{2} - 1 \tag{7.12}$$

$$l = p + q \tag{7.13}$$

During the total internal reflection at the sidewalls, a polarization-dependent negative phase shift occurs as

$$\phi(\theta) = -2\tan^{-1}\left(\frac{\eta\sqrt{n_1^2\sin^2\theta - n_2^2}}{n_1\cos\theta}\right) \tag{7.14}$$

where θ is the incident angle, and η equals 1 and n_1^2/n_2^2 for the TM- and TE-polarized modes, respectively, the phase shift at $\theta = \pi/4$ can be obtained as

$$\phi_0 = -2\tan^{-1}\left(\frac{\eta\sqrt{n_1^2 - 2n_2^2}}{n_1}\right) \tag{7.15}$$

For the high Q WG-like modes with $l \gg 1$ and $m \approx 0$, i.e. $p \approx q \gg 1$, the phase shift θ is close to $\pi/4$, then (7.10) can be rewritten as

$$\lambda \approx 2n_1 a/\sqrt{\left(p + \frac{\phi_0}{\pi}\right)^2 + \left(q + \frac{\phi_0}{\pi}\right)^2} \tag{7.16}$$

and the longitudinal-mode wavelength can be obtained as

$$\lambda_l \approx 2\sqrt{2}n_1 a/\left(l + \frac{4\phi_0}{2\pi}\right) \tag{7.17}$$

The terms $2\sqrt{2}a$ and $4\phi_0$ correspond to the roundtrip length and the phase shift of four-bounced light rays, respectively. So, the corresponding longitudinal-mode interval is

$$\Delta\lambda_l \approx \frac{\lambda^2}{2\sqrt{2}an_g} \tag{7.18}$$

where n_g is the group refractive index. It can be found that the results are consistent with the four-bounced ray model [8]. However, for the high-Q WG-like modes, p and q should have the same parity; thus, the longitudinal-mode number should be an even integer, which means that the high-Q longitudinal-mode spacing is two times of that in (7.18) determined by the cavity roundtrip length.

According to (7.12), the mth-order transverse mode corresponds to $|p-q| = 2m + 2$. Based on (7.16), the transverse-mode wavelength interval between the fundamental and the mth-order transverse modes can be obtained as

$$\Delta\lambda_{0m} \approx \frac{\lambda^3 m(m+2)}{4n_1 n_g a^2} \tag{7.19}$$

Table 7.2 The mode frequencies and Q-factors obtained by FEM in a square microcavity with $a = 3\,\mu m$ and $n_1 = 3.2$.

l	Mode	Representation	f (THz)	Q	l	Mode	Representation	f (THz)	Q
14	$H_z^{7,7}$	B_2	175.059	3.56×10^2	16	$H_z^{8,8}$	A_1	197.068	4.22×10^2
	$H_z^{e,(6,8)}$	A_1	175.908	1.46×10^2		$H_z^{e,(7,9)}$	B_2	198.044	2.23×10^2
	$H_z^{o,(6,8)}$	B_1	176.381	3.21×10^4		$H_z^{o,(7,9)}$	A_2	198.341	3.33×10^4
	$H_z^{e,(5,9)}$	B_2	180.004	1.52×10^2		$H_z^{e,(6,10)}$	A_1	201.371	1.57×10^2
	$H_z^{o,(5,9)}$	A_2	180.008	1.99×10^3		$H_z^{o,(6,10)}$	B_1	201.631	7.89×10^3
	$H_z^{e,(4,10)}$	A_1	185.158	94		$H_z^{e,(5,11)}$	B_2	206.79	2.02×10^2
	$H_z^{o,(4,10)}$	B_1	185.463	1.28×10^3		$H_z^{o,(5,11)}$	A_2	206.857	1.56×10^3
15	$H_z^{7,8}$	E_1	186.342	3.68×10^2	17	$H_z^{8,9}$	E_1	208.349	5.11×10^2
	$H_z^{6,9}$	E_1	188.666	3.85×10^2		$H_z^{7,10}$	E_1	210.494	4.99×10^2
	$H_z^{5,10}$	E_1	193.329	2.89×10^2		$H_z^{6,11}$	E_1	214.691	3.20×10^2

Eqs. (7.16–7.19) are derived under the condition $l \gg 1$, i.e. $n_1 a \gg \lambda$, in such case the transverse-mode interval in (7.19) is much smaller than the longitudinal-mode interval in (7.18), and different transverse modes are assumed to have the similar longitudinal-mode wavelength in (7.17). For the square microcavity with the side length close to the wavelength, the mode interval cannot be expressed as such simple equations.

The mode frequencies and Q-factors of the modes shown in Figure 7.5 obtained by FEM method are listed in Table 7.2 with the longitude mode number l from 14 to 17, and the corresponding irreducible representations of C_{4v} are also marked. The high-Q modes $H_z^{o,(p,q)}$ with even longitude mode numbers l have the highest Q-factors. Especially, the Q-factors of the fundamental WG-like $H_z^{o,(p,q)}$ with $|p-q| = 2$ are about 2 orders of magnitude higher than those of uncoupled modes. The longitude intervals for high-Q modes in the square are twice those in a microdisk with similar size as high-Q WG-like modes only exist with even l, and all the high-Q modes are nondegenerate modes, which make the square microcavity suitable for stable single-mode operation.

Besides the accidentally degenerate modes $F_z^{p,q}$ and $F_z^{q,p}$, the mode coupling can occur between all the modes with the same symmetry, i.e. the field distributions form the same irreducible representations of C_{4v}, when the mode frequencies are close, but the coupling can be neglected for the modes listed in Table 7.2 based on the numerical simulation. The reason is that the external coupling coefficient depends on the modal loss (imaginary part of mode frequency), which is much less than frequency difference in the semiconductor microsquares. For example, $H_z^{7,8}$ and $H_z^{6,9}$ have the mode interval of 2.3 THz, and the imaginary part of the mode frequency for $H_z^{7,8}$ and $H_z^{6,9}$ is about 0.25 THz. The transverse-mode intervals are inversely proportional to the area of the square, which results in closely spaced transverse modes in a large-size square cavity. However, the mode coupling between these transverse modes can still be neglected as the imaginary parts of mode frequencies, which determine the coupling strength, are also inversely proportional to the area of the square.

It should be noted that, in a weakly confined square cavity such as silica square, high modal loss makes mode coupling between different transverse modes possible. The mode coupling between the modes $F_z^{s,s+1}$ and $F_z^{s+2,s-1}$ with a mode number difference of two in both x and y directions results in a high-Q-coupled mode with the field distribution envelope similar to the high-Q fundamental coupled mode $H_z^{o,(p,q)}$ as shown in Figure 7.5b. Thus, high-Q-coupled WG-like modes also exist with odd l, and the longitude mode intervals for high-Q modes are no longer twice of the longitudinal-mode interval.

In the above analysis and numerical simulation, the practical 3D structure is simplified to a 2D structure using the effective index approximation. Compared with real 3D results, the 2D simulation can give accurate mode wavelength [28], but give incorrect high Q factors for the square microcavity with a weak vertical confinement [29, 30], because the vertical radiation loss is not considered in the 2D simulation. However, the 2D analysis and simulation are still powerful tools, because they can give accurate mode structure and predict high-Q WG-like modes. The WG-like modes have a weak field distribution at the center of square; thus, it is also possible to design a square microcavity on a pedestal similar to the microdisk for forming a strong vertical confinement [31].

7.5 Waveguide-Coupled Square Microcavities

Due to the break of rotation symmetry, the mode-field distributions in the square microcavities have multiple angular components and are not homogeneous along the sides of the resonators. Furthermore, the high-Q modes have weak field distribution at the vertices, and the high-Q modes with odd m have weak field distribution at the midpoints of the sidewalls. The square resonator connected with an output waveguide at the place with weak mode-field distribution can still have high-Q modes suitable for realizing directional emission microlasers [24, 32]. The square microcavity with an output waveguide jointed at one vertex or the midpoint of one sidewall as shown in Figure 7.6 is simulated, which is surrounded by a SiN_x layer and the divinylsiloxane bisbenzocyclobutene (DVS-BCB) as fabricated devices. The refractive indices of the InP laser wafer, SiN_x, and BCB are set to be 3.2, 2.0, and 1.54, respectively.

For a square microcavity with a side length of 20 μm and a 1.5-μm-wide waveguide connected to one vertex, the mode-intensity spectrum of TE modes is calculated by the FDTD method and Padé approximation and presented in Figure 7.7a, where the mode numbers are assigned based on the k-space patterns of the mode-field distributions [32]. The modes at 1522.69, 1548.48, and 1575.17 nm are high-Q modes $H_z^{o,(58,60)}$, $H_z^{o,(57,59)}$, and $H_z^{o,(56,58)}$ with Q factors of 1.78×10^5, 1.14×10^5, and 9.15×10^4, respectively, which correspond to the fundamental transverse modes $H_z^{o,(p,q)}$ with even longitudinal-mode number $l = q + p$ and $m = 0$. The corresponding

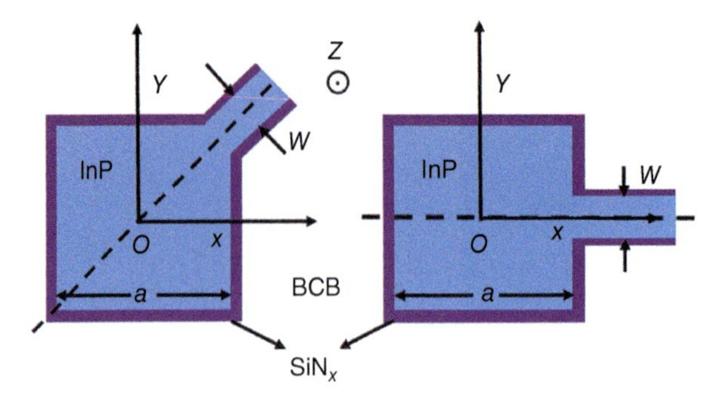

Figure 7.6 Schematic diagram of the square cavity with an output waveguide connecting to one vertex and the midpoint of one sidewall. Source: Yang and Huang [26]. © 2016, IOP Publishing Ltd.

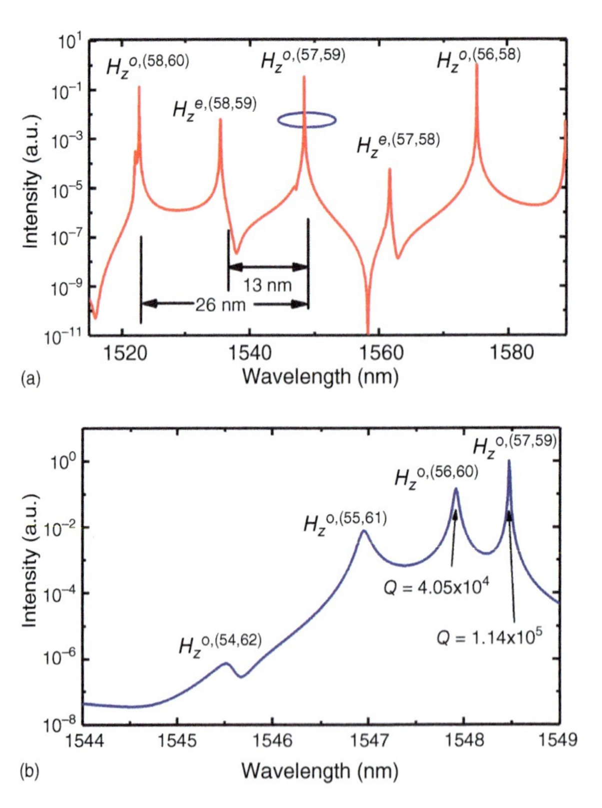

Figure 7.7 (a) Mode-intensity spectrum of TE modes and (b) detailed spectrum for the modes with $l = 116$ in a square microcavity with a side length of 20 μm and a 1.5-μm-wide waveguide connected to one vertex.

longitudinal-mode intervals for these modes around 1550 nm are about 26 nm as those in a perfect square microcavity without the output waveguide. The detailed spectrum for the modes with $l = 116$ around $H_z^{o,(57,59)}$ is shown in Figure 7.7b, where $m = 0$, 1, 2, and 3 correspond to the fundamental, the first-order, the second-order, and the third-order transverse modes $H_z^{o,(57,59)}$, $H_z^{o,(56,60)}$, $H_z^{o,(55,61)}$, and $H_z^{o,(54,62)}$. The mode $H_z^{o,(56,60)}$ at 1547.90 nm has a Q factor of 4.05×10^4, which is much lower than

1.14×10^5 of $H_z^{o,(57,59)}$. However, the corresponding difference between the modal losses of $H_z^{o,(57,59)}$ and $H_z^{o,(56,60)}$ is only $2.3\,\text{cm}^{-1}$. The difference can be enhanced by increasing output waveguide width or decreasing the cavity size, which will make the device more suitable for single-mode operation. The modes at 1535.45 and 1561.69 nm are $H_z^{e,(58,59)}$ and $H_z^{e,(57,58)}$ modes with Q factors of 2.85×10^4 and 2.80×10^4, respectively, which correspond to the symmetric modes $H_z^{e,(p,q)}$ relative to the square diagonal connecting waveguide with an odd longitudinal-mode number. Considering this group of modes, the longitudinal-mode intervals decrease to about 13 nm. The corresponding antisymmetric modes $H_z^{o,(58,59)}$ and $H_z^{o,(57,58)}$are obtained at 1535.44 and 1561.68 nm with Q factors of 8.8×10^3 and 8.4×10^3, respectively, which are much lower than those of symmetric modes.

The field distributions of the high-Q modes are simulated using an excitation source with a narrow bandwidth centered at a resonant wavelength. The field patterns of $H_z^{o,(57,59)}$ and $H_z^{e,(57,58)}$ at the wavelength of 1548.48 and 1561.69 nm are shown in Figure 7.8a,b, respectively. The k-space patterns calculated by equation

$$
\begin{aligned}
U(k_x, k_y) = &\left[\iint F_z(x,y)\cos(k_x x)\cos(k_y y)dxdy\right]^2 \\
&+\left[\iint F_z(x,y)\sin(k_x x)\sin(k_y y)dxdy\right]^2 \\
&+\left[\iint F_z(x,y)\sin(k_x x)\cos(k_y y)dxdy\right]^2 \\
&+\left[\iint F_z(x,y)\cos(k_x x)\sin(k_y y)dxdy\right]^2
\end{aligned}
\tag{7.20}
$$

for $H_z^{o,(57,59)}$ are shown in Figure 7.8c, where the horizontal and vertical axes are $k_x a/\pi - 1$ and $k_y a/\pi - 1$. Because of the evanescent field outside the resonator, the effective side length of square resonators is a little larger than a. Thus, for mode $H_z^{p,q}$, the k-space pattern peak position $(k_x a/\pi - 1,\ k_y a/\pi - 1)$ is expected to be a little smaller than (p, q). The results show that $H_z^{o,(57,59)}$ has two peaks at (56.6, 58.6) and (58.6, 56.6). These high-Q WG-like modes are induced by the mode coupling between two degenerate modes, and have antisymmetric field distributions relative to the square diagonals, which is similar to the case without an output waveguide.

For $H_z^{e,(57,58)}$, the k-space pattern peak is found at (57.1, 57.1), which might be caused by the superposition between several adjacent peaks. Figure 7.8d shows the corresponding k-space pattern calculated by the third term of (7.20), where the strongest and secondary peaks are found at (56.6, 57.6) and (58.5, 55.5). Similarly, from the fourth term of (7.20), two other peaks are obtained at (57.6, 56.6) and (55.5, 58.5). The result indicates two major components $H_z^{57,58}$ and $H_z^{58,57}$, and two minor components $H_z^{56,59}$ and $H_z^{59,56}$. The four components construct the field distribution in Figure 7.8b similar to Figure 7.8a, which is different from the case without an output waveguide. The reason is that the output waveguide introduces a high modal loss for $H_z^{e,(57,58)}$ and $H_z^{e,(56,59)}$, which have strong field distribution at the waveguide-jointed vertex, and results in a mode coupling between them and hence forms a new WG-like coupled mode that does not exist in a perfect square.

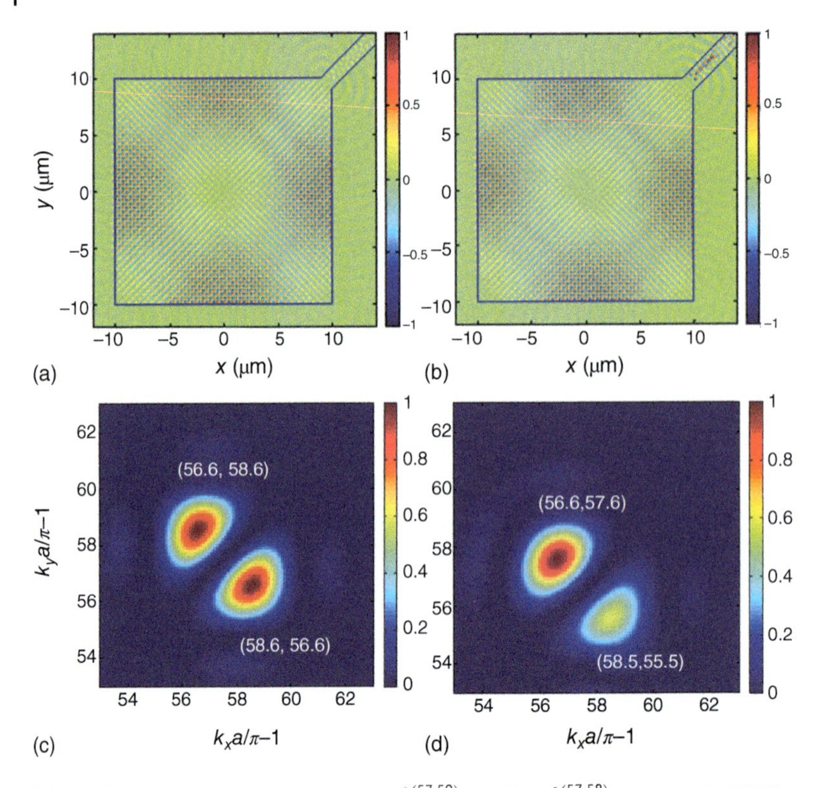

Figure 7.8 Mode-field patterns of (a) $H_z^{o,(57,59)}$ and (b) $H_z^{e,(57,58)}$ obtained by FDTD simulation, and the k-space patterns of (c) $H_z^{o,(57,59)}$ and (d) the third term of (7.20) for $H_z^{e,(57,58)}$ in a square microcavity with a side length of 20 μm and 1.5-μm-wide waveguide connected to one vertex. Source: Long et al. [26]. © 2016, IOP Publishing Ltd.

The mode is marked as $H_z^{e,(57,58)}$ based on its main component for simplification, although the coupled mode has multiple k components. Obvious mode coupling between antisymmetric modes $H_z^{o,(57,58)}$ and $H_z^{o,(56,59)}$ is not observed as they have weak field distribution at the waveguide-jointed vertex; thus, $H_z^{o,(57,58)}$ and $H_z^{o,(56,59)}$ have lower Q factors than the coupled mode combined by $H_z^{e,(57,58)}$ and $H_z^{e,(56,59)}$. The detailed spectra around modes with $l = 115$ are similar to those around $l = 116$, and the high-Q fundamental, first-order, and second-order WG-like modes correspond to the coupled modes induced by ($H_z^{e,(57,58)}$, $H_z^{e,(56,59)}$), ($H_z^{e,(56,59)}$, $H_z^{e,(55,60)}$), and ($H_z^{e,(55,60)}$, $H_z^{e,(54,61)}$), respectively.

For a square microcavity with a side length of 17.8 μm and a 1.4-μm-wide waveguide connected to the midpoint of one sidewall, the mode-intensity spectrum of TE modes is calculated and presented in Figure 7.9a, where the mode numbers are assigned based on the mode-field distributions [33]. The modes at 1530.27, 1560.35, and 1591.64 nm are $H_z^{o,(51,55)}$, $H_z^{o,(50,54)}$, and $H_z^{o,(49,53)}$ modes with Q factors of 3.71×10^4, 2.76×10^4, and 4.22×10^4, respectively, which correspond to the antisymmetric first-order WG-like modes $H_z^{o,(p,q)}$ relative to the square diagonals with l equaling an even number and $|p-q| = 4$. The detailed spectrum for the modes with

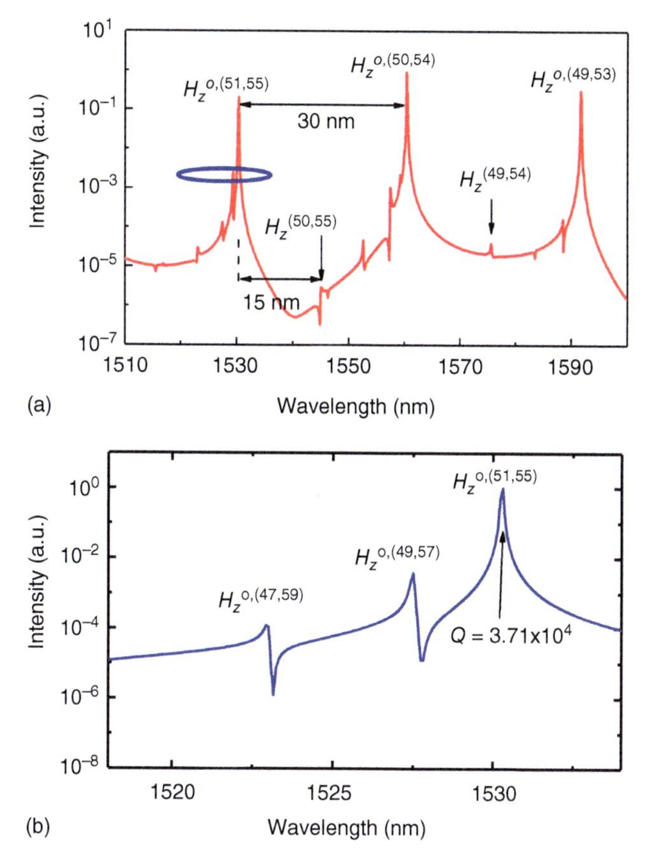

Figure 7.9 (a) Mode-intensity spectrum and (b) magnified spectrum of a group transverse modes for TE modes in a square microresonator with a side length of 17.8 μm and a 1.4-μm-wide waveguide connected to the midpoint of one sidewall.

$l = 106$ around $H_z^{o,(51,55)}$ is shown in Figure 7.9b where $m = 1$, 3, and 5 correspond to the first-order transverse mode $H_z^{o,(51,55)}$, the third-order mode $H_z^{o,(49,57)}$, and the fifth-order mode $H_z^{o,(47,59)}$, respectively. The modes with m equaling to an even number have a strong field distribution at the midpoint and hence a low Q factor. The corresponding longitudinal-mode intervals for these modes around 1550 nm are about 30 nm similar to those in a perfect square microcavity without an output waveguide. The modes at 1545.02 and 1575.65 nm are $H_z^{50,55}$ and $H_z^{49,54}$ modes with the Q factors of 6.76×10^3 and 7.26×10^3, respectively. In the square microcavity with a rounded corner, the Q factors of the modes with even and odd l become similar. Considering this group of modes, the longitudinal-mode intervals decrease to about 15 nm.

The field patterns of $H_z^{o,(51,55)}$ and $H_z^{(50,55)}$ at the wavelengths of 1530.27 and 1545.02 nm are shown in Figure 7.10a,b, respectively. The k-space patterns calculated by (7.20) for $H_z^{o,(51,55)}$ and $H_z^{(50,55)}$ are shown in Figure 7.10c,d, respectively. The results show that $H_z^{o,(51,55)}$ has two peaks at (50.8, 54.8) and (54.8, 50.8). These

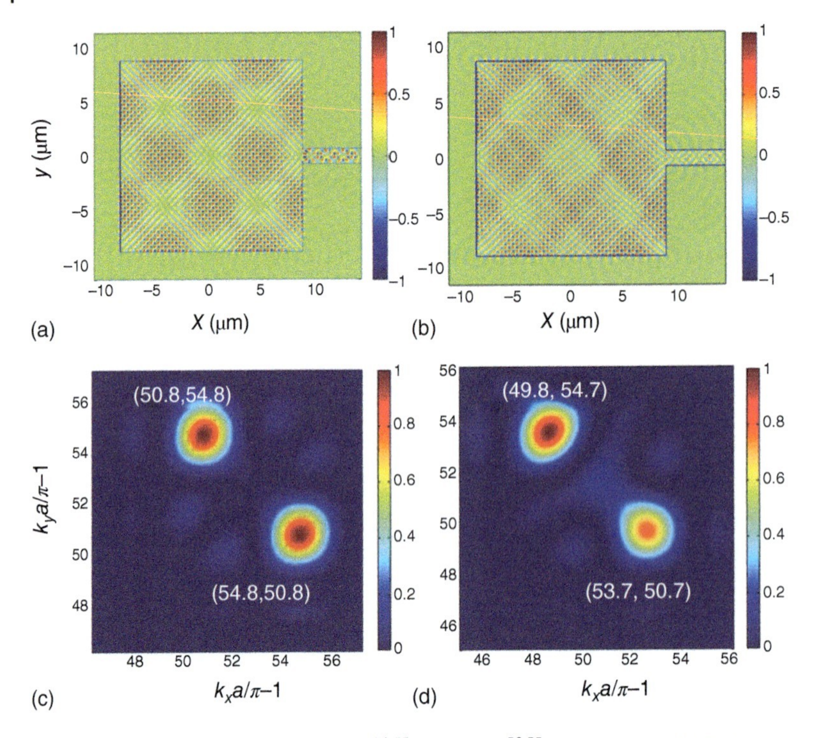

Figure 7.10 Field patterns of (a) $H_z^{o,(51,55)}$ and (b) $H_z^{50,55}$ obtained by FDTD simulation, and the k-space patterns of (c) $H_z^{o,(51,55)}$ and (d) $H_z^{50,55}$. Source: Yang and Huang [26]. © 2016, IOP Publishing Ltd.

high-Q WG-like modes are induced by the mode coupling between two degenerate modes, and have antisymmetric field distributions relative to the square diagonals, which is similar to the case without an output waveguide. For $H_z^{50,55}$, the k-space pattern peaks are found at (49.8, 54.7) and (53.7, 50.7). The result indicates two major components $H_z^{50,55}$ and $H_z^{54,51}$. The two components construct the WG-like mode-field distribution in Figure 7.10d similar to Figure 7.10c with a first-order shaped envelope along sidewall, which is different from the case without an output waveguide. The possible reason is that the output waveguide introduces a high modal loss resulting in an additional mode coupling between the different transverse modes.

From the above simulation results, it can be found that the WG-like modes $H_z^{o,(p,q)}$ in the square microcavity with an output waveguide and l equaling to an even number have the highest Q factors similar to a perfect square microcavity. However, the difference between the Q factors of the modes with even and odd l degrades from 2-order magnitude to less than one order because of the mode coupling between different transverse modes, and the modes with odd l can also have sufficient Q factor for lasing. The output waveguide modifies the mode characteristics besides realizing directional emission, and a careful design of the output waveguide and square cavity will exhibit different mode behaviors for square microcavity lasers. The square microcavity laser with a vertex output waveguide is suitable for low threshold but also possible for multiple transverse-mode lasing, while that with a midpoint waveguide is more suitable for single transverse-mode operation.

7.6 Directional-Emission Square Semiconductor Lasers

Square semiconductor microlasers are fabricated using an AlGaInAs/InP laser wafer grown on n-doped InP substrate by metal–organic chemical vapor deposition. The active region of the laser wafer consists of six compressively strained quantum wells, with 6-nm-thick $Al_{0.24}GaIn_{0.71}As$ quantum wells and 9-nm-thick $Al_{0.44}GaIn_{0.49}As$ barrier layers, confined by the lower cladding layers of 100-nm undoped graded AlGaInAs and 140-nm n-InAlAs, and upper cladding layers of 150-nm undoped graded AlGaInAs and InAlAs. The upper layers are p-InP and p+-InGaAs contacting layers with a total thickness of 1.8 μm.

An 800-nm SiO_2 layer is firstly deposited on the laser wafer by plasma-enhanced chemical vapor deposition (PECVD), and square resonator patterns are transferred onto the SiO_2 layer using standard photolithography and inductively coupled-plasma (ICP) etching techniques. The patterned SiO_2 is used as a hard mask for the ICP etching process to etch the laser wafer with an etching depth of over 4 μm. The scanning electron microscope (SEM) images are shown in Figure 7.11a,b, for the square microcavity lasers with an output waveguide connected to one vertex or the midpoint of one sidewall, respectively. Then, a 200-nm SiN_x layer is deposited on the wafer by PECVD for better adhesion to the following spin-coated DVS-BCB layer for planarization. Afterwards, the BCB film is etched by reactive ion etching (RIE) to expose the top of microsquare cavities. After that, a 450-nm SiO_2 layer is deposited on the wafer and a contact window is opened by photolithography and ICP etching for current injection. A Ti/Pt/Au p-electrode is then deposited by e-beam evaporation and lift-off process, forming a patterned p-electrode. The laser wafer is mechanically lapped down to a thickness of about 120 μm, and an Au–Ge–Ni metallization layer is deposited as n-electrode. The microscopic pictures of fabricated square microlasers with output waveguide connected to the vertex and the midpoint of one sidewall are shown in Figure 7.11c,d, respectively [32].

For a square semiconductor microlaser with a side length of 20 and 1.5 μm-wide waveguide connected to a vertex, the threshold current is 6 mA and the maximum output power coupled into a multiple-mode fiber is 101 μW at the continuous wave (CW) current of 43 mA. The lasing spectra are measured by an optical spectrum analyzer with a resolution of 0.02 nm and plotted in Figure 7.12a at the continuous currents of 17, 35, 47, and 64 mA. At 17 mA, the dominant lasing modes have a longitudinal-mode interval of 11.8 nm around 1550 nm, which corresponds to a group index n_g of 3.6 calculated from the mode wavelength interval (7.18).

Ignoring the existence of two main transverse modes, near single longitudinal-mode operations at 17 and 47 mA are observed with the side-mode suppression ratios (SMSR) of about 30 dB. However, three groups of longitudinal modes are observed at 35 and 64 mA. The lasing mode characteristics are in good agreement with the numerical results. The main lasing modes at 17 and 47 mA are WG-like modes $H_z^{o,(p,q)}$ with an even longitudinal-mode number, and the main lasing

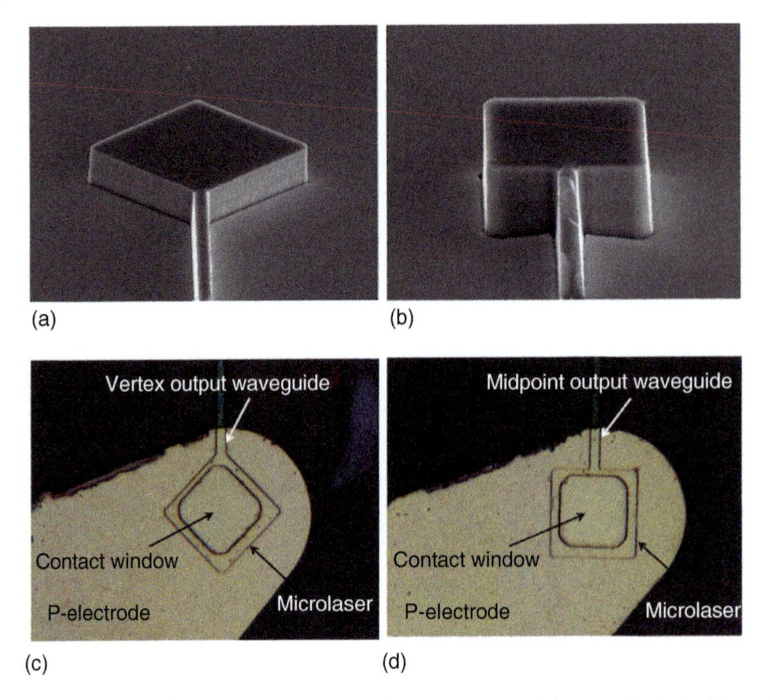

Figure 7.11 SEM images for square microresonators with output waveguide connected to (a) one vertex and (b) the midpoint of one sidewall after ICP etching, and microscopic images for fabricated square microlasers with output waveguide connected to (c) one vertex and (d) the midpoint of one sidewall. Source: Long et al. [32].

modes at 35 and 64 mA are WG-like modes $H_z^{e,(p,q)}$ with an odd longitudinal-mode number. Multiple longitudinal-mode lasing is observed at 35 and 64 mA because the WG-like modes with odd l have much smaller Q factors than those with even l. Furthermore, the high Q factors of 1.14×10^5 and 4.05×10^4 for the fundamental and first-order transverse modes in Figure 7.7 result in the lasing of two transverse modes, because the two modes may have similar modal loss according to the material absorption loss but different spatial field distributions according to the numerical simulation. Figure 7.12b shows the lasing spectrum at the CW current of 20 mA. Four transverse modes around the main lasing mode are clearly observed at the wavelengths of 1544.47, 1545.66, 1546.77, and 1547.33 nm. The corresponding transverse-mode wavelength intervals are 2.86, 1.67, and 0.56 nm, which agree well with the analytical values of 3.0, 1.6, and 0.60 nm, respectively, extracted from Eq. (7.19). Furthermore, two-port square microlasers were demonstrated by connecting two waveguides to the vertices of the square microcavity [34].

For a square microlaser with a side length of 20 and 1.5-μm-wide waveguide connected to the midpoint of one sidewall, single-mode operations with the lasing mode wavelengths of 1533.60, 1546.54, 1560.83, and 1577.49 nm and the SMSRs of 36, 33, 42, and 37 dB are realized at the CW injection currents of 13, 33, 53, and 80 mA, respectively, as shown in Figure 7.13. The longitudinal-mode interval of 11.7 nm is obtained around 1550 nm, which corresponds to a group index of 3.6 from Eq. (7.18).

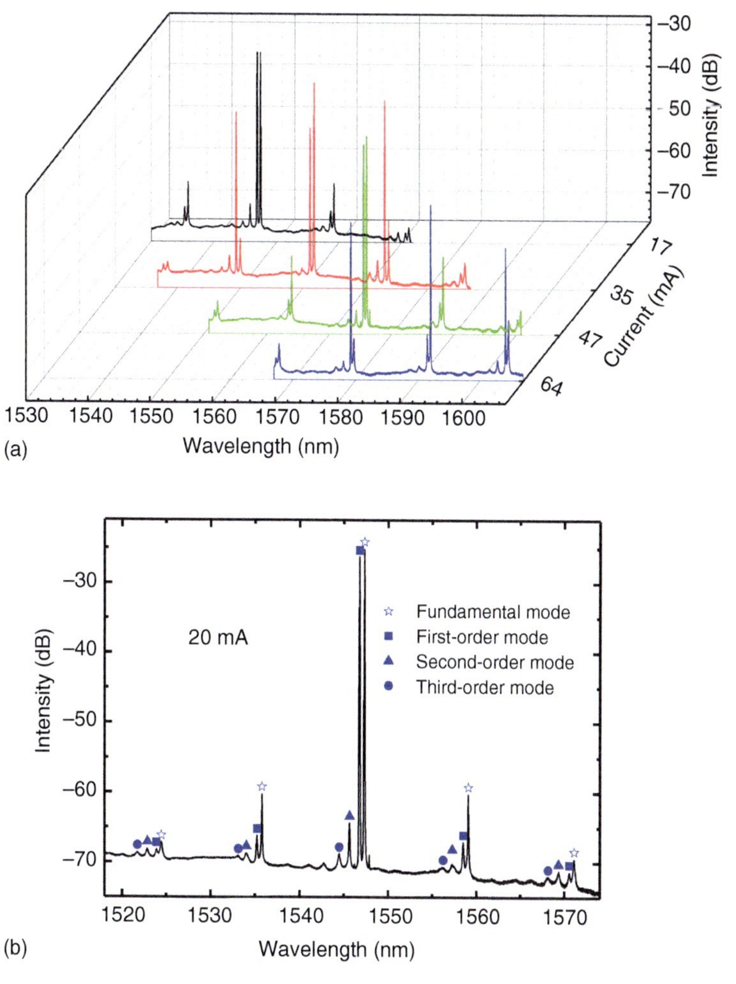

Figure 7.12 (a) Lasing spectra at the continuous currents of 17, 35, 47, and 64 mA, and (b) lasing spectrum at 20 mA for a microsquare laser with $a = 20\ \mu m$ and vertex connected waveguide width $w = 1.5\ \mu m$. The fundamental, first-order, second-order, third-order modes are marked by stars, squares, triangles, and circles, respectively. Source: Long et al. [32]. © 2014, IEEE.

The square microlaser with a midpoint waveguide is easy to realize in single-mode operation, which accords with the simulation results [24, 25].

7.7 Dual-Mode Lasing Square Lasers with a Tunable Interval

From the above experimental results, it can be found that dual-transverse-mode lasing is easy to realize in the square microlaser with a vertex waveguide. Different transverse modes have different spatial distributions, which make use of spatial

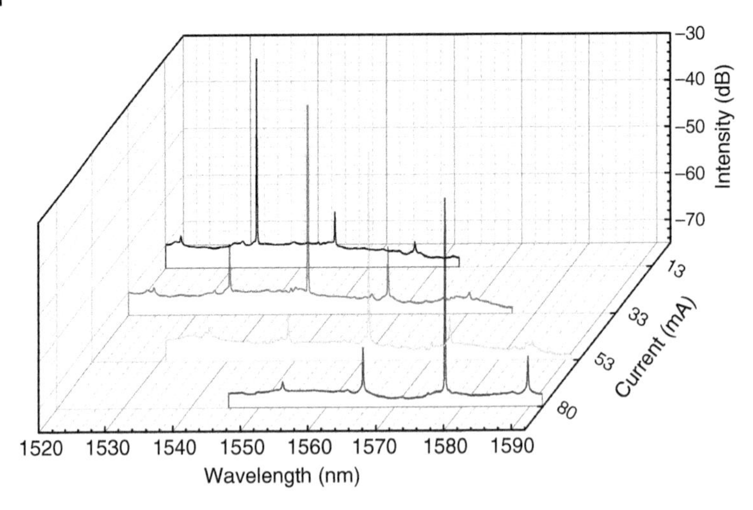

Figure 7.13 Lasing spectra at the CW currents of 13, 33, 53, 80 mA, a microsquare laser with a side length $a = 20\,\mu\text{m}$ and a midpoint connected waveguide width $w = 1.5\,\mu\text{m}$. Source: Long et al. [32]. © 2014, IEEE.

injection to modulate the lasing mode possible [35]. The fundamental transverse WG-like mode $H_z^{o,(86,88)}$ and the first-order transverse WG-like mode $H_z^{o,(85,89)}$ are found at 1553.85 and 1553.58 nm in the square microcavity with a side length of 30 μm with the mode Q factors of 1.46×10^5 and 5.52×10^4, respectively. Accounting for the absorption loss, the two modes can have near mode Q factors for realizing dual-mode lasing. The corresponding mode-field patterns $|H_z|$ are presented in Figure 7.14a,b. Along the lines connecting the midpoints of the adjacent sides, the fundamental mode has the strongest distribution, while the first-order transverse mode has a weak distribution. So the refractive index step Δn, as shown in Figure 7.14c, may result in a tunable transverse-mode wavelength interval between the fundamental and the first-order transverse modes.

The mode wavelengths of the square microcavity with a refractive index step Δn are simulated. The mode wavelength interval $\Delta\lambda$ between $H_z^{o,(86,88)}$ and $H_z^{o,(85,89)}$ vs. Δn is plotted as circular symbols in Figure 7.14d for the microsquare resonator with $a = 30\,\mu\text{m}$, $w_g = 2.5\,\mu\text{m}$, and $W = 4\,\mu\text{m}$. As Δn increases from −0.004 to 0.005, the mode wavelength interval $\Delta\lambda$ increases from 0.04 to 1.36 nm. The simulated mode wavelength interval $\Delta\lambda$ between $H_z^{o,(57,59)}$ and $H_z^{o,(56,60)}$ for the microsquare resonator with $a = 20\,\mu\text{m}$, $w_g = 1.5\,\mu\text{m}$, and $W = 2\,\mu\text{m}$ is also calculated and plotted as square symbols. As Δn increases from −0.004 to 0.005, $\Delta\lambda$ increases from 0.28 to 1.33 nm. The results indicate that the transverse-mode wavelength intervals may be modulated by the refractive index step Δn in the injection window area caused by the injection current. Furthermore, the mode wavelength interval $\Delta\lambda$ can be adjusted by changing the side length of the square microcavity according to Eq. (7.19). The wavelength intervals $\Delta\lambda$ are 1.86 and 0.27 nm around 1550 nm as $a = 10$ and 30 μm and $\Delta n = 0$, which corresponds to the frequency differences of 238 and 34 GHz.

Figure 7.15a shows the microscopic image of a fabricated 30-μm-side-length microsquare laser with a 2.5-μm-width output waveguide. The wavelength interval

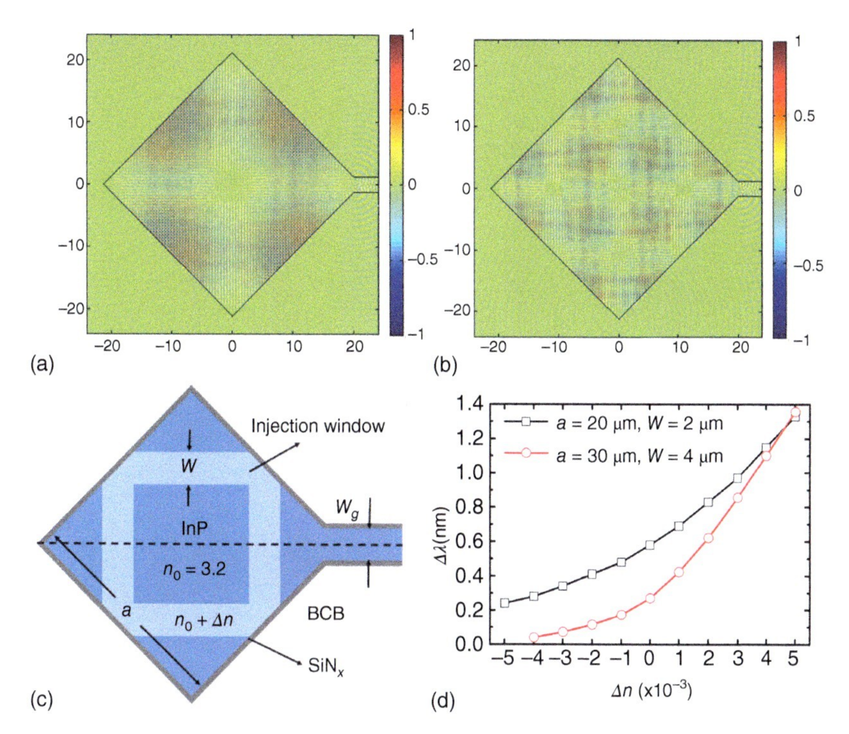

Figure 7.14 Mode-field patterns $|H_z|$ for (a) the fundamental and (b) the first-order transverse mode in the square microcavity. (c) Schematic diagram of the microsquare resonator with a square-ring current injection window. (d) Simulated mode wavelength intervals vs. Δn for the microsquare resonators with $a = 20\,\mu m$, $w_g = 1.5\,\mu m$, $W = 2\,\mu m$ (square symbols), and $a = 30\,\mu m$, $w_g = 2.5\,\mu m$, $W = 4\,\mu m$ (circular symbols). Source: Long et al. [35]. © 2015, Optical Society of America.

and the intensity ratio of the dual mode are measured and plotted in Figure 7.15b as the functions of the injection current. Dual-mode lasing with the intensity ratio less than 2.5 dB is realized from 89 to 108 mA, and the corresponding wavelength interval increases from 0.25 to 0.37 nm, which corresponds to the frequency difference from 31 to 46 GHz.

The output power of the dual-mode microsquare laser is coupled into a tapered single-mode fiber (SMF) and amplified by an erbium-doped fiber amplifier (EDFA). The amplified output is filtered by a tunable bandpass optical filter, and fed into a 32-GHz high-speed photodetector for the microwave generation. The microwave signal is measured using a 43-GHz bandwidth electric spectrum analyzer (ESA). The optical spectra measured at a resolution of 0.02 nm are plotted in Figure 7.15c at the injection currents of 90, 95, 100, and 105 mA, where the optical spectra for 95, 100, and 105 mA are vertically shifted by 50, 100, and 150 dB for clarity, respectively. From the detailed lasing spectrum at 90 mA, two four-wave mixing (FWM) peaks at 1562.44 and 1563.19 nm are observed, which are generated by the two lasing peaks at 1562.69 and 1562.94 nm. Two other wide peaks at 1561.29 and 1562.08 nm are the second-order and third-order transverse modes.

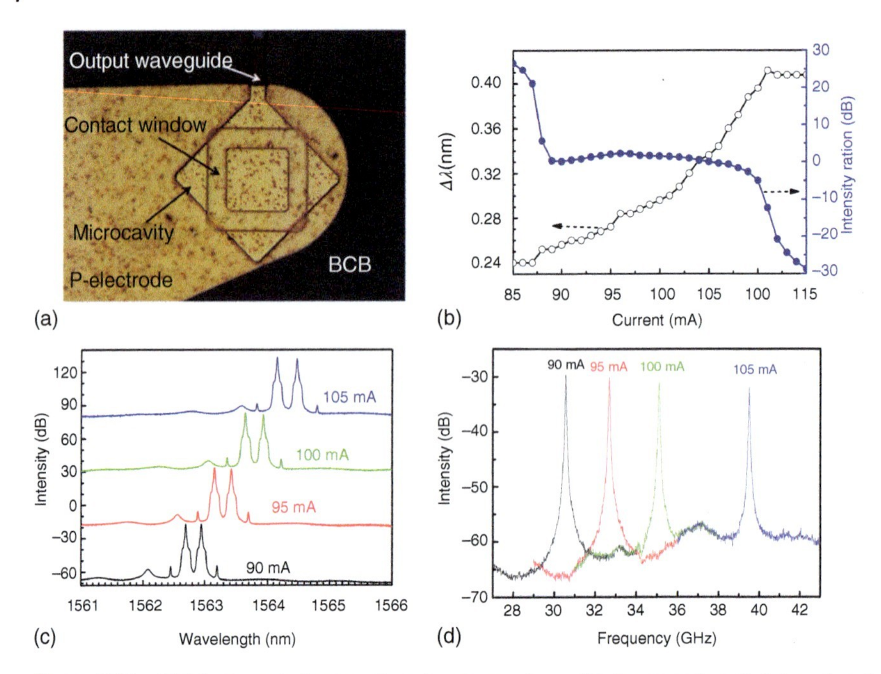

Figure 7.15 (a) Microscopic image of a microsquare laser, (b) the wavelength interval and intensity ratio of the dual-transverse-mode around 1563 nm vs. the injection current. (c) Optical spectra and (d) microwave signal spectra generated using the dual-mode microsquare laser at the injection currents of 90, 95, 100, and 105 mA. Source: Long et al. [35]. © 2015, Optical Society of America.

The corresponding microwave signals are presented in Figure 7.15d with the frequencies of 30.56, 32.70, 35.12, and 39.51 GHz and the 3-dB linewidths of 47, 53, 54, and 47 MHz, respectively, at the injection currents of 90, 95, 100, and 105 mA. Tunable microwave signals are realized based on the dual-mode square microlaser, and the linewidth of the microwave signals is mainly limited by the lasing mode linewidth of the microlaser. The microwave signals can also be directly detected from the electrode of a microlaser under dynamic states due to external optical injection [36].

7.8 Application of Dual-Mode Square Microlasers

Based on the dual-mode lasing square microcavity lasers, the signal frequency can be extended from tens of GHz to the THz range through mode beating. Sub-THz wave is generated based on a dual-mode square microcavity laser and the uni-traveling-carrier photodiode (UTC-PD), the obtained signal frequency is mainly limited by the response bandwidth of the UTC-PD [37].

An 18-μm-side-length square microcavity laser with a 1.5-μm-wide vertex output waveguide was used in the experiment of sub-THz signal generation. Figure 7.16a shows the schematic diagram of the experimental setup with the dual-mode square

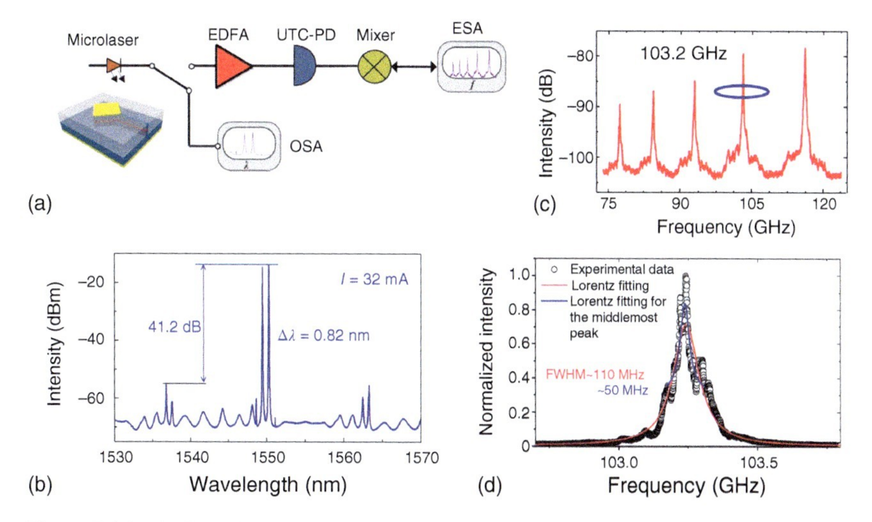

Figure 7.16 (a) Experimental setup for THz wave generation using a dual-mode square microlaser. (b) Lasing spectrum at an injection current of 32 mA. (c) Original mixed frequency signal. (d) Detailed electrical spectrum around 103.2 GHz. Source: From Yang et. al. [38]. © 2018, IOP Publishing Ltd.

microcavity laser. The lasing output from the square microcavity laser was coupled into a tapered SMF and amplified by an EDFA with a gain of 32 dB and then detected by a UTC-PD and a harmonic mixer. The output electrical signal from the mixer was fed into an ESA. Figure 7.16b shows the optical spectrum of the square microcavity laser under the injection current of 32 mA. The two main lasing peaks have a wavelength interval of 0.82 nm and an intensity ratio of about 0.5 dB. The lasing spectrum indicates the lasing of the fundamental and first-order transverse modes with an SMSR of about 41.2 dB. Figure 7.16c shows the generated sub-THz signal at the UTC-PD mixer with the dual-mode square microcavity laser at 32 mA, which indicates a center frequency of 103.2 GHz for the beating signal. The other peaks are induced by the local oscillator (LO) in the frequency mixer. Figure 7.16d shows the detailed electrical spectrum for the generated sub-THz waves signal around 103.2 GHz. A full-width at the half-maximum (FWHM) of about 110 MHz is obtained according to a Lorentz fitting. However, the signal linewidth is broadened due to the system instability and/or signal drifting. Using the Lorentz fit with a smaller range for the middlemost peak, the linewidth is obtained about 50 MHz as shown in Figure 7.16d, which is supposed to be the actual linewidth [38].

Dual-mode lasers can also be used as the seeding source to initiate the FWM effect in the highly nonlinear fiber (HNLF), thus generating wideband OFCs and optical pulses [39–41]. The experimental setup for the OFC generation is shown in Figure 7.17a. The lasing output from the dual-mode square microcavity laser is firstly preamplified by an $EDFA_0$. Then, the amplified signal goes through an optical band-pass filter (BPF) and is boosted by a high-power $EDFA_1$, which can initiate the FWM effect in the $HNLF_1$ and produce lots of FWM sidebands. Similar to the first stage,

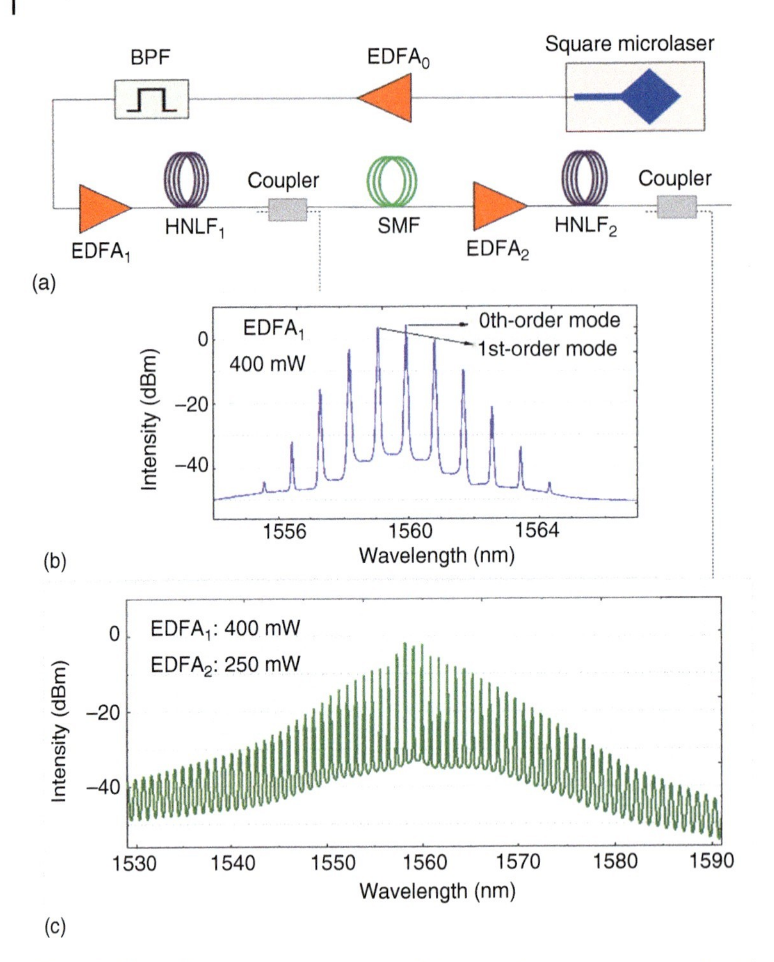

Figure 7.17 (a) Experimental setup for OFC generation based on a dual-mode square microcavity laser. Generated OFC spectra after the (b) first stage and (c) second stage of HNLF. Source: Based on Weng et al. [41]. © 2019, IEEE.

the EDFA$_2$ and HNLF$_2$ are exploited for the cascaded FWM spread-spectrum process. Before injecting to the second amplifier, a 50-m-long standard SMF is added to compress the chirped waveform produced in HNLF$_1$. Each HNLF has a length of 500 m and a nonlinear coefficient of 10 W^{-1} km^{-1}. A total insertion loss of 7 dB is estimated for the whole transmission system.

An optical spectrum analyzer with a resolution of 0.02 nm is used to measure the spectra of the OFCs and the optical signal-to-noise ratios (OSNRs) of the sidebands. The OFC spectrum after the first stage of HNLF$_1$ is shown in Figure 7.17b, where 11 teeth are found at the EDFA$_1$ pump power of 400 mW. Then, the multiwavelength source generated in the first stage is broadened to a Gaussian-shaped comb spectrum in the second spread-spectrum stage, with a bandwidth exceeding 100 nm at the EDFA$_2$ pump power of 250 mW. There are 70 FWM lines with OSNR >10 dB in the range from 1529 to 1591 nm as shown in Figure 7.17c, covering the C-band of dense wavelength division multiplexing (DWDM) systems. 26 FWM lines with OSNR >20 dB are observed in the range from 1548 to 1572 nm. The bandwidth and flatness of the comb can further be improved by compressing the laser linewidth, suppressing the $1/f$ frequency noise and stimulated Brillouin backscattering in the system [39]. The results indicate that dual-mode square microcavity lasers can replace the separated lasers for the ongoing research on dual-pumped fiber-based OFC generation. Considering the simple fabrication process, the strong correlation between two lasing modes, and the high repetition rates, we expect the OFCs based on the dual-mode square microcavity lasers can be good candidates for WDM systems.

Furthermore, strong mode correlation is experimentally demonstrated by comparing the lasing mode linewidths of dual-mode square microlaser and the linewidth of the beating microwave signal [42]. The beating microwave at 35.9GHz with a linewidth of 12MHz was measured for a dual-mode lasing square microlaser with a side length of 20 μm and a 1.5-μm-wide vertex output waveguide. The microwave signal linewidth of 12 MHz is about one order less than the lasing mode linewidths of 88 and 102 MHz, which indicate the phase noises of the two lasing modes are partly synchronism caused by the carrier fluctuation.

7.9 Lasing Spectra Controlled by Output Waveguides

The square microcavity laser with an output waveguide connecting the midpoint of a sidewall is suitable for single transverse-mode lasing [33]. The simulated mode Q factors vs. the midpoint waveguide width w are plotted in Figure 7.18 for the

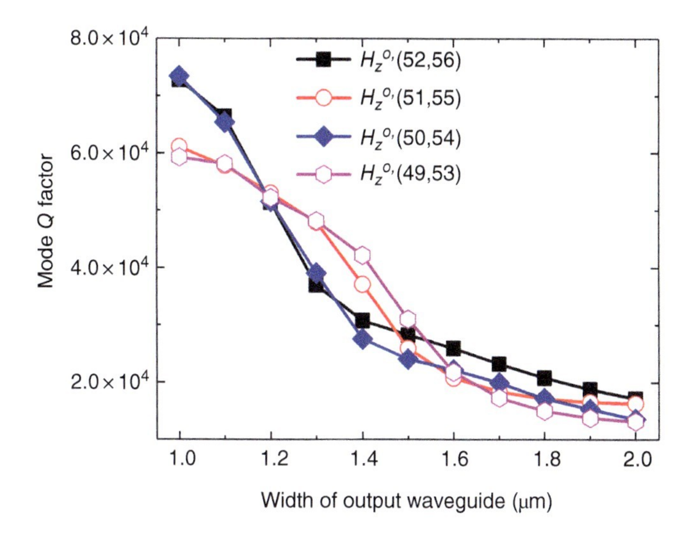

Figure 7.18 Mode Q factors vs. the output waveguide width for $H_z^{o,(52,56)}$, $H_z^{o,(51,55)}$, $H_z^{o,(50,54)}$, and $H_z^{o,(49,53)}$ in the square microcavities with a side length of 17.8 μm. Source: Tang et al. [33]. © 2015, Optical Society of America.

high-Q modes $H_z^{o,(52,56)}$, $H_z^{o,(51,55)}$, $H_z^{o,(50,54)}$, and $H_z^{o,(49,53)}$ in square microcavities with a side length of 17.8 μm. The mode-field distributions of $H_z^{o,(52,56)}$ and $H_z^{o,(50,54)}$ modes are symmetrical to the middle line of the output waveguide, while those of $H_z^{o,(51,55)}$ and $H_z^{o,(49,53)}$ modes are antisymmetrical, due to the even and odd mode numbers. The mode Q factors of the symmetrical and antisymmetrical modes degrade with the increase of the waveguide width at different rates. The mode Q factors of the symmetrical modes decrease more quickly with the increase of w than the antisymmetrical modes as $w < 1.4$ μm. The symmetrical modes have higher Q factors as $w < 1.2$ μm, but the antisymmetrical modes have higher Q factors as 1.2 μm < w < 1.47 μm. As 1.6 μm < w < 1.8 μm, the mode Q factors of the symmetrical modes are slightly higher than those of the antisymmetrical modes again. Therefore, by adjusting the output waveguide width, we can even have the high Q symmetrical or antisymmetrical modes with the wavelength interval as four times as the longitudinal-mode interval, which is important for the square microlaser with a wide continuous-wavelength tuning range. In the following part, we fabricate square microlasers based on the simulated results.

The lasing spectra of a square microlaser with a = 17.8 μm and w = 1.8 μm are measured by an optical spectrum analyzer at a resolution of 0.06 nm and presented in Figure 7.19a at 298 K and I = 6, 15, 20, 30, 40, 50, and 58 mA, where the adjacent spectra are relatively shifted by 10 dB for clarity. Evident resonance peaks at 1516.48, 1529.39, 1542.37, and 1555.54 nm marked by A, B, C, and D are observed at 6 mA. The wavelength intervals 12.91, 13.01, and 13.17 nm can be fitted by Eq. (7.18) with the group index n_g ranging from 3.57 to 3.62. As the current increases from 15 to 20 mA, the lasing mode jumps 26.75 nm over two longitudinal-mode intervals to mode D. This indicates that mode D has a higher-mode Q factor than mode C and the mode interval between the high Q modes is twice of the longitudinal-mode interval, which is in good agreement with the simulation results. Furthermore, the coupling efficiency to the SMF is measured, which is defined as the power coupled to the SMF to that measured by a 5-mm-diameter detector 2 mm away from the cleaved facet. The measured coupling efficiency is about 5.9% as $I < 10$ mA and 21% as $I > 20$ mA. Based on the simulated and experimental results, the increase of the coupling efficiency corresponds to the transition of the lasing modes from the antisymmetric mode to the symmetric mode. As the injection current continues to increase, the dominant lasing mode keeps constant at mode D due to the high Q factor of mode D.

The lasing characteristics are also studied for a square microlaser with a side length of 17.8 μm and the output waveguide width of 1.4 μm. The lasing spectra at I = 5, 15, 25, 35, 45, 55, and 65 mA are presented in Figure 7.19b at 291 K, where the adjacent spectra are relatively shifted by 10 dB for clarity. Three evident peaks at 1519.33, 1532.06, and 1544.95 nm are observed at I = 5 mA, and marked as A, B, and C. The measured coupling efficiency to the SMF is only 5.86%, so the lasing mode should be an antisymmetric mode relative to the output waveguide, which agrees well with the simulation results in Figure 7.18 as the output waveguide width is 1.4 μm. Due to the high mode Q factor of the mode B and the large mode interval between the adjacent high Q antisymmetric modes, the lasing mode keeps

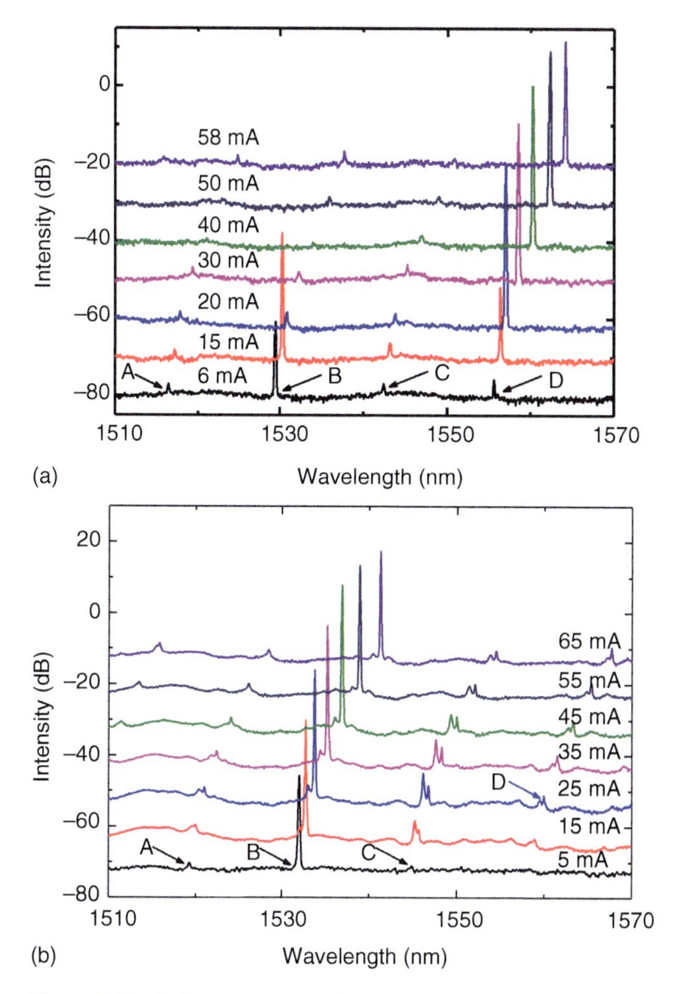

Figure 7.19 Lasing spectra at different currents for the square microlaser with a side length of 17.8 μm and the output waveguide width of (a) 1.8 μm and (b) 1.4 μm. Source: Tang et al. [33]. © 2015, Optical Society of America.

at the mode B with a continuous-wavelength tuning range of 9.26 nm with the SMSR >26 dB from 5 to 65 mA.

The square microlasers were also fabricated with four output waveguides connected to four vertices of the square resonator [43]. For a microsquare laser with a side length of 20 μm and four output waveguides with a width of 1.5 μm, the output powers coupled into an SMF and a multiple-mode fiber and applied voltage were measured at 288 K and plotted as functions of injection current in Figure 7.20a, where the inset of a microscopic image is present. The threshold current of 11 mA is much larger than that with a single-output waveguide due to the lower mode Q factor for the microcavity with four output waveguides. The output powers measured through the SMF and multiple-mode fiber are 29 and 206 μW at 63 mA, respectively. The kinks of the output power–current curve are mainly caused by lasing mode

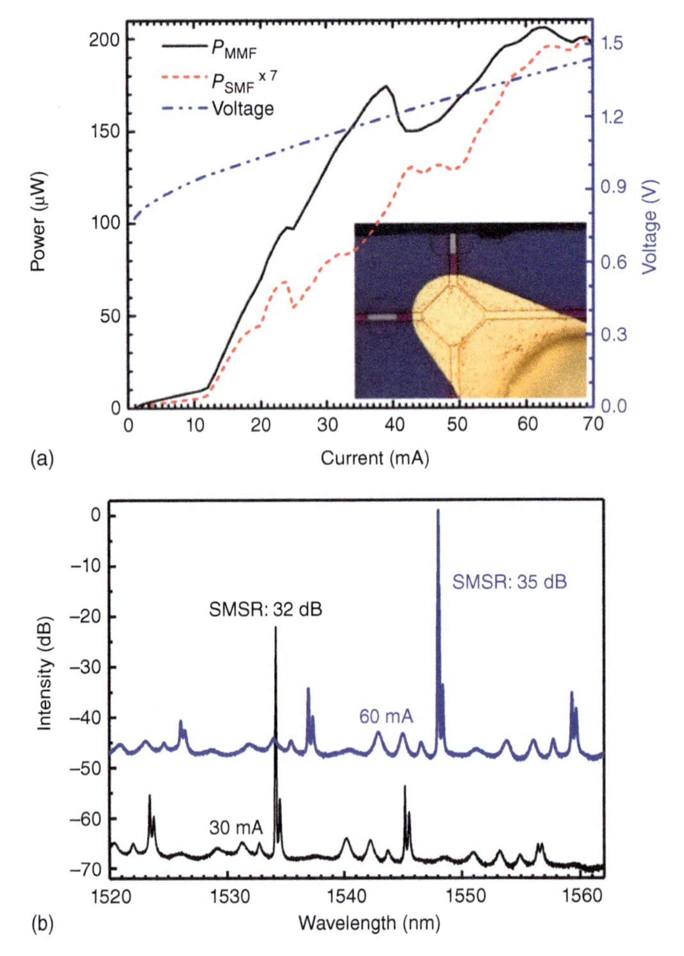

Figure 7.20 (a) Output powers coupled into single-mode fiber and multiple-mode fiber and applied voltage vs. CW injection current at 288 K, (b) and lasing spectra at currents of 30 mA and 60 mA, for a square microcavity laser with a side length of 20 μm and four vertex waveguides with a width of 1.5 μm.

jumping as the device temperature rises with the injection current. The lasing spectra at 30 and 60 mA were measured and are shown in Figure 7.20b with those at 60 mA shifted 20 dB vertically for clarity. Single-mode operations with the SMSRs of 32 and 35 dB were realized at 30 and 60 mA. The results indicate that the square microlasers are suitable as a light source for photonic integrated circuits, and especially can provide multiple sources from a microlaser.

7.10 Circular-Side Square Microcavity Lasers

To further enhance the wavelength interval of the dual-transverse-mode square microcavity lasers, we have proposed and demonstrated the deformed square

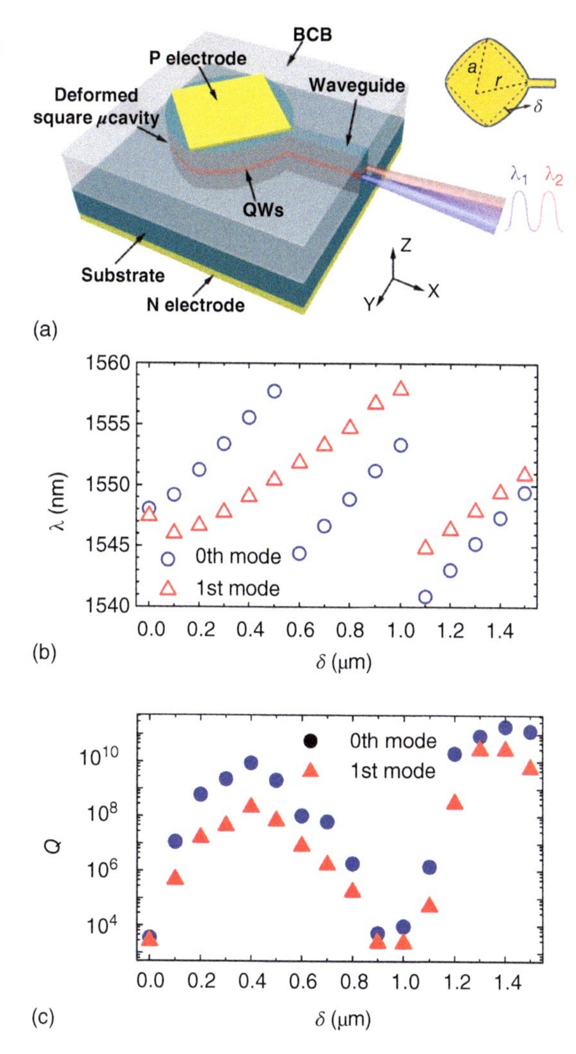

Figure 7.21 (a) Schematic diagram of waveguide-coupled circular-side square microcavity laser. (b) Mode wavelengths and (c) corresponding mode Q factors for the WGMs around 1550 nm vs. the deformation amplitude δ. Source: From Yang et. al. [38]. © 2018, IOP Publishing Ltd.

microcavities with the flat sides being replaced by circular sides [44]. Figure 7.21a shows the schematic diagram of a deformed square microcavity laser with the flat sides replaced by circular arcs, where a is the side length of square, δ is the deformation parameter, and r is the radius of the circular arcs. Considering the mode-field distributions of the fundamental and first-order transverse modes, the effective round-trip length of the fundamental mode will be longer than that of the first-order mode in the circular-side square microcavity. Thus, with the increase of deformation parameter δ, the wavelength of the fundamental mode should increase much faster than that of the first-order mode, and the wavelength interval should be enhanced greatly.

The high-Q whispering-gallery modes (WGMs) in the circular-side square microcavities are numerically simulated by the 2D FEM. The TE WGMs with symmetric magnetic field (H_z) distributions relative to the middle line of the output waveguide are considered here. For the circular-side square microcavity with $a = 16\,\mu\text{m}$ and an output waveguide with width $w = 1.5\,\mu\text{m}$, the mode wavelengths and Q factors vs. δ obtained by FEM are plotted in Figure 7.21b,c for the fundamental (0th) and first-order (1st) transverse modes. The mode wavelengths of the fundamental transverse modes redshift about 17 nm as δ increases 0.1 μm due to the increase of the effective cavity size (or the effective round-trip length), which is close to the longitudinal-mode wavelength interval $\Delta\lambda_l$. The mode wavelengths of the fundamental and first-order transverse modes, shown in Figure 7.21b, are kept around 1550 nm by choosing different longitudinal modes. For the WGMs belonging to the same longitudinal-mode number, the transverse-mode interval $\Delta\lambda_t$ increases gradually with the deformation parameter δ due to the increase of effective round-trip length difference. However, if $\Delta\lambda_t > \Delta\lambda_l/2$, the wavelength interval between the nearby fundamental and first-order transverse modes becomes $\Delta\lambda = \Delta\lambda_l - \Delta\lambda_t$. Then, the wavelength interval will decrease with δ increasing when $\delta > 0.6\,\mu\text{m}$ as shown in Figure 7.21b. The mode Q factors of the fundamental and first-order order transverse modes are 8.7×10^9 and 2.1×10^8 at $\delta = 0.4\,\mu\text{m}$, and are 1.9×10^{11} and 2.7×10^{10} at $\delta = 1.4\,\mu\text{m}$, respectively, which are much larger than those in a square microcavity. The mode Q factors are low around $\delta = 0.95\,\mu\text{m}$, because there are no long-lived light rays according to the light ray simulation [44].

The mode-field distributions of $|H_z|$ are presented in Figure 7.22a–d at $\delta = 0, 0.5, 0.9$, and 1.3 μm for the 0th- and 1st-order transverse modes, respectively. The high-Q mode field patterns in Figure 7.22b,d have narrow transverse field distributions with near-constant width and flat-phase planes similar to the plane wave. The modes take ultra-high Q factors as the field patterns are well confined in the transverse direction with near-zero radiation loss at the vertices. In contrast, the main transverse-mode field patterns in Figure 7.22c are focused too much by the circular sides as concave mirrors. Strong focused effect causes additional field beams with a large radiation loss, and the ultra-high Q modes exist as the deformation magnitudes are smaller and larger than the strong focused case.

The circular-side square microcavity lasers are fabricated on the AlGaInAs/InP MQWs laser wafer with the parameters of $a = 16\,\mu\text{m}$ and $w = 1.5\,\mu\text{m}$ as used in Figure 7.21. The inset of Figure 7.23a shows the top-view optical-microscope image of a fabricated circular-side square microcavity laser. Figure 7.23a shows the lasing spectrum for a circular-side square microcavity lasers with $\delta = 1.1\,\mu\text{m}$. Dual-mode lasing with a wavelength interval of 3.43 nm is obtained with an SMSR of 26 dB and intensity ratio of about 0.4 dB. The fundamental and first-order transverse modes are marked by circle and triangle symbols, respectively. Different from the nondeformed square microcavity laser, the fundamental lasing mode is at the short-wavelength side as the two lasing modes have different longitudinal-mode numbers. Figure 7.23b shows the lasing mode wavelength interval $\Delta\lambda$ vs. the deformation δ for the circular-side square microcavity lasers. The lasing fundamental transverse mode is at long-wavelength side when $\delta < 0.7\,\mu\text{m}$, $\Delta\lambda$ first increases with δ and reaches 7.6 nm at $\delta = 0.7\,\mu\text{m}$ as shown in Figure 7.23b. At $\delta = 0.9\,\mu\text{m}$, single transverse-mode lasing is realized due to the lower-mode Q factor. As δ increases beyond 1.1 μm, dual-mode lasing appears again because the enhancement of the mode Q factors, and $\Delta\lambda$ decreases because the lasing fundamental transverse mode is at short-wavelength side and moves toward the first-order transverse mode as the simulated results in Figure 7.21.

The experimental results indicate that dual-mode lasing can be easily realized in the circular-side square microcavity with ultrahigh-Q modes, and the transverse-mode interval can be adjusted by varying the deformation parameter δ. However, the wavelength interval between the lasing modes is also limited by the longitudinal-mode interval due to the lasing mode competition. In addition, the mode-field patterns are mainly determined by the mode numbers p and q of wave-mode numbers along the directions of the square sides instead of the longitudinal- and transverse-mode numbers. The nearby high Q confined modes usually have totally different values of p and q; even they have the same longitudinal-mode number or transverse-mode number. For example, the fundamental and the first-order transverse modes with the same longitudinal-mode number will have the mode numbers of $(p, q) = (p, p+2)$ and $(p-1, p+3)$, and the intensity overlap is much weaker than that in a Fabry–Pérot cavity. So the square microcavity lasers satisfy the stable condition of dual-mode stationary solution based on nonlinear gain analysis. Furthermore, the ultra-high Q modes with mode-field patterns in Figure 7.22 are mainly determined by the longitudinal- and transverse-mode numbers again, instead of the aforementioned mode numbers p and q. But stable dual-mode lasing condition can satisfy for circular-side square resonator microlasers in most deformation cases [44].

A lot of high-order transverse-mode peaks appear in the lasing spectra of Figure 7.23a due to the ultrahigh mode Q factor. To suppress high-order transverse modes, we design output waveguide-connecting angle for controlling mode Q factor. Based on numerical simulation, circular-side square microlasers were fabricated with different δ, $a = 16\,\mu m$, and a 1.5-μm-wide output waveguide. Figure 7.24a shows the SEM image of a deformed circular side square microresonator with a

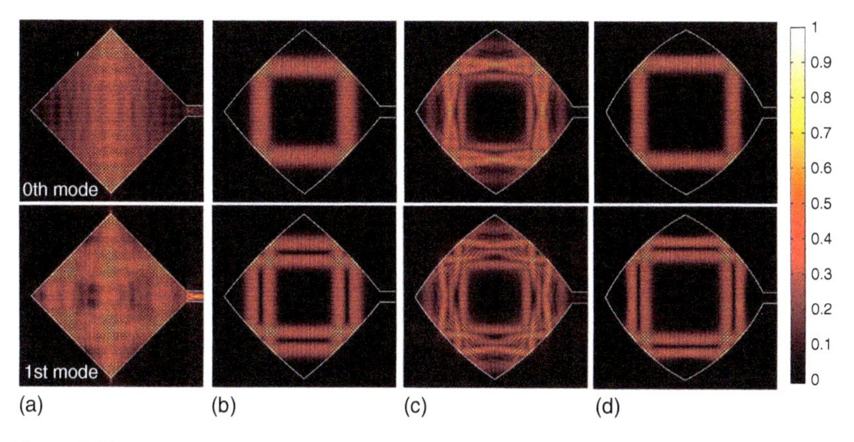

Figure 7.22 The distributions of $|H_z|$ for the 0th (upper) and 1st (lower) symmetric transverse modes at (a) $\delta = 0$, (b) 0.5 μm, (c) 0.9 μm, and (d) 1.3 μm, respectively. Source: Weng et al. [44]. © 2017, American Physical Society.

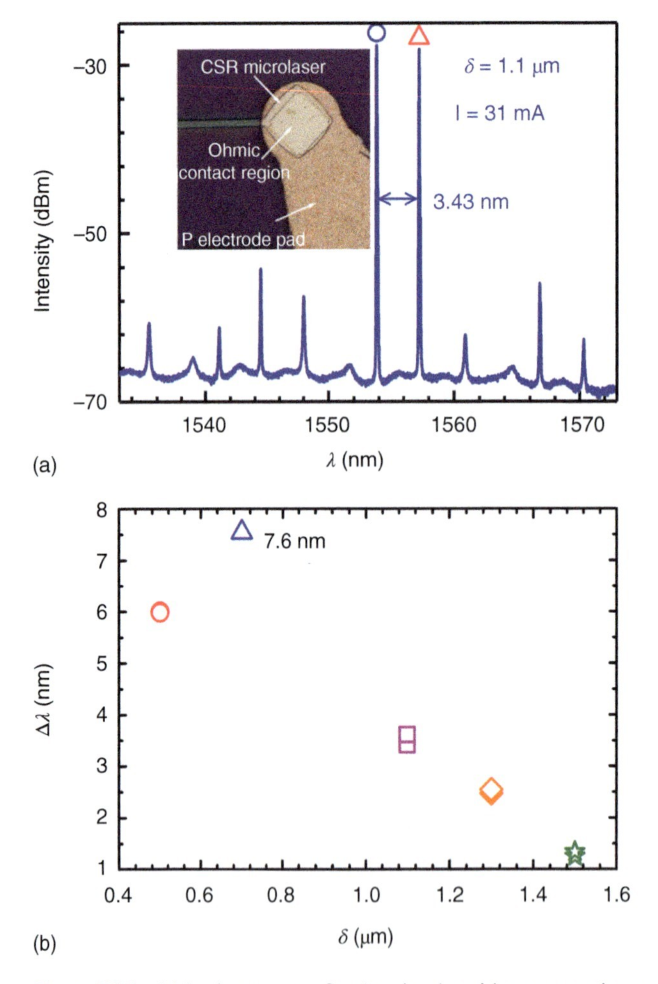

Figure 7.23 (a) Lasing spectra for the circular-side square microcavity lasers with $\delta = 1.1$ micrometer. Insets: optical microscope image of a circular-side square microlaser. (b) Wavelength intervals of the circular-side square microcavity lasers with different δ. Source: From Yang et. al. [38]. © 2018, IOP Publishing Ltd.

tangential output waveguide after ICP etching. The tangential output waveguide can have a higher coupling efficiency with the confined mode as it is nearly parallel to the transmission direction of the mode light ray. Figure 7.24b shows the optical microscope image of a fabricated microlaser.

By butt-coupling a multimode fiber to the cleaving facet of the output waveguide to collect the output light, the microlasers are measured at the heatsink temperature of 288 K. The lasing spectrum of a microlaser with $\delta = 2.7\,\mu$m at an injection current of 36 mA is shown in Figure 7.25a, which indicates a uniform dual-mode lasing at 1566.91 and 1561.51 nm corresponding to the fundamental transverse mode and the first-order transverse mode, respectively. Figure 7.25b shows the lasing spectra for the circular-side square microlasers with δ from 1.9 to 2.5 μm,

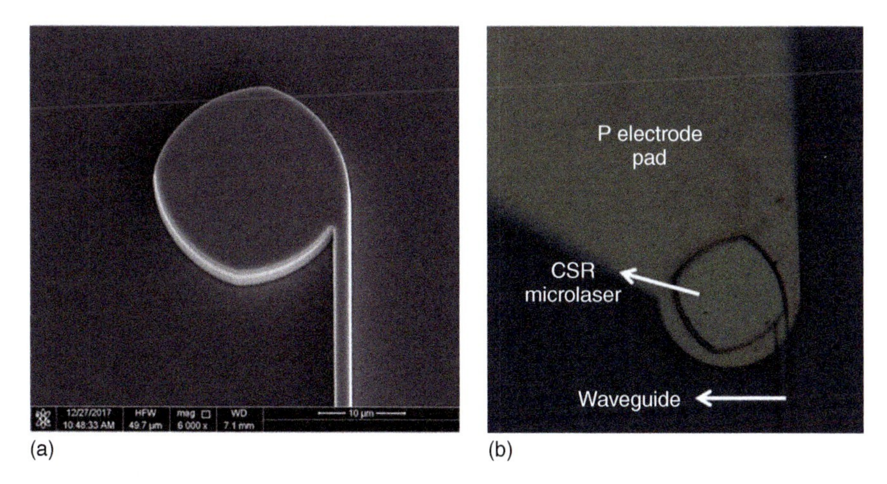

Figure 7.24 (a) SEM image of a circular-side square microresonator with a tangential output waveguide after ICP etching. (b) Optical microscope image of a fabricated microlaser. Source: Weng et al. [45].

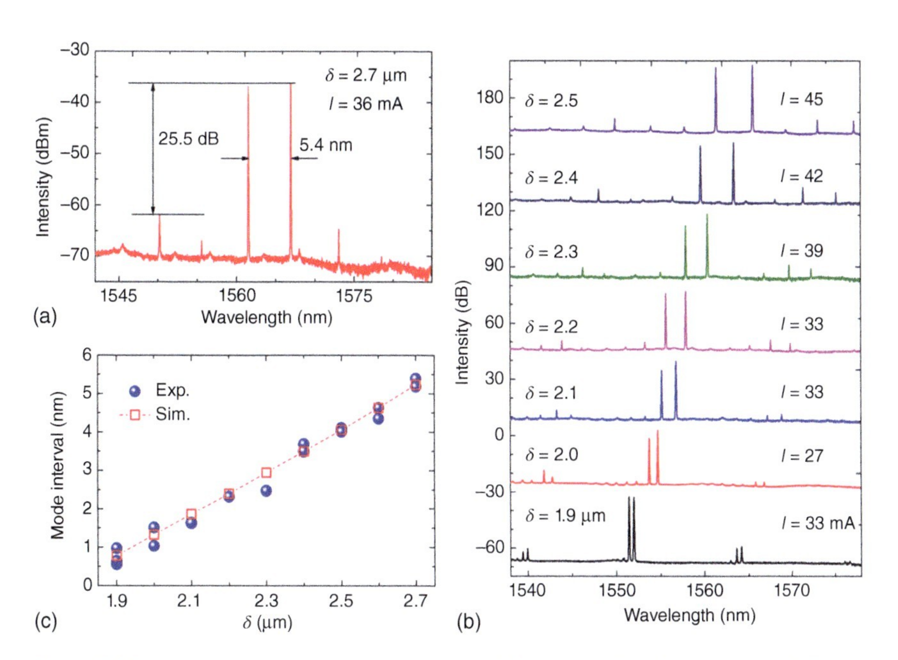

Figure 7.25 (a) Lasing spectrum of the circular-side square microlaser with $\delta = 2.7\ \mu m$ at 36 mA. (b) Lasing spectra of the microlasers with different deformation value δ. (c) Mode interval between the fundamental transverse mode and the first-order transverse mode vs. δ for comparing experimental and simulated results. Source: Weng et al. [45]. © 2018, Optical Society of America.

which indicate a purer dual-mode lasing spectrum than those in Figure 7.23. Both the lasing wavelengths and the mode interval gradually increase with the increase of δ, which are consistent with the simulation results. The measured lasing-mode-wavelength intervals vs. δ are depicted in Figure 7.25c as solid circles. The measured transverse-mode intervals are varied with δ at the rate of 5.9 nm μm^{-1} for the microlasers as $1.9 \leq \delta \leq 2.5\,\mu m$, which agree well with the numerical result of 5.5 nm μm^{-1} depicted by open squares [45].

7.11 Summary

In summary, a quasi-analytical model similar to Marcatili's scheme is used to describe the confined modes in 2D square microcavities, and far-field emission method based on the quasi-analytical near-field distributions is applied to estimate the mode Q factors. Furthermore, the group theory is used to analyze the symmetry of the confined modes in the microsquare, and the mode coupling between the modes with the same irreducible representations is not forbidden. The mode coupling between the modes with even longitude mode number would happen and results in high-Q-coupled WG-like modes. The WG-like modes have a sinusoidal-modulated envelope along the sidewall, which make directly connecting an output waveguide to the position with weak field distribution possible. InP-based waveguide-coupled directional-emission square microcavity lasers are fabricated and electrically injected lasing is realized at room temperature. Dual-mode lasing with tunable intervals is demonstrated by using a spatial current injection to modulate the lasing modes as different transverse modes have distinct field distributions, and a tunable microwave signal is obtained using the dual-mode microlaser. The dual-mode square microcavity lasers are applied to generate sub-THz waves and OFCs. By properly designing waveguide width to control the mode Q factor difference between different longitudinal modes, single-mode square microlaser with a wide continuous tunable wavelength range is demonstrated. Deformed square microcavities with the flat sidewalls replaced by circular arcs are proposed and demonstrated for enhancing the mode confinement and increasing the dual-mode interval from tens of GHz to THz range. In addition, the ultra-high mode Q factors for the deformed square microcavity may have merit on the application of optical sensor.

The WG-like mode-field distributions in square are obviously different from the WGMs in the microdisks. The WG-like modes typically distribute over the whole cavity with sinusoidal-modulated envelope along the sidewall, and spatial distributions of different transverse modes have a large offset, which make using output waveguide and spatial injection to modulate and control lasing mode possible. The WG-like modes also promise high injection efficiency and avoid effect of carrier hole burning and diffusion on high-speed direct modulation. Different functional square semiconductor microcavity lasers have been demonstrated. We believe that the laser performance can be improved and the practical application in photonics integration and optical interconnects can be realized in the future.

References

1 Li, C., Ma, N., and Poon, A.W. (2004). Waveguide-coupled octagonal microdisk channel add-drop filters. *Opt. Lett.* 29: 471–473.

2 Yang, Y.D. and Huang, Y.Z. (2007). Symmetry analysis and numerical simulation of mode characteristics for equilateral-polygonal optical microresonators. *Phys. Rev. A* 76: Art. no. 023822.

3 Chen, Q., Yang, Y.D., and Huang, Y.Z. (2007). Finite-difference time-domain analysis of deformed square cavity filters with a traveling-wave-like filtering response by mode coupling. *Opt. Lett.* 32: 967–969.

4 Yang, Y.D., Huang, Y.Z., Che, K.J. et al. (2009). Equilateral-triangle and square resonator semiconductor microlasers. *IEEE J. Sel. Top. Quantum Electron.* 15: 879–884.

5 Fang, H.H., Ding, R., Lu, S.Y. et al. (2013). Whispering-gallery mode lasing from patterned molecular single-crystalline microcavity array. *Laser Photonics Rev.* 7: 281–288.

6 Xu, C.X., Dai, J., Zhu, G.P. et al. (2014). Whispering-gallery mode lasing in ZnO microcavities. *Laser Photonics Rev.* 8: 469–494.

7 Huang, Y.Z., Guo, W.H., and Wang, Q.M. (2000). Influence of output waveguide on mode quality factor in semiconductor microlasers with an equilateral triangle resonator. *Appl. Phys. Lett.* 77: 3511–3513.

8 Guo, W.H., Huang, Y.Z., Lu, Q.Y., and Vu, L.J. (2003). Whispering-gallery-likel modes in square resonators. *IEEE J. Quantum Electron.* 39: 1106–1110.

9 Guo, W.H., Huang, Y.Z., Lu, Q.Y., and Yu, L.J. (2003). Modes in square resonators. *IEEE J. Quantum Electron.* 39: 1563–1566.

10 Yang, Y.D. and Huang, Y.Z. (2007). Mode analysis and Q-factor enhancement due to mode coupling in rectangular resonators. *IEEE J. Quantum Electron.* 43: 497–502.

11 Lv, X.M., Huang, Y.Z., Yang, Y.D. et al. (2014). Influences of carrier diffusion and radial mode field pattern on high speed characteristics for microring lasers. *Appl. Phys. Lett.* 104: Art. no. 161101.

12 Guo, W.H., Huang, Y.Z., Lu, Q.Y., and Yu, L.J. (2004). Comparison of free spectral range and quality factor for two-dimensional square and circular microcavities. *Chin. Phys. Lett.* 21: 79–80.

13 Poon, A.W., Courvoisier, F., and Chang, R.K. (2001). Multimode resonances in square-shaped optical microcavities. *Opt. Lett.* 26: 632–634.

14 Moon, H.J., An, K., and Lee, J.H. (2003). Single spatial mode selection in a layered square microcavity laser. *Appl. Phys. Lett.* 82: 2963–2965.

15 Boriskina, S.V., Benson, T.M., Sewell, P., and Nosich, A.I. (2005). Optical modes in 2-D imperfect square and triangular microcavities. *IEEE J. Quantum Electron.* 41: 857–862.

16 Huang, Y.Z., Chen, Q., Guo, W.H., and Yu, L.J. (2005). Experimental observation of resonant modes in GaInAsP microsquare resonators. *IEEE Photonics Technol. Lett.* 17: 2589–2591.

17 Hattori, H.T., Liu, D.Y., Tan, H.H., and Jagadish, C. (2009). Large square resonator laser with quasi-single-mode operation. *IEEE Photonics Technol. Lett.* 21: 359–361.

18 Liu, D.Y., Hattori, H.T., Fu, L. et al. (2010). The temperature dependence of InGaAs single-wavelength quantum well and multi-wavelength quantum dot square resonator microlasers. *J. Phys. D: Appl. Phys.* 43: Art. no. 135102.

19 Che, K.J., Yang, Y.D., and Huang, Y.Z. (2010). Mode characteristics for square resonators with a metal confinement layer. *IEEE J. Quantum Electron.* 46: 414–420.

20 Ma, R.M., Oulton, R.F., Sorger, V.J. et al. (2011). Room-temperature sub-diffraction-limited plasmon laser by total internal reflection. *Nat. Mat.* 10: 110–113.

21 Bittner, S., Bogomolny, E., Dietz, B. et al. (2013). Experimental observation of localized modes in a dielectric square resonator. *Phys. Rev. E* 88: Art. no. 062906.

22 Lee, C.W., Wang, Q., Lai, Y.C. et al. (2014). Continuous-wave InP-InGaAsP microsquare laser-a comparison to microdisk laser. *IEEE Photonics Technol. Lett.* 26: 2442–2445.

23 Guo, C.C., Xiao, J.L., Yang, Y.D. et al. (2016). Lasing characteristics of wavelength-scale aluminum/silica coated square cavity. *IEEE Photonics Technol. Lett.* 28: 217–220.

24 Zhao, W. and Huang, Y.Z. (2007). Analysis of directional emission in square resonator lasers with an output waveguide. *Chin. Opt. Lett.* 5: 463–465.

25 Huang, Y.Z., Che, K.J., Yang, Y.D. et al. (2008). Directional emission InP/GaInAsP square-resonator microlasers. *Opt. Lett.* 33: 2170–2172.

26 Yang, Y.D. and Huang, Y.Z. (2016). Mode characteristics and directional emission for square microcavity lasers. *J. Phys. D: Appl. Phys.* 49: Art. no. 253001.

27 Wiersig, J. (2006). Formation of long-lived, scarlike modes near avoided resonance crossings in optical microcavities. *Phys. Rev. Lett.* 97: Art. no. 253901.

28 Chen, Q., Huang, Y.Z., Guo, W.H., and Yu, L.J. (2005). Analysis of modes in a freestanding microsquare resonator by 3-D finite-difference time-domain. *IEEE J. Quantum Electron.* 41: 997–1001.

29 Yang, Y.D., Huang, Y.Z., and Chen, Q. (2007). Comparison of Q-factors between TE and TM modes in 3-D microsquares by FDTD simulation. *IEEE Photonics Technol. Lett.* 19: 1831–1833.

30 Li, J., Yang, Y.D., and Huang, Y.Z. (2010). Mode simulation for midinfrared microsquare resonators with sloped sidewalls and confined metals. *IEEE Photonics Technol. Lett.* 22: 459–461.

31 Chen, Q. and Huang, Y.Z. (2006). Investigation of mode characteristics for a square cavity with a pedestal by a three-dimensional finite-difference time-domain technique. *J. Opt. Soc. Am. B: Opt. Phys.* 23: 1287–1291.

32 Long, H., Huang, Y.Z., Yang, Y.D. et al. (2014). Mode characteristics of unidirectional emission AlGaInAs/InP square resonator microlasers. *IEEE J. Quantum Electron.* 50: 981–989.

33 Tang, M.Y., Sui, S.S., Yang, Y.D. et al. (2015). Mode selection in square resonator microlasers for widely tunable single mode lasing. *Opt. Express* 23: 27739–27750.

34 Che, K.J., Lin, J.D., Huang, Y.Z. et al. (2010). Two-port InGaAsP/InP square resonator microlasers. *Electron. Lett* 46: 585–U62.

35 Long, H., Huang, Y.Z., Ma, X.W. et al. (2015). Dual-transverse-mode microsquare lasers with tunable wavelength interval. *Opt. Lett.* 40: 3548–3551.

36 Liu, B.W., Huang, Y.Z., Long, H. et al. (2015). Microwave generation directly from microsquare laser subject to optical injection. *IEEE Photonics Technol. Lett.* 27: 1853–1856.

37 Weng, H.Z., Wada, O., Han, J.Y. et al. (2017). Sub-THz wave generation based on a dual wavelength microsquare laser. *Electron. Lett.* 53: 939–940.

38 Yang, Y.D., Weng, H.Z., Hao, Y.Z. et al. (2018). Square microcavity semiconductor lasers. *Chin. Phys. B* 27: Art. no. 114212.

39 Weng, H.Z., Han, J.Y., Li, Q. et al. (2018). Optical frequency comb generation based on the dual-mode square microlaser and a nonlinear fiber loop. *Appl. Phys. B* 124: Art. no. 91.

40 Weng, H.Z., Huang, Y.Z., Ma, X.W. et al. (2018). Optical frequency comb generation in highly nonlinear fiber with dual-mode square microlasers. *IEEE Photonics J.* 10: Art. no. 7102009.

41 Weng, H.Z., Yang, Y.D., Wu, J.L. et al. (2019). Dual-mode microcavity semiconductor lasers. *IEEE J. Sel. Top. Quantum Electron.* 25: Art. no. 1501408.

42 Weng, H.Z., Huang, Y.Z., Ma, X.W. et al. (2017). Spectral linewidth analysis for square microlasers. *IEEE Photonics Technol. Lett.* 29: 1931–1934.

43 Che, K.J., Yao, Q.F., Huang, Y.Z. et al. (2011). Multiple-port InP/InGaAsP square-resonator microlasers. *IEEE J. Sel. Top. Quantum Electron.* 17: 1656–1661.

44 Weng, H.Z., Huang, Y.Z., Yang, Y.D. et al. (2017). Mode Q factor and lasing spectrum controls for deformed square resonator microlasers with circular sides. *Phys. Rev. A* 95: Art. no. 013833.

45 Weng, H.Z., Yang, Y.D., Xiao, J.L. et al. (2018). Spectral engineering for circular-side square microlasers. *Opt. Express* 26: 9409–9414.

8

Hexagonal Microcavity Lasers and Polygonal Microcavities

8.1 Introduction

Optical microcavities have attracted considerable research interests in both fundamental physics studies and device applications [1]. To achieve high-quality (Q) factor and small mode volume (V) simultaneously, the mode light should be confined inside the microcavity with a near-unity reflectivity, which can be achieved by the photonic forbidden band in photonic crystal microcavities, or by the total internal reflection (TIR) in whispering-gallery-mode (WGM) optical microcavities. Since photonic integration has become an important research topic nowadays, the WGM optical microcavities with monolithic integration capability and low-processing complexity have attracted significant attention [2, 3]. As the most representative cavity shapes, the WGM microcavities with circularly rotational symmetries in different geometries have been widely studied for realizing low-threshold microlasers [4, 5]. Although the circular WGM microcavities can have extremely high Q factors, one major shortcoming appears as the isotropic emission along the cavity rim, which causes a significant difficulty in efficient collection of the output light from the circular microlasers.

Besides the circular WGM microcavities, regular polygonal microcavities supporting high-Q whispering-gallery-like (WG-like) modes (denoted as WGMs in the following for short) have also been studied widely in the past decades [6–10]. The regular polygonal microcavities have distinct mode properties compared with the circular microcavities when the side number is not large [11]. For the WGMs in the equilateral-triangular and square microcavities with integrable internal dynamics, quasi-analytical solutions were obtained as presented in Chapters 6 and 7. The dynamics is not integrable but instead pseudo-integrable in the regular polygonal microcavities with the side number larger than four, which results in the pseudo-integrable leakage loss for the WGMs [12]. High-Q WGMs can still exist in the regular polygonal microcavities as the superscar states along periodic orbits (POs) with weak field distributions at the corners [13]. In this chapter, theoretical analyses and experimental results are presented for the regular polygonal optical microcavities, especially the hexagonal microcavity lasers. This chapter is organized as follows. In Section 8.2, the mode characteristics of the regular polygonal microcavities are presented. In Section 8.3, the WGMs in hexagonal microcavities are studied

Microcavity Semiconductor Lasers: Principles, Design, and Applications, First Edition.
Yong-zhen Huang and Yue-de Yang.

by ray dynamics and cavity symmetry analyses, and the numerical simulations. In Section 8.4, the experimental results of waveguide-coupled unidirectional emission hexagonal microcavity lasers are presented. In Section 8.5, lasing characteristics are presented for an octagonal microlaser. Finally, a summary is given in Section 8.6.

8.2 Mode Characteristics of Regular Polygonal Microcavities

8.2.1 Symmetry Analyses Based on Group Theory

In this section, the symmetry characteristics of the WGMs in regular polygonal microcavities are analyzed based on the group theory. By assuming a given mode-field distribution in the vertical direction, a practical three-dimensional (3D) polygonal microcavity can be simplified to a two-dimensional (2D) polygon with an effective refractive index. The 2D model is widely used because it can give accurate mode structure and predict high-Q WGMs efficiently, although it may give incorrect Q factors sometimes. An illustration of a 2D regular polygon is shown in Figure 8.1. The center of the cavity O is the coordinate origin, n_1 and n_2 are the refractive indices of the inner and external regions, R_1 and R_2 are the maximum and minimum radii of the resonators, respectively. The symmetries of the regular polygonal microcavity can be described by the point group C_{Nv}. The symmetry operations of C_{Nv} include rotation operations and mirror operations σ. The rotation operation C_N^l means rotating an angle of $2l\pi/N$ in counterclockwise (CCW) direction, where N is the side number of the polygon, $l = 0, 1, \ldots, N-1$, and $l = 0$ correspond to the identity operation. All these rotation elements constitute the invariant subgroup C_N of the point group C_{Nv}. σ_1 and σ_2 are the symmetry operations on the OA and OB mirror surfaces, respectively, where A is one of the vertices and B is the midpoint of one side. The other $N-2$ mirror symmetry elements are not shown as they belong to the same category of σ_1 and σ_2. In addition, σ_1 and σ_2 also belong to the same category when N is odd.

The WGMs in a 2D polygonal microcavity can be separated into transverse-electric (TE) modes and transverse-magnetic (TM) modes based on their polarization. The TM (TE) WGMs are defined as the modes that have magnetic (electric) fields in r–φ plane and electric(magnetic) fields perpendicular to the plane. By omitting the time dependence $\exp(-i\omega t)$, the z-direction magnetic field H_z of the TE modes and the electric field E_z of the TM modes in the regular polygonal microcavity are expressed as $F_z(\vec{r})$. The z-direction field distribution $F_z(\vec{r})$ satisfies the 2D Helmholtz equation

$$\nabla^2 F_z(\vec{r}) + n^2(\vec{r}) k_0^2 F_z(\vec{r}) = 0 \tag{8.1}$$

where k_0 is the wavenumber in vacuum, $n(\vec{r})$ is the refractive index distribution and equals to n_1 and n_2 in the inner and external regions of the microcavity, respectively. In the polar coordinates, Eq. (8.1) is written as

$$\left(\frac{\partial^2}{\partial r^2} + \frac{1}{r}\frac{\partial}{\partial r} + \frac{1}{r^2}\frac{\partial^2}{\partial \varphi^2}\right) F_z(r,\varphi) + n^2(r,\varphi) k_0^2 F_z(r,\varphi) = 0 \tag{8.2}$$

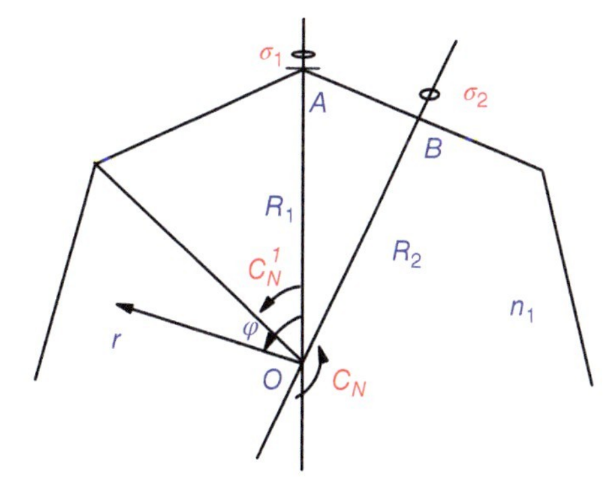

Figure 8.1 An illustration of a 2D regular polygonal microcavity. The operators of the point group C_{Nv} as well as the coordinates are indicated. Source: Modified from Yang and Huang [11].

where r and φ are the radius and angle in the polar coordinates, respectively. To compare with the circular microcavities, the z-direction field distributions in the regular polygonal microcavities can be expanded based on the angular components as

$$F_z(r,\varphi) = \sum_{\nu=-\infty}^{\infty} f_\nu(r)\exp(i\nu\varphi) = \sum_{\nu=0}^{\infty}[g_{1\nu}(r)\cos(\nu\varphi) + g_{2\nu}(r)\sin(\nu\varphi)] \tag{8.3}$$

where ν is the angular wavenumber, and f and g are the field distributions in the r direction for traveling- and standing-wave representations, respectively, and satisfy

$$\begin{cases} g_{1\nu}(r) = f_\nu(r) + f_{-\nu}(r) \\ g_{2\nu}(r) = i[f_\nu(r) - f_{-\nu}(r)] \end{cases} \tag{8.4}$$

The mode-field distributions can be regarded as the superposition of traveling waves or standing waves. It should be noted that the standing-wave representation is more standard due to the mirror symmetry and the break of the circular rotational symmetry. When the refractive index distribution is independent of position angle, the solution of Eq. (8.2) can be expressed as Bessel function or Hankel function, so the field distributions satisfy

$$\begin{cases} f_\nu(r) = a_\nu J_\nu(n_1 kr) & (r < R_2) \\ f_\nu(r) = b_\nu H_\nu^{(1)}(n_2 kr) & (r > R_1) \end{cases} \tag{8.5}$$

where J_ν and $H_\nu^{(1)}$ are the Bessel function and the Hankel function of the first kind.

Assuming that the field distribution $F_z(r,\varphi)$ satisfies Eq. (8.2), it can be kept under all the symmetry operations of $C_{N\nu}$ according to the symmetry theory. The field distribution of its degenerate mode is obtained by symmetry transformation

$$F_z^C(r,\varphi) = CF_z(r,\varphi) \tag{8.6}$$

where C is a symmetry operation of the point group $C_{N\nu}$. Under the symmetry operation, r is invariant. The transformation of the field distribution is equivalent to the inverse transformation of the coordinates, so Eq. (8.6) can be rewritten as

$$CF_z(r,\varphi) = F_z(r, C^{-1}\varphi) \tag{8.7}$$

Table 8.1 Character table of the point group C_{Nv} for even N.

C_{Nv}	E	C_N^l	$\frac{N}{2}\sigma_1$	$\frac{N}{2}\sigma_2$
A_1	1	1	1	1
A_2	1	1	−1	−1
B_1	1	$(-1)^l$	1	−1
B_2	1	$(-1)^l$	−1	1
E_n	2	$2\cos(2nl\pi/N)$	0	0

Table 8.2 Character table of the point group C_{Nv} for odd N.

C_{Nv}	E	C_N^l	$N\sigma_1$
A_1	1	1	1
A_2	1	1	−1
E_n	2	$2\cos(2nl\pi/N)$	0

The symmetry operation satisfies $C_N^1\varphi = \varphi + 2\pi/N$, $\sigma_1\varphi = -\varphi$, and all the other operations in the group can be generated by the two elements ($C_N = [C_N^1, (C_N^1)^2, \ldots, (C_N^1)^N]$, $C_{Nv} = [C_N, \sigma_1 C_N]$).

The $2N$ degenerate field distributions $CF_z(r, \varphi)$ are taken as the basis vector, then each element of the point group C_{Nv} can be represented as a $2N \times 2N$ matrix, but this matrix is reducible. It is very complicated to simplify a $2N \times 2N$ matrix, so the irreducible representations of the point group C_{Nv} are directly given in Tables 8.1 and 8.2, and the corresponding angular components in Eq. (8.3) can be classified into the irreducible representations accordingly. Tables 8.1 and 8.2 are the characters of the point group C_{Nv} with even and odd N, respectively [14].

When N is even, the characters of point group C_{Nv} are listed in Table 8.1, where $1 \leq l \leq N-1$ and $1 \leq n \leq N/2 - 1$. The character table is a $(N+2)\times(N+2)$ matrix, and includes 4 1D representations and $N/2 - 1$ 2D representations. For the following field distributions:

$$A_{1z}(r,\varphi) = \sum_{m=0}^{\infty} g_{1mN}(r)\cos(mN\varphi) \tag{8.8}$$

$$A_{2z}(r,\varphi) = \sum_{m=0}^{\infty} g_{2mN}(r)\sin(mN\varphi) \tag{8.9}$$

$$B_{1z}(r,\varphi) = \sum_{m=0}^{\infty} g_{1(m+1/2)N}(r)\cos((m+1/2)N\varphi) \tag{8.10}$$

$$B_{2z}(r,\varphi) = \sum_{m=0}^{\infty} g_{2(m+1/2)N}(r)\sin((m+1/2)N\varphi) \tag{8.11}$$

it can be obtained under the symmetry operations that

$$C_N^1 A_{1z} = A_{1z},\ C_N^1 A_{2z} = A_{2z},\ C_N^1 B_{1z} = -B_{1z},\ C_N^1 B_{2z} = -B_{2z}$$
$$\sigma_1 A_{1z} = A_{1z},\ \sigma_1 A_{2z} = -A_{2z},\ \sigma_1 B_{1z} = B_{1z},\ \sigma_1 B_{2z} = -B_{2z} \tag{8.12}$$

Compared with Table 8.1, it can be found that A_{1z}, A_{2z}, B_{1z}, and B_{2z} form the A_1, A_2, B_1, and B_2 representations of point group C_{Nv}, respectively. Furthermore, for the mode-field distributions

$$F_{nz}(r, \varphi) = \sum_{m=-\infty}^{\infty} f_{mN+n}(r) \exp(i(mN + n)\varphi) \tag{8.13}$$

$$F_{nz}^*(r, \varphi) = \sum_{m=-\infty}^{\infty} f_{mN+n}(r) \exp(-i(mN + n)\varphi) \tag{8.14}$$

it can be obtained under the symmetry operations that

$$C_N^1 F_{nz} = \exp(-i2n\pi/N) G_{nz},\quad C_N^1 F_{nz}^* = \exp(i2n\pi/N) F_{nz}^*,$$
$$\sigma_1 F_{nz} = F_{nz}^*,\quad \sigma_1 F_{nz}^* = F_{nz} \tag{8.15}$$

Then, the symmetry operations C_N^1 and σ_1 can be expressed based on F_{nz} and F_{nz}^* as

$$C_N^1 = \begin{pmatrix} \exp(-i2n\pi/N) & 0 \\ 0 & \exp(i2n\pi/N) \end{pmatrix} \tag{8.16}$$

$$\sigma_1 = \begin{pmatrix} 0 & 1 \\ 1 & 0 \end{pmatrix} \tag{8.17}$$

Compared with Table 8.1, F_{nz} and F_{nz}^* form the E_n representations with $n = 1, 2, 3, \ldots, N/2 - 1$.

When N is odd, the characters of point group C_{Nv} are listed in Table 8.2, where $1 \leq l \leq N - 1$ and $1 \leq n \leq (N - 1)/2$. The character table is a $(N + 1) \times (N + 1)$ matrix, and includes 2 1D representations and $(N - 1)/2$ 2D representations. Similar to even N, the mode-field distributions A_{1z} and A_{2z} form the A_1 and A_2 representations of point group C_{Nv}, respectively. Furthermore, the mode-field distributions F_{nz} and F_{nz}^* form the E_n representations with $n = 1, 2, 3, \ldots, (N - 1)/2$. Basically, the one-dimensional (1D) representations A_1, A_2, B_1, and B_2 correspond to the nondegenerate modes, and the 2D representations E_n correspond to double-degenerate modes in the microcavities.

The double-degenerate modes with traveling-wave distributions F_{nz} and F_{nz}^* as in Eqs. (8.13, 8.14) can also be expressed as standing-wave modes G_{nz} and G_{nz}^* with field distributions

$$G_{nz1}(r, \varphi) = \frac{F_{nz}(r, \varphi) + F_{nz}^*(r, \varphi)}{\sqrt{2}} \tag{8.18}$$

$$G_{nz2}(r, \varphi) = \frac{F_{nz}(r, \varphi) - F_{nz}^*(r, \varphi)}{\sqrt{2}} \tag{8.19}$$

It can also be obtained under the symmetry operations that

$$C_N^1 G_{nz1} = \cos(2n\pi/N) G_{nz1} + \sin(2n\pi/N) G_{nz2},$$

$$C_N^1 G_{nz2} = \cos(2n\pi/N)G_{nz2} - \sin(2n\pi/N)G_{nz1}$$
$$\sigma_1 G_{nz1} = G_{nz1}, \qquad \sigma_1 G_{nz2} = -G_{nz2} \tag{8.20}$$

Then, the symmetry operations C_N^1 and σ_1 can be expressed based on G_{nz1} and G_{nz2} as

$$C_N^1 = \begin{pmatrix} \cos(2n\pi/N) & -\sin(2n\pi/N) \\ \sin(2n\pi/N) & \cos(2n\pi/N) \end{pmatrix} \tag{8.21}$$

$$\sigma_1 = \begin{pmatrix} 1 & 0 \\ 0 & -1 \end{pmatrix} \tag{8.22}$$

The characters of C_N^1 and σ_1 are $2\cos(2n\pi/N)$ and 0, respectively; thus, G_{nz1} and G_{nz2} also form the E_n representations compared with Tables 8.1 and 8.2. In fact, the traveling- and standing-wave modes are different representations of the double-degenerate modes, which correspond to the eigenstates of group operations C_N^1 and σ_1, respectively. If and only if the modes can be expressed as the eigenstates of both C_N^1 and σ_1, the modes are nondegenerate modes, such as the modes with the field distributions in Eqs. (8.8–8.11).

The regular polygonal microcavities with side number $N \to \infty$ are similar to the circular microcavities, in which all WGMs are double-degenerate modes except the mode with angular wave number $\nu = 0$. In the regular polygonal microcavities, the double-degenerate modes will split into two nondegenerate modes when the corresponding angular wave number $\nu = mN$ for odd N and $\nu = mN/2$ for even N, which is observed in the equilateral-triangular and square microcavities as presented in Chapters 6 and 7. The field distributions with the angular components of wave numbers that are congruent modulo N can form the same irreducible representations and cannot be divided by the symmetries, and then the modes in regular polygonal microcavities should have multiple angular components as those in Eqs. (8.8–8.13). The phenomena can also be ascribed to the scattering due to the nonuniform refractive index distribution in the $R_2 < r < R_1$ region. In addition, the symmetry analysis presented above suits all the cavities with the symmetries of the point group $C_{N\nu}$ [11].

8.2.2 Numerical Simulations of WGMs in Regular Polygonal Microcavities

To examine the symmetry characteristics of the WGMs in the polygonal microcavities, the mode-field distributions and the quality (Q) factors of the WGMs are simulated by the finite-difference time-domain (FDTD) method as presented in Chapter 3. The TM modes are used as examples in the simulations.

The WGMs in circular microcavities are double-degenerate modes, and are denoted with an angular mode number of m_ν and a radial mode number of m_r. When circular microcavities are deformed to regular polygonal microcavities, these modes transform to corresponding confined modes [15]. Here, the confined WGMs in the polygonal microcavities are also marked as TM_{m_ν,m_r} with the angular and radial mode numbers m_ν and m_r. To compare the WGMs in different polygonal microcavities and circular microcavities, the definition of the mode numbers is slightly different from the triangular and square microcavities presented in Chapters 6 and 7.

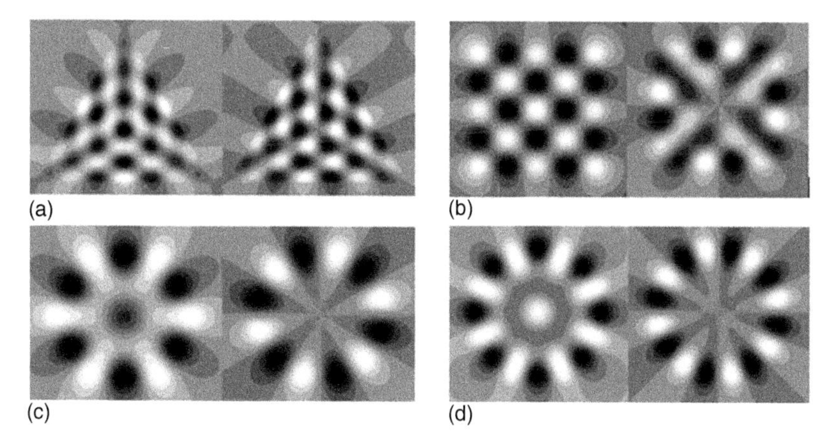

Figure 8.2 The electric field patterns for (a) $TM_{9,1}$, (b) $TM_{8,1}$, (c) $TM_{6,1}$, and (d) $TM_{8,1}$ modes in the regular triangular, square, hexagonal, and octagonal microcavities with side lengths of 3, 2, 1, and 0.8 μm, respectively. The left and right field distributions are A_1 and A_2 representations, respectively [11].

The nondegenerate modes with standing-wave distribution in the polygonal microcavities with a refractive index of 3.2 surrounded by air are simulated by the FDTD method. Symmetric and antisymmetric sources relative to the σ_1 plane are used to excite the modes with the corresponding symmetry. The obtained E_z field distributions are plotted in Figure 8.2 for (a) $TM_{9,1}$, (b)$TM_{8,1}$, (c) $TM_{6,1}$, and (d) $TM_{8,1}$ modes in the regular triangular, square, hexagonal, and octagonal microcavities with side lengths of 3, 2, 1, and 0.8 μm, respectively. The left and right field distributions correspond to A_1 and A_2 irreducible representations of point group C_{Nv}, respectively. The field distributions forming A_1 representations have a peak value at the center of the microcavities, which correspond to the angular component of $\nu = 0$. Those mode fields forming A_2 representations have zero intensity at the center and corners. The mode frequencies and Q factors obtained by the FDTD simulation are listed in Table 8.3. The differences between the mode frequencies of the two nondegenerate modes are very small, because they are split from the same WGMs. The Q factors of the modes of A_2 representations are larger

Table 8.3 Mode frequencies and Q factors of $TM_{9,1}$, $TM_{8,1}$, $TM_{6,1}$, and $TM_{8,1}$ modes in the triangular, square, hexagonal, and octagonal microresonators, respectively.

	A_1 representation		A_2 representation	
	f (THz)	Q-factors	f (THz)	Q-factors
Triangular	194.79	2.5×10^3	194.81	5.5×10^3
Square	148.79	1.2×10^2	150.28	3.2×10^3
Hexagonal	142.42	3.4×10^3	143.62	4.4×10^3
Octagonal	166.26	4.6×10^2	166.34	9.1×10^4

than those of the modes of A_1 representations due to the lower energy loss from the corners. The microdisk on a pedestal is usually used to obtain strong vertical waveguiding in such microcavity. Then, different field distributions in the center region can result in a mode selection. The modes with field distributions of A_1 representations will be suppressed in the microresonators with a pedestal, so the regular polygonal microcavities can be used to realize real single-mode operation.

It is difficult to distinguish the double-degenerate traveling-wave modes in the FDTD simulation, because these modes have same mode frequency and Q factor. Therefore, the field distributions of the standing-wave modes are first simulated by the FDTD method using the corresponding symmetry condition, and then the traveling-wave field distributions can be obtained as

$$F_{nz}(r,\varphi) = G_{nz1}(r,\varphi) + iG_{nz2}(r,\varphi)\sqrt{\frac{\iint |G_{nz1}(r,\varphi)|^2 r\ dr\ d\varphi}{\iint |G_{nz2}(r,\varphi)|^2 r\ dr\ d\varphi}} \tag{8.23}$$

where the square root term is used to normalize the two standing-wave field distribution $G_{nz1}(r, \varphi)$ and $G_{nz2}(r, \varphi)$. Considering the time dependence $\exp(-i\omega t)$, the time-varying field distributions can be obtained. For the TM modes in the regular polygonal microcavities, the electric field distributions $E_z(r, \varphi, t)$ can be expanded by $\exp(i\nu\varphi)$ as

$$E_z(r,\varphi,t) = \sum_{\nu=-\infty}^{\infty} f_\nu(r)\exp[i(\nu\varphi - \omega t)] \tag{8.24}$$

And then the energy of each angular component can be obtained approximately as

$$f(\nu) = \int |f_\nu(r)|^2 r\ dr \tag{8.25}$$

Equation (8.25) can be used to represent the energy ratio of different angular components.

The angular component distributions of the $TM_{11,1}$ traveling-wave modes in regular triangular, square, hexagonal, and octagonal microcavities are calculated and plotted in Figure 8.3, where the sum of all angular component energy is normalized to 1. It can be found that all angular components with an integer multiple of N are not zero. For the regular hexagonal and octagonal microcavities, one dominant component can be found at 11. However, the proportions of multiple components are similar in the regular triangular and the square microcavities, which indicate that field distributions are much different from those in a circular microcavity. The negative and positive angular wave numbers represent the components propagating clockwise (CW) and CCW, respectively. The field components propagating in the opposite direction are large in the equilateral-triangular and square microcavities, that is, the traveling-wave mode contains the components propagating in both directions. This is also the reason why the traveling-wave mode cannot be excited by a simple traveling-wave excitation source. As mentioned above, the scattering in the $R_2 < r < R_1$ region of the regular polygonal microcavities results in their distinct mode properties compared with the circular microcavities. It can be expected that

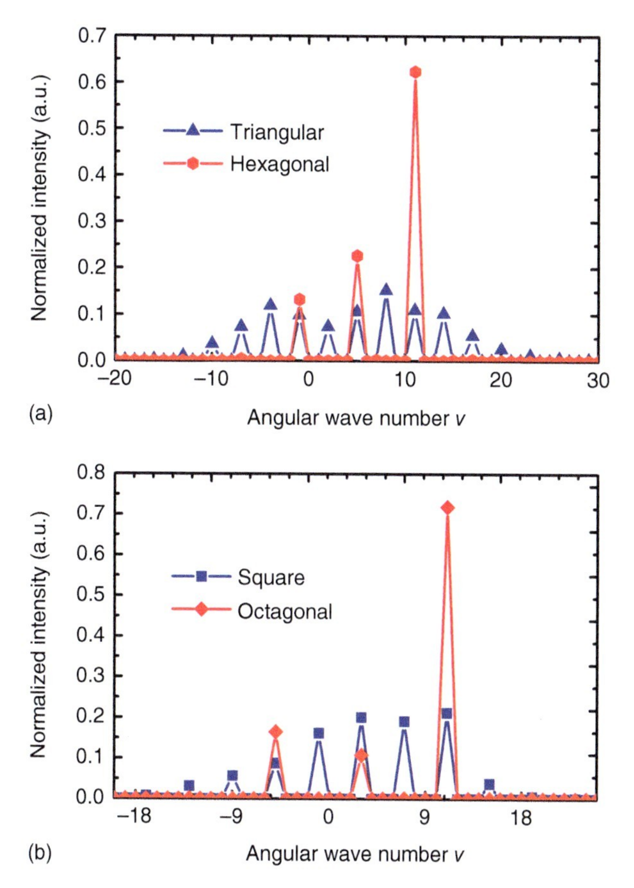

Figure 8.3 Angular component distributions of $TM_{11,1}$ traveling wave modes in regular (a) triangular and hexagonal, and (b) square and octagonal microcavities.

there is no dominant angular component for the WGMs with the angular mode number being much larger than the side number, because the mode fields of WGMs with high angular mode number mostly distribute in the rim region and will be greatly affected by the corner region.

8.2.3 Circular-Side Polygonal Microcavities

Apart from the regular polygonal microcavities, circular-side polygonal microcavities (CSPMs) have been proposed to enhance the mode Q factors and control the transverse mode interval simultaneously, and suppress the undesired high-order transverse modes [16]. The circular sides can work as concave mirrors to enhance the mode confinement, and introduce an additional degree of freedom for manipulating the WGMs in CSPMs.

A schematic diagram of two adjacent sides of a CSPM is shown in Figure 8.4. For the sake of simplification, the other sides are not presented in the figure owing to the rotational symmetry of the CSPM. The cavity geometry is determined by the

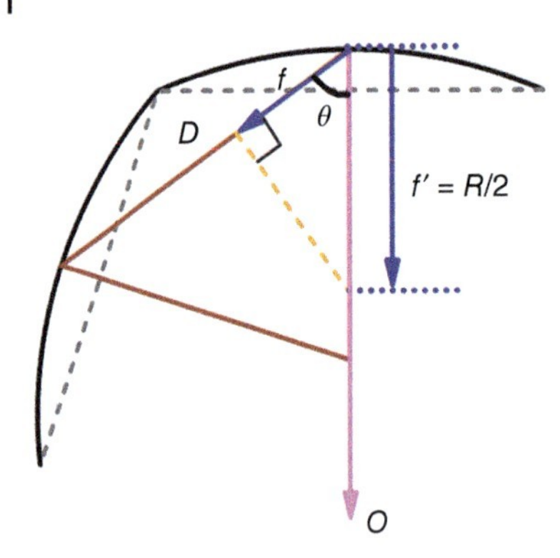

Figure 8.4 Schematic diagram of two adjacent sides of a CSPM. Source: Tang et al. [16]. © 2019, American Physical Society.

following three parameters, the side number of the regular polygonal microcavity N, the distance between the midpoints of adjacent sides D, and the deformation degree d defined as

$$d = D/f \tag{8.26}$$

where f is the focal length along the line connecting the midpoints of adjacent sides as shown in Figure 8.4. The focal length f can be obtained as $f = f' \cos\theta = R\cos\theta/2$, where f' and R are the paraxial focal length and the radius of circular sides, respectively, and θ is the incident angle of the light ray propagating along the line connecting the midpoints of adjacent sides. The CSPM is a regular polygonal microcavity with $d = 0$ and is a circular microcavity with $d = 4$. In the following discussion, we restrict the deformation degree d in the range from 0 to 4 covering all the shapes deformed from the regular polygonal microcavity to the circular microcavity.

Basically, varying the deformation degree d will not change the symmetry properties of the CSPMs. The confined modes can be still classified into irreducible representations of the point group $C_{N\nu}$, which gives the general rules of the degeneracy of the WGMs as presented in Section 8.2.1. Thus, part of the double-degenerate modes will split into two nondegenerate standing-wave modes with different mode frequencies and Q factors. However, the converging effect of the circular-side concave mirrors will reduce the light scattering at the vertices greatly, and decrease the differences in the mode frequencies and Q factors between the split degenerate modes compared with the regular polygonal microcavities.

Ray optics are introduced to describe the WGMs in the CSPMs. High Q WGMs typically correspond to the inherent POs that connect the midpoints of the sides, which appear as fixed points in the Poincaré surface of sections (SOSs). The POs of the CSPMs with different side numbers of 3, 4, 5, 6, and 8 are shown in Figure 8.5, where the orbits reflected between the opposite sides for the CSPMs with even N are neglected because these light rays cannot be totally reflected. Different from the regular polygonal microcavities, which have multiple PO families, the inevitable fixed orbits in the CSPMs are the light rays connected the midpoints of the circular sides.

To demonstrate the ray dynamics, the Poincaré SOSs of the CSPMs with $d = 1.5$, and $N = 3$, 4, 11, and 12 are calculated and shown in Figure 8.6a–d, respectively, where χ is the incident angle of the light ray on the boundary, S is the distance from

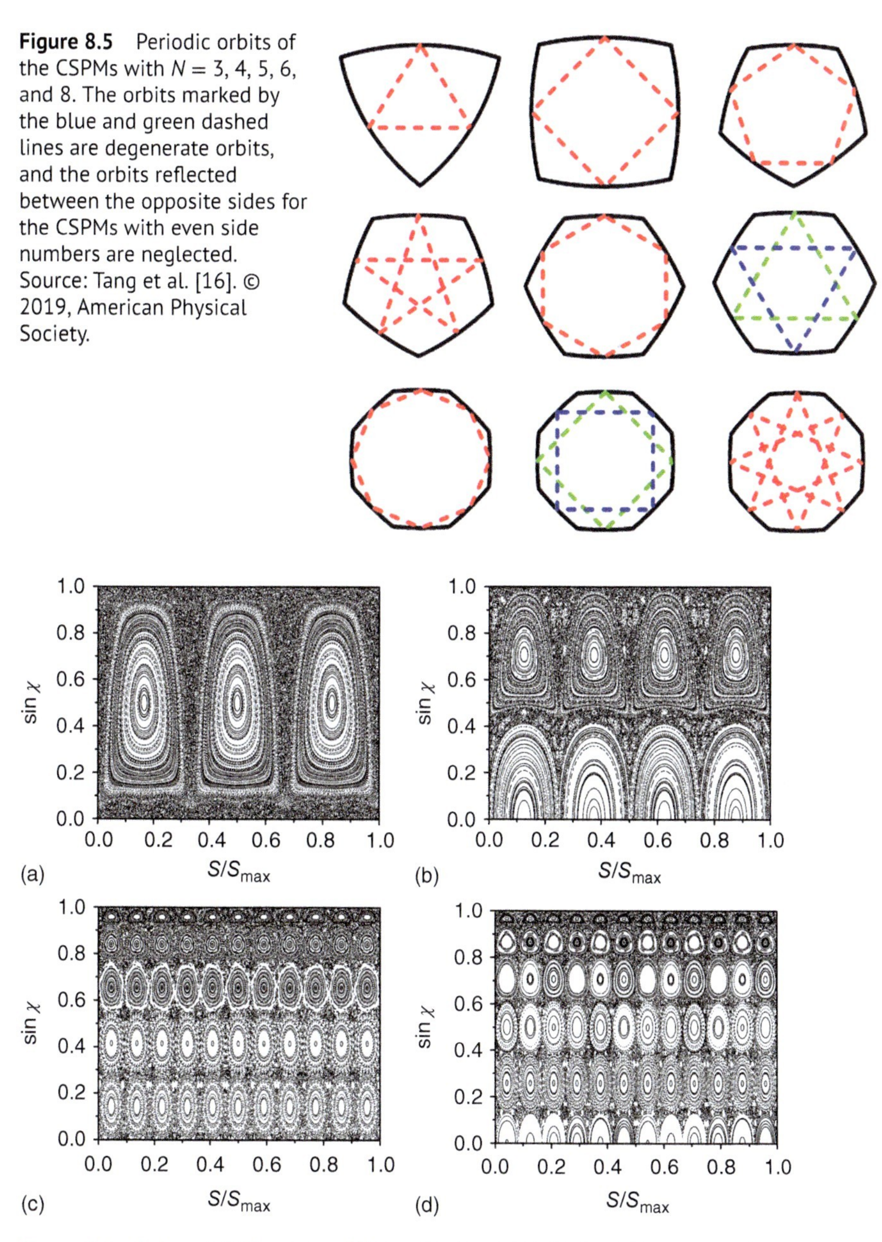

Figure 8.5 Periodic orbits of the CSPMs with $N = 3, 4, 5, 6$, and 8. The orbits marked by the blue and green dashed lines are degenerate orbits, and the orbits reflected between the opposite sides for the CSPMs with even side numbers are neglected. Source: Tang et al. [16]. © 2019, American Physical Society.

Figure 8.6 Poincaré SOSs of the CSPMs with $N =$ (a) 3, (b) 4, (c) 11, and (d) 12. The deformation degrees of the CSPMs are 1.5. Source: Tang et al. [16]. © 2019, American Physical Society.

one of the vertex along the boundary in CCW direction, and S_{max} is the perimeter of the CSPM. In the calculation of the Poincaré SOS, 200 light rays are used with random initial conditions in the phase space and the first 500 reflections are recorded for each ray. Only the upper half of the SOSs are shown in the figures as the lower half will have similar structure due to the symmetry of the CSPMs.

In general, the phase space of a regular polygonal microcavity is constructed by segmented horizontal lines, and that of a circular microresonator is constructed by horizontal lines. In the SOSs of the CSPMs, these orbits evolved into many regular islands arranged in an array, and chaotic sea appears around the islands. Figure 8.6a,c show the SOSs of the CSPMs with N equaling prime numbers of 3 and 11, where 1 and 5 kinds of nondegenerate orbits are observed and the islands on the same horizontal line have similar structure. For the case $N = 4$ as shown in Figure 8.6b, there are a nondegenerate orbit reflected by the adjacent sides and double-degenerate orbits reflected by the opposite sides. For the case $N = 12$ as shown in Figure 8.6d, the islands on the same horizontal line have a variety of structures and the degeneracies from the top islands to the bottom islands are 1, 2, 3, 4, 1, and 6, respectively.

When the side number is fixed, the cavity shape can be changed by varying the deformation degree d, and the ray dynamics will be modulated accordingly. In the following discussion, the light ray reflections between the adjacent sides of the CSPMs are considered, as these light rays have the largest incident angle benefiting for TIR. A 2×2 monodromy matrix $T(d)$ for the reflections between the adjacent sides around the midpoints under the first-order paraxial approximation is obtained as (the axis chooses the line connecting the midpoints of the adjacent sides)

$$T(d) = \begin{pmatrix} \dfrac{\delta\phi_{j+1}}{D} \\ \dfrac{\delta\theta_{j+1}}{\sin(\pi/N)} \end{pmatrix} = \begin{pmatrix} \dfrac{d}{2} - 1 & -\sqrt{2} \\ \dfrac{d}{\sqrt{2}} - \dfrac{d^2}{4\sqrt{2}} & \dfrac{d}{2} - 1 \end{pmatrix} \begin{pmatrix} \dfrac{\delta\phi_j}{D} \\ \dfrac{\delta\theta_j}{\sin(\pi/N)} \end{pmatrix} \tag{8.27}$$

where $\delta\phi_j$ and $\delta\theta_j$ are the position offset and the incident angle offset, respectively. The determinant of $T(d)$ equals 1. Similar to the steady condition in the coaxial spherical cavity, $T(d)$ should satisfy the same stability condition such that the half of the sum of the matrix diagonals is between −1 and 1. Thus, the stable condition can be obtained as $0 < d < 4$, which means that the CSPM considered here is always a stable cavity for the paraxial light rays under the first-order approximation. The mapping relation in Eq. (8.27) results in concentric ellipses in the SOS except some specific deformations. Hence, the circular side can enhance the light confinement in most cases.

One interesting phenomenon is the generation of fixed points under the evolution of the ray trajectories. The condition $(T(d))^r = I$ can give a sequence deformation degree d_r with $r \geq 3$, where I is the identity matrix and r is an integer. The corresponding d_r can be obtained as

$$d_r = 2\cos(2q\pi/r) + 2 \tag{8.28}$$

where q is a positive integer satisfying $q < r/2$. Then, the incident angle offsets and the position offsets will loop after r times of reflections. Each regular island will degenerate to discrete fixed points with a number of $r/(r, N)$ for a single ray trajectory, where (r, N) is the maximum common divisor of r and N. Unstable fixed points appear in the Poincaré SOSs of the CSPMs with specific deformations that corresponding to a small r, e.g. $r = 3$. The light rays around the island center can directly couple to chaotic sea in the phase space through the high-order terms in the transmission matrix. Such phenomena are found in the CSPMs with different side numbers. The unstable fixed points and surrounding ray trajectories form quasi-stable "star islands" with discrete fixed points around the island center based on the ray dynamics analyses. Although the light rays in the star island are not really stable, the life times are relatively long, which can support moderate-high Q modes in the CSPMs.

The results of ray dynamic analyses indicate that the introduction of the circular sides can enhance the optical confinement of the WGMs in most cases for the polygonal microcavities. The concave mirror effect can confine the light rays in a stable island and eliminate the optical losses for realizing high Q WGMs. The quasi-stable ray dynamics at some specific deformations give another degree for controlling the mode Q factors. In addition, the CSPMs introduce an additional degree of freedom for manipulating the mode structures of the WGMs and may lead to novel applications of these microcavities.

8.3 WGMS in Hexagonal Microcavities

In Section 8.2, the characteristics of the WGMs in the polygonal microcavities are studied by symmetry analyses and numerical simulations. Although the Q factors of the WGMs are relatively lower compared with the circular microcavities, their novel properties lead to advantages in mode control and directional emission. The WGMs in equilateral-triangular and square microcavities are presented in Chapters 6 and 7. Another interesting polygonal structure is the regular hexagonal microcavities, because the ZnO- or nitride-based semiconductor materials for ultraviolet laser diodes generally have a wurtzite crystal structure and typically present a natural hexagonal cross section [10, 17]. The bottom-up synthesized structures allow smooth surface faceting and hence reduce the scattering loss effectively. WGM lasing has been successfully demonstrated in the synthesized hexagonal micro/nanocavity lasers. On the other hand, deformation can be easily introduced and controlled in the top-down fabrication processes, which allows the manipulation of the lasing mode and waveguide-coupled unidirectional emission [18–20]. These works show that hexagonal microcavities have received great attention. In this section, a comprehensive study on the mode characteristics of WGMs in the regular hexagonal microcavities is presented, including the analyses of ray dynamics and cavity symmetry, and the numerical simulations of WGMs inside the hexagonal cavities.

8.3.1 Periodic Orbits in Hexagonal Microcavities

Figure 8.7a shows a simplified 2D hexagonal cavity with a side length of a in x–y plane, where Ω_i and Ω_o are the internal and external regions of the hexagon with refractive indices of n_i and n_o, respectively, and the center of the hexagon is placed at O. Both TM- and TE-polarized WGMs are considered in the analyses, here the TM (TE) WGMs are defined as the modes have magnetic (electric) fields in x–y plane

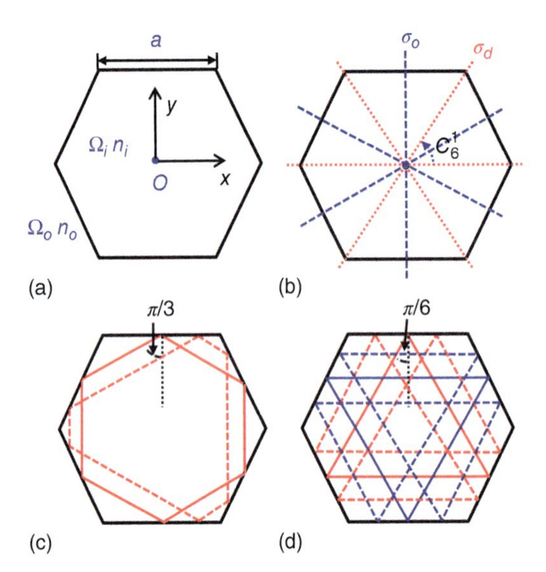

Figure 8.7 (a) Schematic diagram and (b) symmetry operators of a 2D hexagonal microcavity. (c) The hexagonal POs and (d) the triangular POs in the hexagonal microcavity. The solid lines and the dashed lines indicate the ray trajectories connecting the midpoints of the sides and the other ray trajectories in the same orbit family with the same incident angles, respectively. Source: Yang et al. [20]. © 2019, Chinese Laser Press.

and electric (magnetic) fields perpendicular to the plane. In the practical 3D case, the polarization of the mode is defined based on its dominant electromagnetic field components. Figure 8.7b shows the symmetry operators of the hexagonal microcavity.

Due to the nonintegrable internal dynamics, there is a lack of analytical solution in the hexagonal microcavity. To give an intuitive description of the WGMs, the semiclassical ray model is introduced into the hexagonal microcavity with the dimension much larger than the optical wavelength.

The light can be confined along the hexagonal POs with the same incident angle of $\pi/3$ at the six sides as shown in Figure 8.7c. For the hexagonal microcavity with a refractive index ratio $n_i/n_o > 2/\sqrt{3}$, these light rays can be totally reflected at the cavity sides. The solid line indicates the ray trajectory connecting the midpoints of adjacent sides, and the dashed line indicates one ray trajectory within the same orbit family with equal orbit length and incident angle. The WGMs corresponding to the hexagonal POs are the dominant modes in the hexagonal microcavity with a weak refractive index contrast, i.e. a small value of n_i/n_o. However, the triangular POs with the incident angle of $\pi/6$ as shown in Figure 8.7d should be considered in the hexagonal microcavity with $n_i/n_o > 2$ as these light rays can also be totally reflected. The solid lines in Figure 8.7d indicate two isolated equilateral-triangular ray trajectories connecting the midpoints of secondary neighbor sides with the shapes of up triangle "Δ" and down triangle "∇." The dashed lines indicate the ray trajectories in the same orbit family with the lengths twice those of the solid lines. The triangular POs in the hexagonal microcavity are very similar to the POs in regular triangular microcavity. The difference is that the triangular POs in the hexagonal microcavity are doubly degenerated.

In addition, the light rays reflected by the opposite sides with incident angle of 0 also form triply degenerate POs, which are neglected in this study because these light rays cannot be totally reflected. Overall, there are three families of POs in the hexagonal microcavities with the degeneracies of 1, 2, and 3. The numbers and degeneracies of the POs in general polygonal microcavities are given in [16].

In the hexagonal cavities with finite sizes, the incident angles at the sides cannot be exactly $\pi/3$ or $\pi/6$ considering the transverse distribution of the POs. For the light ray trajectory with an initial incident angle θ_b other than 0, $\pi/6$, and $\pi/3$, the subsequent incident angles θ_i cannot remain unchanged but only take a few finite values. Here, the light rays in the CW and CCW directions are not distinguished. Then, θ_i is restricted in the range from 0 to $\pi/2$, and all the possible incident angle θ_i is given by

$$\theta_i = \begin{cases} \theta_b, 2\pi/3 - \theta_b, \theta_b - \pi/3; & \pi/3 < \theta_b < \pi/2 \\ \theta_b, \pi/3 - \theta_b, 2\pi/3 - \theta_b; & \pi/6 < \theta_b < \pi/3 \\ \theta_b, \pi/3 - \theta_b, \pi/3 + \theta_b; & 0 < \theta_b < \pi/6 \end{cases} \tag{8.29}$$

In general, there are three different incident angles θ_1, θ_2, and θ_3 within the ranges of (0, $\pi/6$), ($\pi/6$, $\pi/3$), and ($\pi/3$, $\pi/2$), respectively, satisfying $\theta_1 + \theta_2 = \pi/3$ and $\theta_2 + \theta_3 = 2\pi/3$. For the light rays near the hexagonal POs with a small offset angle of $\Delta\theta$ from $\pi/3$, the three incident angles are $\Delta\theta$ and $\pi/3 \pm \Delta\theta$. The small incident angle of $\Delta\theta$ below TIR condition results in pesudointegrable leakage. For the light rays near the triangular POs with a small offset angle of $\Delta\theta$ from $\pi/6$, the three incident angles are $\pi/2 - \Delta\theta$ and $\pi/6 \pm \Delta\theta$. All the three incident angles are above the TIR condition in the hexagonal microcavity with $n_i/n_o > 2$. Another type of loss is boundary-wave leakage [8]. However, for the high-Q superscar modes in the regular polygonal microcavities, the boundary-wave leakage is relatively weak due to the destructive interference between the light waves with different incident angles at the boundaries, similar to that in the equilateral-triangular and square microcavities. For the hexagonal POs and the triangular POs, the destructive interference results from the light waves with the two incident angles of $\pi/3 \pm \Delta\theta$ and $\pi/6 \pm \Delta\theta$, respectively. Although the boundary-wave leakage of the light with incident angle near $\pi/6$ should be larger than that with the angle near $\pi/3$, the destructive interference can minimize the boundary-wave leakage. In addition, the width of rectangle formed by triangular POs is wider than the one formed by hexagonal POs, which leads to a smaller offset angle $\Delta\theta$ and lower scattering loss at the corners for the triangular POs [9]. Taking the pesudointegrable leakage into account, the WGMs corresponding to the hexagonal POs do not necessarily have higher Q than those corresponding to the triangular POs; thus, both kinds of WGMs should be considered in the analyses and simulations.

For the hexagonal microcavity with the dimension much larger than the optical wavelength, the resonance conditions of the above two kinds of WGMs can be derived under the semiclassical approximation. During the TIR at the sidewalls, a polarization-dependent negative phase shift occurs as

$$\delta(\theta) = -2\tan^{-1}\left(\frac{\beta\sqrt{n_i^2\sin^2\theta - n_o^2}}{n_i\cos\theta}\right) \tag{8.30}$$

where θ is the incident angle, and β equals 1 and n_i^2/n_o^2 for the TM- and TE-polarized modes, respectively. For the WGMs corresponding to the hexagonal POs (denoted as hexagonal WGMs for short), the round-trip length is $3\sqrt{3}a$ as shown in Figure 8.7c, and then the resonance condition can be written as

$$\text{Re}(3\sqrt{3}n_i ka) + 6\delta\left(\frac{\pi}{3}\right) = 2l\pi \tag{8.31}$$

where k is the wavenumber in vacuum and l is the longitudinal mode number. For the WGMs corresponding to the triangular POs (denoted as triangular WGMs for short), the ordinary round-trip length is $9a$ as the dashed lines shown in Figure 8.7d. The resonance condition of the triangular POs is generally written as

$$\text{Re}(9n_i ka) + 6\delta\left(\frac{\pi}{6}\right) = 2l\pi. \tag{8.32}$$

The transverse mode number and the longitudinal mode number of the triangular WGMs should have the same parity. Otherwise, it will lead to a trivial solution with all-zero fields in the cavities similar to that in the equilateral-triangular microcavities as presented in Chapter 6 [21]. Thus, the mode interval of the fundamental transverse triangular WGMs should be twice that obtained in Eq. (8.32) corresponding to an effective round-trip length of $9a/2$ similar to the solid lines shown in Figure 8.7d.

8.3.2 Symmetry Analyses and Mode Coupling

The group theory was used to analyze the symmetry of the modes in Section 8.2.1 for the hexagonal microcavities. The symmetry of a regular hexagonal microcavity can be described by the point group C_{6v} as shown in Figure 8.7b. The point group C_{6v} includes a rotational subgroup $C_6 = \{E, C_6^1, C_6^2, C_6^3, C_6^4, C_6^5\}$, three mirror elements relative to the line connecting the midpoints of opposite sides σ_0, and three mirror elements relative to the diagonals σ_d (σ_2 and σ_1 in Figure 8.1). The character of the point group C_{6v} can be obtained from Table 8.1 as $N = 6$. The point group C_{6v} has four 1D and two 2D irreducible representations. The mode-field distributions in the hexagonal microcavity should be classified into the irreducible representations of the point group.

The degeneracy of the WGMs should strictly fulfill the symmetry of the hexagonal microcavity. The WGMs in the hexagonal microcavity typically have multiple angular components with a difference of 6. The WGMs with the angular components of $6m$ and $6m+3$ are nondegenerate standing-wave modes form A and B representations, respectively, where m is an integer. The WGMs with the angular components of $6m+1$ and $6m+5$ ($6m+2$ and $6m+4$) are double-degenerate modes and form E_1 (E_2) representation. The double-degenerate modes forming E_1 (E_2) representation can be expressed as two traveling-wave modes with the angular components of

$6m + 1$ or $6m + 5$ ($6m + 2$ or $6m + 4$), and can also be expressed as two standing-wave modes with the angular components of both $6m + 1$ and $6m + 5$ ($6m + 2$ and $6m + 4$).

For the WGMs corresponding to the hexagonal POs, the modes obviously follow the symmetry analyses above. For the WGMs corresponding to the triangular POs, the situation will be slightly complicated, as these modes exhibit quadruple degeneracy according to the light rays shown in Figure 8.7d. Part of the degeneracy results from the propagation directions of the CW and CCW similar to the WGMs in the equilateral-triangular microcavity, and the other part of the degeneracy results from the double-degenerate "Δ" and "∇" POs in the hexagonal microcavity. According to the group theory analyses, the quadruple-degenerate modes become four nondegenerate standing-wave modes forming A_1, A_2, B_1, and B_2 representations, or two groups of double-degenerate modes forming E_1 and E_2 representations, depending on the angular components of the modes. The nondegenerate modes will have different mode Q factors.

The physical mechanism of the removal of the degeneracy resulting from the double-degenerate "Δ" and "∇" POs can be understood by the mode-coupling theory. In the scheme of ray optics, the triangular POs "Δ" with a small offset angle $\Delta\theta$ is slowly diverging from the stable POs. After some time, it will cross the corner to the adjacent side, and then become the triangular POs "∇." Thus, considering the offset of the incident angle, the two triangular POs are no longer isolated with each other, which results in coupling between the two kinds of triangular WGMs. In fact, the mode coupling always happens for the modes with leakage unless it is forbidden by the cavity symmetry. The external coupling will lead to two modes with enhanced and reduced Q factors corresponding to the destructive and constructive interference of the loss channels.

8.3.3 Numerical Simulation of WGMs in Hexagonal Microcavities

For the 2D hexagonal microcavities with a refractive index distribution of $n(x, y)$, Maxwell's equations for the confined optical field can be replaced by the scalar-wave equation similar to Eq. (8.1)

$$-\nabla^2\psi(x,y) = n^2(x,y)\frac{\omega^2}{c^2}\psi(x,y), \tag{8.33}$$

where ω is the angular frequency, c is the light speed in vacuum, and ψ represents the field distribution. The WGMs in the hexagonal microcavities are simulated by the finite-element method (FEM) (a commercial software: COMSOL Multiphysics 5.0) for revealing their mode characteristics. In the 2D microcavities, the TE and TM modes are simulated separately. A perfectly matching layer is used to absorb the outgoing waves terminating the simulation window. The obtained eigenvalues (ka) in the hexagonal microcavities are complex numbers. The real parts of ka give the mode frequencies, and the Q factors can be obtained as $Q = \mathrm{Re}(k)/2|\mathrm{Im}(k)|$.

Figure 8.8a shows the simulated mode Q factors and normalized frequencies $\mathrm{Re}(ka)$ for the TE modes in a hexagonal microcavity with $n_i/n_o = 3.2/1$. The transverse mode order is defined as the node number of the field distribution envelope along one sidewall similar to that in the equilateral-triangular and square microcavities for the high-Q superscar WGMs. The fundamental transverse (0th) WGMs corresponding to the hexagonal POs have the Q factors ranging from 1200 to 2300. The magnetic-field amplitude distribution of one hexagonal WGM (A in Figure 8.8a) is shown in Figure 8.8b. There are a lot of modes having Q factors much higher than the hexagonal WGMs. Based on the field distributions, these modes are denoted as triangular WGMs as they propagate along the triangular POs. As shown in Figure 8.8a, the zeroth triangular WGMs have the relatively high Q factors, and the mode with the highest Q is marked as mode B. Figure 8.8c shows the magnetic-field amplitude distribution of mode B, which indicates mixed fields of the "Δ" and "∇" triangular WGMs due to the mode coupling. The mode interval of the zeroth triangular WGMs is about 15% larger than that of the hexagonal WGMs consistent with the interval ratio of $2\sqrt{3}/3$, which is derived from the reciprocal of a half of the round-trip length ratio owing to the absence of odd longitudinal modes for the zeroth triangular WGMs. Two high Q modes are found at $ka = 39.85$ and 41.16, where the degeneracy of the modes are fully removed, and the WGMs are nondegenerate standing-wave modes with the angular components of $3m$. The other modes appear as two double-degenerate mode pairs with two different Q factors because of the mode coupling between the "Δ" and "∇" triangular WGMs, which remove the degeneracy induced by the double-degenerate POs [9]. For the triangular WGMs with the angular components of $3m$, the degeneracy resulting from the propagation directions of CW and CCW is also removed due to the cavity geometry-induced coupling or scattering between the two directions. The numerical simulation results are consistent with the symmetry analyses.

In the hexagonal microcavity, the first-order (1st) triangular WGMs can also have much higher Q factors than the hexagonal WGMs as shown in Figure 8.8a. The first triangular WGMs appear at the center between two zeroth triangular WGMs, as they have odd and even longitudinal mode numbers, respectively. Three groups of nondegenerate standing-wave modes are found at $ka = 39.24$, 40.55, and 41.86 with high Q factors. Figure 8.8d shows the magnetic-field amplitude distribution of a nondegenerate first triangular WGM (C in Figure 8.8a). Considering the first triangular WGMs, the mode interval of the triangular WGMs is much smaller than that of hexagonal WGMs, which agrees well with Eqs. (8.31, 8.32) with a round-trip length ratio of $3/\sqrt{3}$.

To further demonstrate the mode properties of the WGMs in the hexagonal microcavities, the cavities with different refractive index ratio n_i/n_o are considered. Figure 8.9a shows the simulated mode Q factors and normalized frequencies Re(ka) of the TE modes in a hexagonal microcavity with $n_i/n_o = 3.2/1.54$. The zeroth hexagonal WGMs have the highest Q ranging from 1300 to 2200, which is almost the same as the hexagonal WGMs in the microcavity with $n_i/n_o = 3.2/1$. The mode Q factors of the triangular WGMs are lower than 1000 due to the high boundary-wave leakage as the incident angle $\pi/6$ is only slightly higher than the TIR angle of 0.16π.

. Figure 8.9b shows the simulated mode Q factors of modes A (shown in Figure 8.8a,b) and B (shown in Figure 8.8a) as functions of n_i/n_o. The Q factor of the hexagonal WGM (mode A) almost does not vary with the refractive index ratio indicating the dominant loss is pesudointegrable leakage, which is insensitive

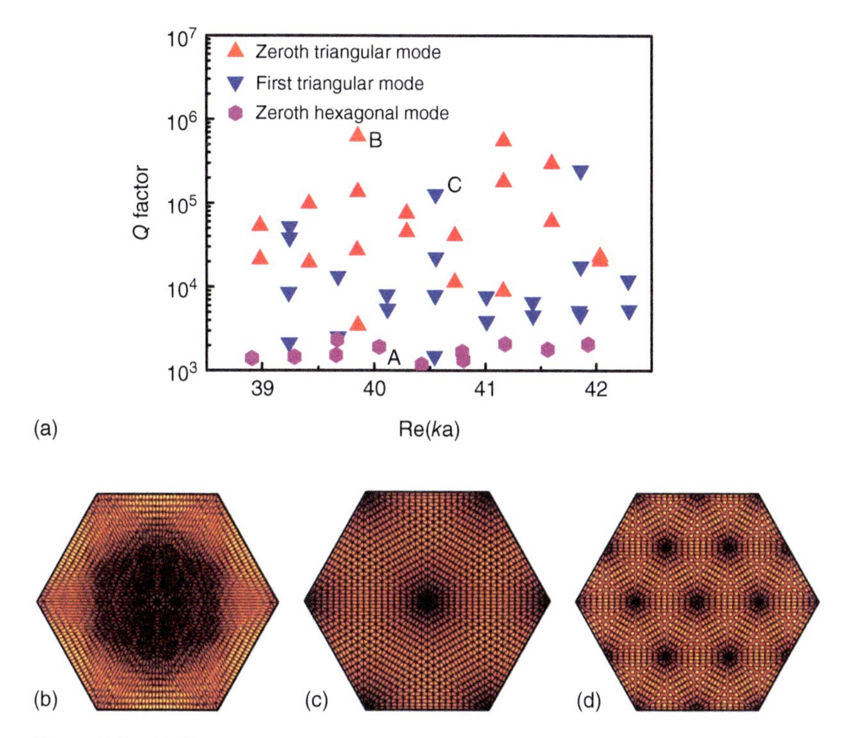

Figure 8.8 (a) Simulated TE modes in hexagonal microcavity with refractive indices of 3.2/1. (b)–(d) The magnetic-field amplitude distributions of the modes marked by A, B, C. Source: Yang et al. [20]. © 2019, Chinese Laser Press.

to refractive index. The existence of pesudointegrable leakage loss limits the Q factors of the hexagonal WGMs. On the contrary, the triangular WGMs have much higher Q factors in the hexagonal microcavity with high refractive index ratio due to the absence of pesudointegrable leakage. The numerical simulation results are consistent with the analyses of ray optics. The Q factor of the triangular WGM (mode B) shows an exponential decrease as $n_{i/}n_o < 2.6$. The possible reason is that with the decrease of $n_{i/}n_o$, the boundary-wave leakage increases, and the destructive interference between the two boundary waves with $\pi/6 \pm \Delta\theta$ also becomes worse as the smaller incident angle $(\pi/6 - \Delta\theta)$ is close to the TIR angle. The results show that whether the hexagonal WGM or the triangular WGM dominates depends on the refractive index ratio of the hexagonal microcavity.

8.3.4 WGMs in Wavelength-Scale Hexagonal Microcavities

In the wavelength-scale hexagonal microcavities, the WGMs cannot be described by the ray model but can be numerically simulated. The mode wavelengths obtained by Eq. (8.31) will be significantly larger than the actual values, because the propagation constants corresponding to transverse distributions are not taken into account. We consider both the TE and TM modes, as they have distinct mode properties in the wavelength-scale microcavities. Every WGM in the hexagonal

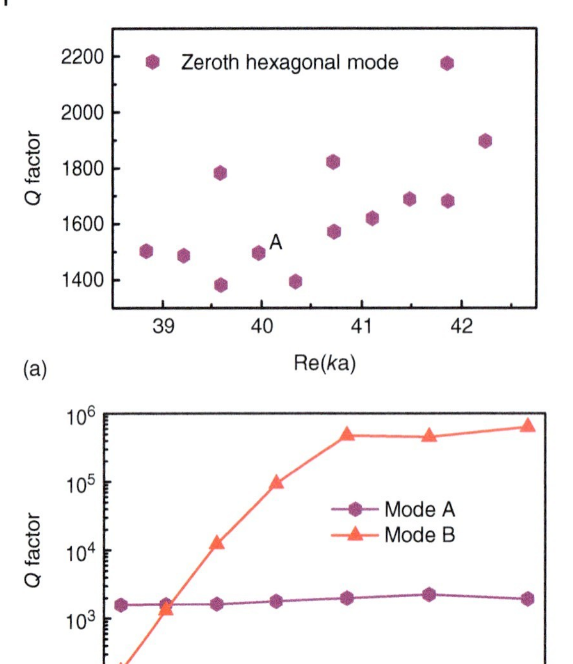

Figure 8.9 (a) Mode Q factors vs. normalized frequency for TE modes in the hexagonal microcavity with refractive indices of 3.2/1.54. (b) Simulated mode Q factors of modes A and B as functions of $n_{i/}n_o$. Source: Yang et al. [20]. © 2019, Chinese Laser Press.

microcavity has one main angular component, and hence is denoted as $TE_{m_\nu,1}$ or $TM_{m_\nu,1}$ with the angular mode number m_ν corresponding to the main component [11]. Figure 8.10a,b shows the mode Q factors of the $TE_{m_\nu,1}$ and $TM_{m_\nu,1}$ modes in the wavelength-scale hexagonal microcavity with refractive indices of 3.2/1 as functions of angular mode number ν, respectively. The symmetries of the modes are defined based on the symmetries of the electric-(magnetic-)field amplitude distributions relative to one diagonal for the TM (TE) modes. The Q factors of the WGMs in the wavelength-scale hexagonal microcavity show a peak at $m_\nu = 6$. The nondegenerate antisymmetric $TE_{6,1}$ or $TM_{6,1}$ modes have the highest Q factor. The secondary high Q mode appears at $m_\nu = 5$ and 7 for the TE and TM modes, respectively. The insets show the magnetic- and electric-field amplitude distributions of the TE and TM modes with $m_\nu = 5$, 6, and 7. The TE modes with $m_\nu > 7$ and the TM mode with $m_\nu > 9$ have the Q factors about 1 order of magnitude lower than the peak value. The TM modes generally have higher Q factors than TE modes, and the modes $TM_{m_\nu+1,1}$ and $TE_{m_\nu,1}$ have close resonant frequencies.

To characterize the WGMs in the hexagonal microcavity fabricated with different material, Figure 8.11a,b shows mode Q factors of TE and TM modes with $m_\nu = 5$, 6, and 7 in the wavelength-scale hexagonal microcavity as functions of $n_{i/}n_o$, respectively. The analytical Q factors of TE and TM modes in a circular microcavity are presented as a comparison. With the decrease of $n_{i/}n_o$, the modes with $m_\nu = 7$ become the highest Q mode among the four modes shown in Figure 8.11. For the TE modes,

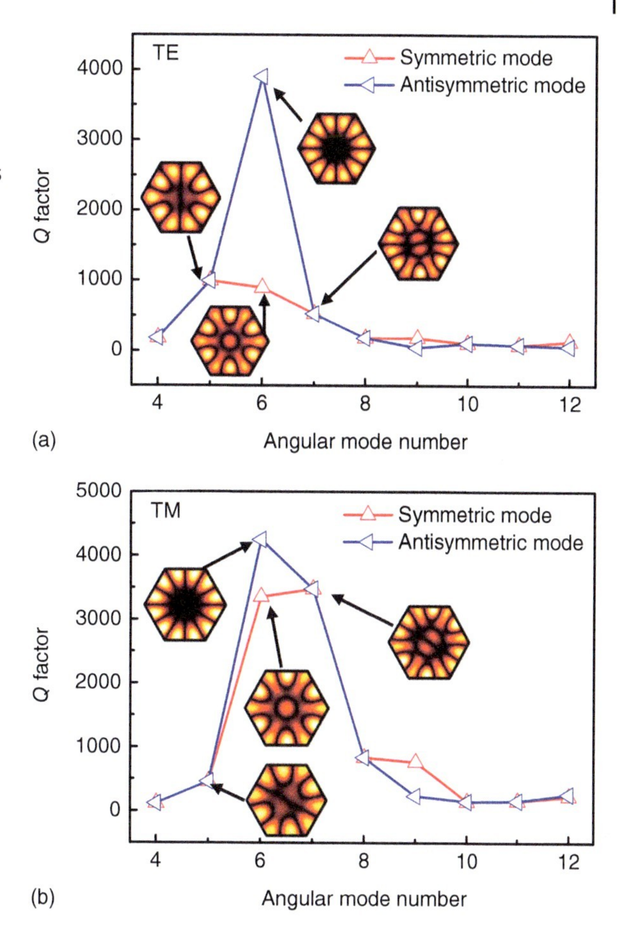

Figure 8.10 Mode Q factors of (a) TE and (b) TM modes in the wavelength-scale hexagonal microcavity with refractive indices of 3.2/1 as functions of angular mode number. The insets show the magnetic- and electric-field amplitude distributions of the TE and TM modes, respectively. Source: Yang et al. [20]. © 2019, Chinese Laser Press.

the Q factors of the $TE_{5,1}$ and antisymmetric $TE_{6,1}$ modes are about 10% and 20% lower than those in the circular microcavity. For the TM modes, the Q factor of antisymmetric $TM_{6,1}$ agrees very well with that in the circular microcavity. The Q factors of the WGMs in the hexagonal cavity with small angular mode number are comparable with that in the perfectly circular cavities, which indicate that the hexagonal cavities are suitable for the demonstration of wavelength-scale microcavity lasers with the diagonal length close to the light wavelength.

8.4 Unidirectional Emission Hexagonal Microcavity Lasers

The hexagonal microcavity lasers can be formed by both the bottom-up grown and top-down fabrication processes. To exactly control the sizes and geometries of the hexagonal microcavity lasers for practical applications, the top-down process is typically required. The GaN [22] and molecular single-crystalline [23] hexagonal microcavity arrays were fabricated using the top-down processes. In this section,

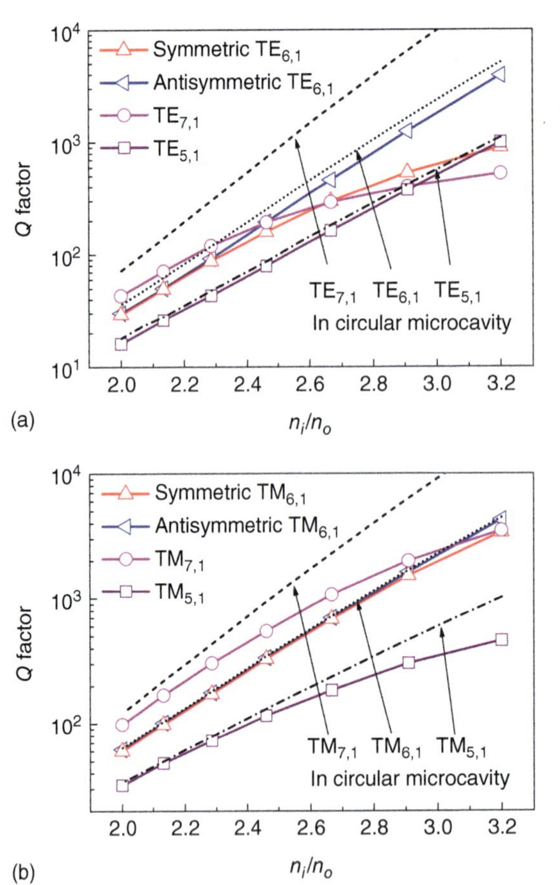

Figure 8.11 Mode Q factors of (a) TE and (b) TM modes in the wavelength-scale hexagonal microcavity as functions of n_i/n_o. Source: Yang et al. [20]. © 2019, Chinese Laser Press.

lasing characteristics of waveguide-coupled unidirectional emission hexagonal microcavity lasers are presented [18, 19], which are suitable for the photonic integrated circuits and optical interconnects.

8.4.1 Waveguide-Coupled Hexagonal Microcavity Lasers

The deformed hexagonal resonator microcavity, connecting a vertex output waveguide confined by bisbenzo-cyclobutene (BCB), is simulated using 2D FEM for the symmetric and antisymmetric TE modes. The perfectly magnetic conductor and perfectly electric conductor condition are set at the symmetry axis of the output waveguide and hexagonal resonator for calculating the two degenerate modes with different symmetry. The perfectly matched layer with a thickness of 1 μm is used to terminate the computation region. The effective refractive indices of the hexagonal microcavity and surrounding BCB cladding layer are taken to be 3.2 and 1.54, respectively. For a hexagonal resonator with a side length $a = 10\,\mu\text{m}$ and the width of the output waveguide $w = 1.5\,\mu\text{m}$, the calculated longitudinal mode wavelength interval is 14.5 nm around 1550 nm. The mode field distributions of $|H_z|$ for the fundamental and first-order transverse modes are calculated and depicted in Figure 8.12a,b at

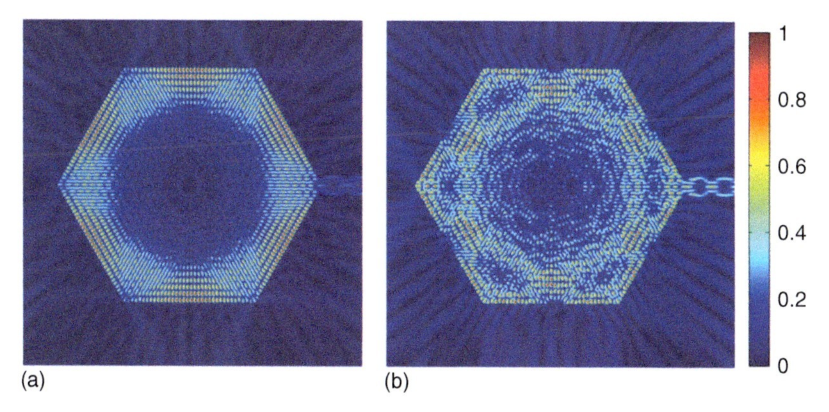

Figure 8.12 Mode-field distributions of $|H_z|$ for symmetric TE modes of (a) the fundamental transverse mode at 1543.7 nm and (b) the first-order transverse mode at 1539.7 nm, respectively, in a hexagonal microresonator with $a = 10\,\mu m$ and $w = 1.5\,\mu m$. Source: Wang et al. [19]. © 2018, Optical Society of America.

the mode wavelengths of 1543.7 and 1539.7 nm, respectively, and the corresponding mode Q factors are only 1.53×10^3 and 3.97×10^2. The mode-field patterns indicate strong radiation loss for the first-order transverse mode.

The AlGaInAs/InP hexagonal microcavity lasers with an output waveguide butt-coupled to one vertex of the hexagon were fabricated on multiple-quantum-well (MQW) epitaxial laser wafer using standard contact photolithography and inductively coupled plasma (ICP) dry etching process [24]. The inset of Figure 8.13a shows the microscope image of a fabricated hexagonal microcavity laser with a side length of 10 μm and an output waveguide width of 1.5 μm. The hexagonal microcavity lasers were mounted on a thermoelectric cooler (TEC) to control the substrate temperature. Figure 8.13a shows the multimode fiber-coupled light power and the applied voltage vs. the continuous-wave injection current measured at a TEC temperature of 288 K. The threshold current is about 6 mA corresponding to a current density of 2.3 kA cm^{-2}. Figure 8.13b shows the lasing spectra at injection currents of 6, 30, and 55 mA, where single-mode operation was achieved with an side-mode suppression ratio (SMSR) of 42.8 dB at the injection current of 30 mA.

Although the hexagonal microcavity laser exhibits excellent single-transverse-mode properties as shown in Figure 8.13b, the Q factors of WGMs in a hexagonal microcavity are much lower than those in a circular microcavity with similar size, especially for hexagonal microcavity with a low refractive index. Center holes and round corners were introduced for improving the mode Q factors [19]. Mode Q factor modification effect of a center hole with a radius of R_h is simulated for the hexagonal microresonator with a side length $a = 10\,\mu m$, a vertex output waveguide width $w = 1.5\,\mu m$, and rounding vertices with a radius of 1.5 μm. The rounding vertices usually occur after dry etching and wet chemical etching processes. The mode Q factors and wavelengths for symmetric and antisymmetric fundamental TE modes are calculated and presented in Figure 8.14 as a function of the center hole radius R_h. The mode Q factors remain near constant of about 2.81×10^3 as

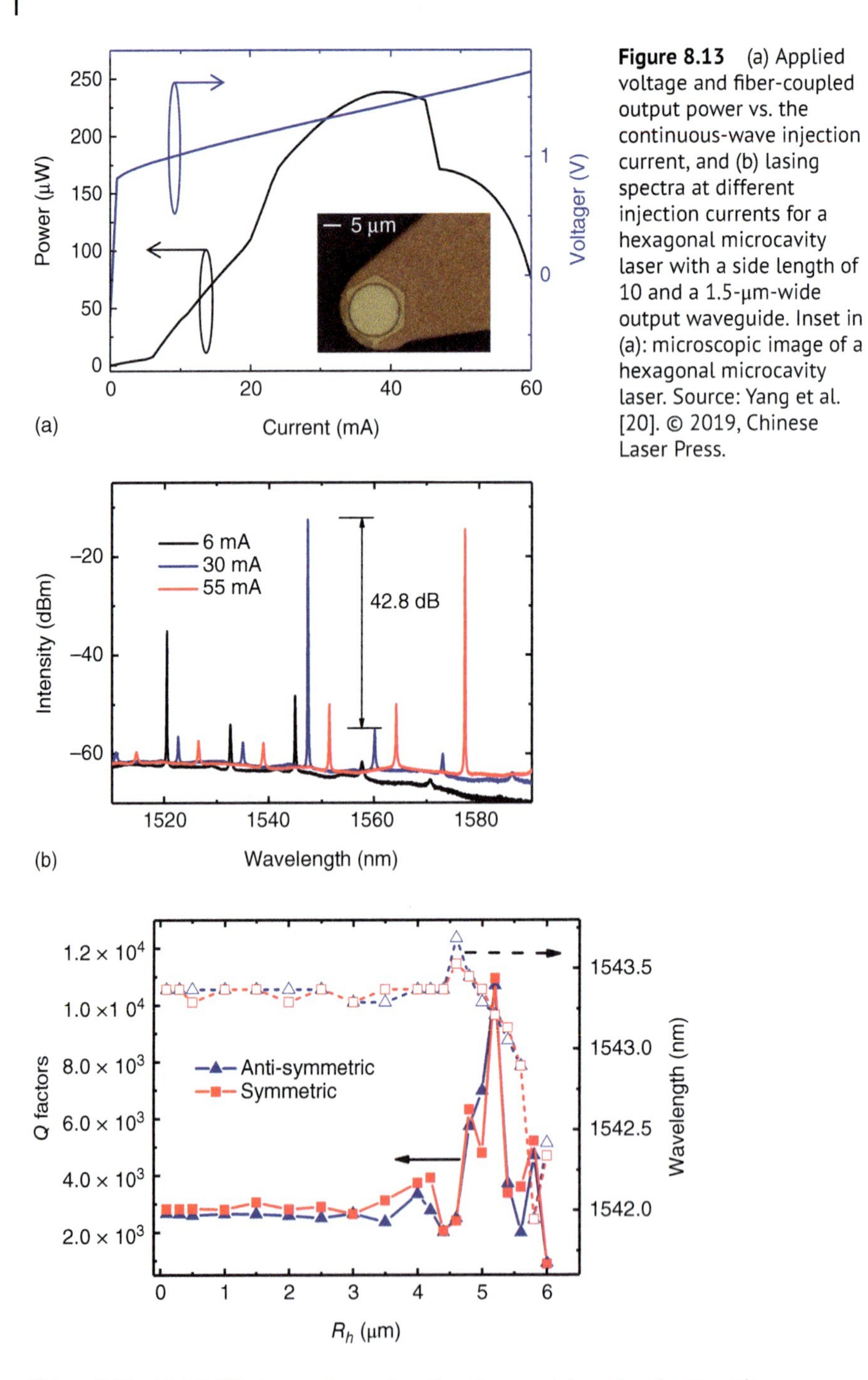

Figure 8.13 (a) Applied voltage and fiber-coupled output power vs. the continuous-wave injection current, and (b) lasing spectra at different injection currents for a hexagonal microcavity laser with a side length of 10 and a 1.5-μm-wide output waveguide. Inset in (a): microscopic image of a hexagonal microcavity laser. Source: Yang et al. [20]. © 2019, Chinese Laser Press.

Figure 8.14 Mode Q factors and wavelengths of symmetric and antisymmetric fundamental mode vs. the center hole radius R_h for a deformed hexagonal microcavity with $a = 10\,\mu\text{m}$, $w = 1.5\,\mu\text{m}$, and a radius of 1.5 μm for round vertices. Source: Wang et al. [19]. © 2018, Optical Society of America.

$R_h < 3.5\,\mu\text{m}$ due to nearly zero field distribution at the center, and then oscillate with enhanced values of 1.09×10^4 for symmetric modes, and 1.07×10^4 for antisymmetric modes, respectively, as $R_h = 5.2\,\mu\text{m}$. Further increase of R_h will lead to serious deformation and destruction of the mode field distributions, and result in the sharp decline of the mode Q factors as $R_h > 5.8\,\mu\text{m}$. The different mode Q factors between the symmetric and antisymmetric modes in hexagonal resonators are beneficial for stable single-mode operation. In addition, the rounding vertices result in the higher mode Q factor of 2.81×10^3 than the original value of 1.53×10^3 for the mode in Figure 8.12a.

8.4.2 Circular-Side Hexagonal Microcavity Lasers

The mode Q factors of WGMs in square microcavities can be enhanced by replacing the flat sides with circular arcs, and the lasing spectra can be engineered by tuning the deformation parameters. The waveguide-coupled deformed hexagonal microcavity lasers with the flat sides replaced by circular arcs were also proposed for enhancing the Q factors of WGMs [18]. Low-threshold lasing was demonstrated while maintaining single-transverse-mode property.

A schematic diagram of a proposed circular-side hexagonal resonator (CSHR) is shown in Figure 8.15a, which is surrounded by SiN_x and BCB layers similar to following device. The mode characteristics of the CSHR are simulated by 3D FDTD method. The cross-sectional view of the structure is shown in Figure 8.15b. The refractive indices of AlGaInAs quantum wells, InP, SiN_x, and BCB are set to 3.4, 3.17, 2.0, and 1.54, respectively, and the thicknesses of both AlGaInAs layer and SiN_x layer are set to 200 nm. The top-view schematic diagram of the CSHR is shown in Figure 8.15c, where a, R, d are the side length of the hexagon, the deformation

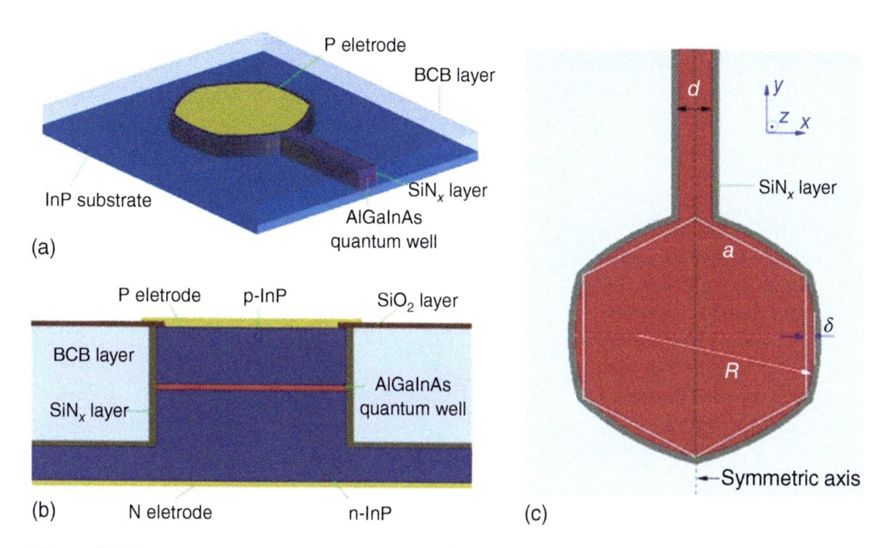

Figure 8.15 (a) Schematic diagram of a CSHR, (b) cross-sectional, and (c) lateral schematic diagrams of the CSHR. Source: Xiao et al. [18]. © 2017, Optical Society of America.

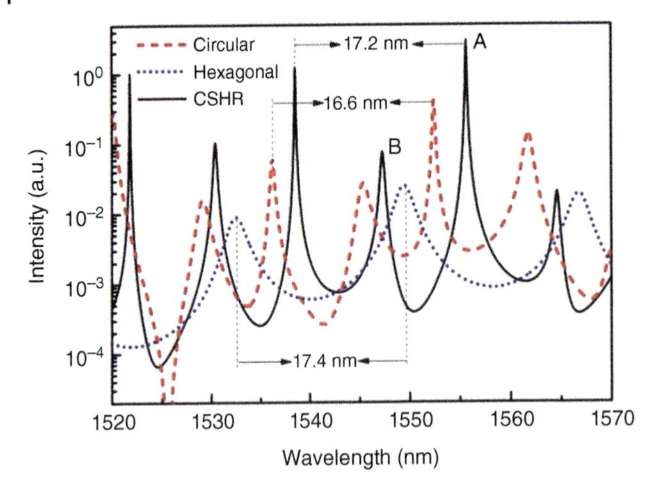

Figure 8.16 Intensity spectra for symmetric TE mode in circular resonator with radius of 7.5 μm, a hexagonal resonator with side-length of 7.5 μm, and a CSHR with $a = 7.5$ μm, $R = 14$ μm. The resonators are connected to a 2-μm-wide output waveguide. Source: Xiao et al. [18]. © 2017, Optical Society of America.

curvature radius, and the width of the output waveguide, respectively, and δ is a deformation amplitude determined by a and R.

In the simulation, the spatial steps Δx, Δy, and Δz are set to be 30 nm, and the time step Δt is set to be 5.62×10^{-17} seconds according to the Courant condition. A Gaussian-modulated cosine impulse of (3.48) is used as the exciting source inside the resonator. Symmetric or antisymmetric exciting resources relative to the midline of the output waveguide are applied to simulate the modes with different symmetries. The mode wavelengths are calculated from the intensity spectra obtained by Padé approximation from the time-domain FDTD output. The simulated intensity spectra of the symmetric TE modes are plotted in Figure 8.16 as dashed, dotted, and solid lines for a circular resonator with radius of 7.5 μm, a hexagonal resonator with $a = 7.5$ μm, and a CSHR with $a = 7.5$ μm and $R = 14$ μm respectively, all connecting a 2-μm-wide output waveguide. For the circular resonator, Q factors of 1.47×10^3 and 2.77×10^3 are obtained for modes at the wavelengths of 1545.3 and 1552.7 nm with a longitudinal mode interval of 16.6 nm. For the hexagonal resonator, mode Q factor is only 7.7×10^2 at the wavelength of 1549.5 nm, which is too low for low-threshold lasing. For the CSHR, two sets of transverse modes are observed as for circular resonator. The mode Q factors of 1.50×10^4 and 3.76×10^3 are obtained for the fundamental and first-order transverse modes at 1555.7 and 1547.3 nm. The longitudinal mode interval around 1550 nm is about 17.2 nm, and the mode Q factor of 1.50×10^4 is 5.4 times of the high Q mode in the circular resonator. The results indicate that the CSHR is suitable for realizing low-threshold single-mode operations.

Figure 8.17a shows the SEM image of a waveguide-coupled deformed hexagonal microcavity laser after ICP etching of laser cavity. Figure 8.17b shows the microscopic image of a fabricated deformed hexagonal microcavity laser. The devices were cleaved over the output waveguide leaving a length of ~10 μm and bonded on an AlN submount with a thin-film resistance of 30 Ω in a series to match the impedance to ~50 Ω. For the waveguide-coupled deformed hexagonal microcavity laser with a side length $a = 7.5\,\mu m$, circular arc radius $R = 14\,\mu m$, and an output waveguide width $d = 2\,\mu m$, the multimode fiber-coupled output power and the applied voltage vs. the CW injection current are measured at a temperature of 288 K and plotted in Figure 8.17c. The threshold current is about 2.5 mA with a current density of 1.5 kA cm^{-2}, and the maximum coupled output power is 63 μW at 21 mA. Compared with the nondeformed hexagonal microcavity laser, the lasing threshold current density is much lower owing to the enhanced mode Q factors. Figure 8.17d shows the lasing spectra at injection currents of 2.5, 14, and 23 mA, where three fundamental transverse modes are found around 1535.54, 1550.95, and 1566.67 nm, and a very wide first-order transverse mode is found around 1560 nm at 14 mA. Single-mode operation is demonstrated with an SMSR of 43 dB at 14 mA. Figure 8.17e shows the small-signal modulation responses of the waveguide-coupled deformed hexagonal microcavity laser at different bias currents. The 3-dB bandwidth increases from 8.2 to 13.0 GHz as the bias current is increased from 6 to 21 mA, and the resonance peak height decreases from 3.0 to below 0.2 dB showing a very flat response curve as the bias current is above 14 mA.

Apart from single-mode lasing, the deformation provides another parameter to control the WGMs inside microcavities. The transverse mode intervals can be controlled by the deformation parameters as different transverse modes have different round-trip length. In addition, the wavelength interval can be tuned by adjusting the injection current as different modes have different field distributions similar to those in a square microcavity. Nonlinear dynamics, including chaos, four-wave mixing, and high-order oscillations states, were observed from deformed hexagonal microlasers, and random bits were generated based on the laser output without external optical injection or feedback [25].

8.5 Octagonal Resonator Microlasers

In this section, we present lasing characteristics for an AlGaInAs/InP octagonal resonator microlaser with a side length of 10.8 μm and a 2-μm-wide vertex output waveguide [26]. Octagonal resonator microlasers were fabricated with a ring pattern of p-electrode for matching the mode-field pattern. Based on 2D FDTD method, mode intensity distribution $|H_y|^2$ of the fundamental TE mode is obtained and shown in Figure 8.18a, which is symmetric relative to the output waveguide. The intensity in the region outside the microcavity and that in the vertex waveguide are amplified 64 and 36 times for clarity, respectively. The calculated mode Q-factor and wavelength are 1.0×10^4 and 1512 nm, respectively. The simulation results indicate that the octagonal resonator has a large scattering loss at the vertices, which remains an obstacle for the realization of high-output efficiency microlasers. The lasing spectra measured at TEC temperature of 287 K are plotted in Figure 8.18b at an injection current of 40 mA, with main lasing mode wavelength of 1534.15 nm and a side-mode suppression ratio of 42 dB.

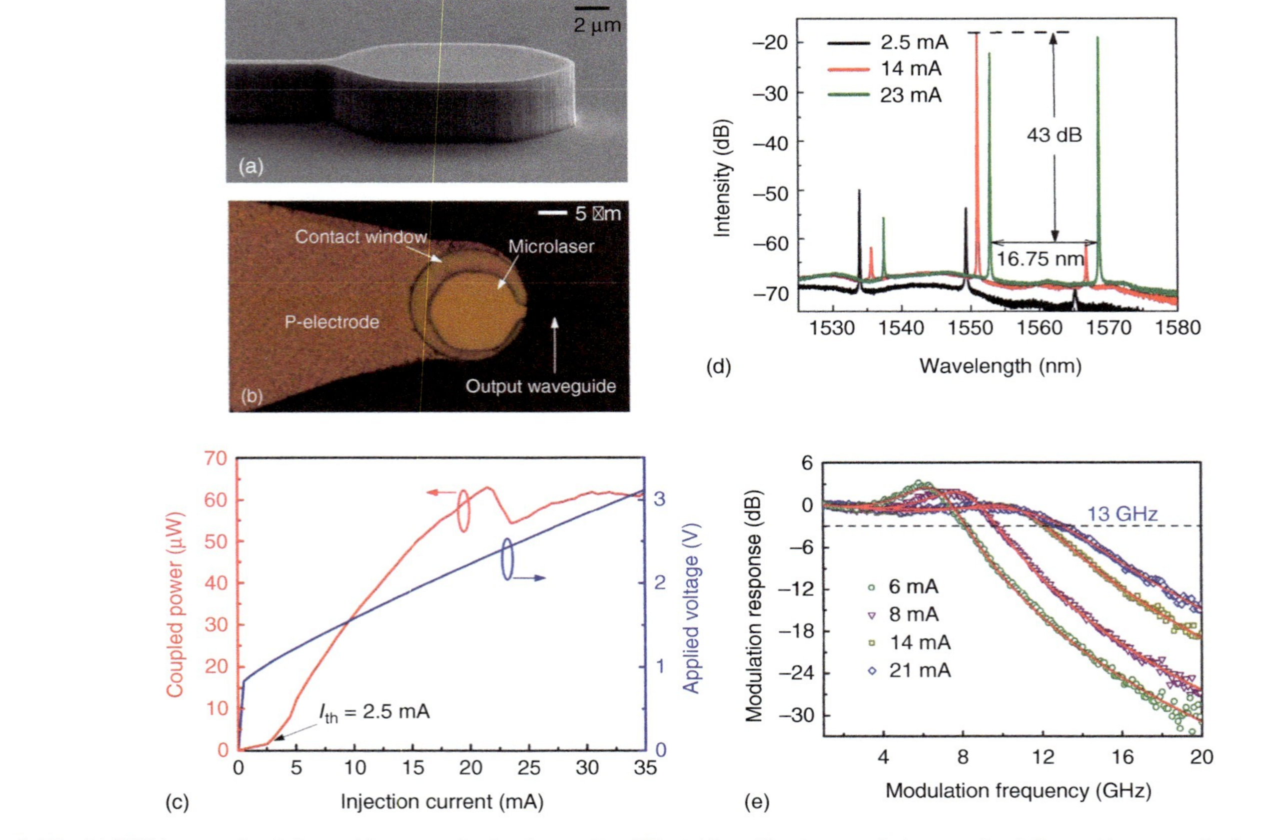

Figure 8.17 (a) SEM image of a deformed hexagonal microlaser after ICP etching, (b) microscopic image of a deformed hexagonal microcavity laser. (c) Applied voltage and fiber-coupled output power vs. the continuous-wave injection current. (d) Lasing spectra at the injection currents of 2.5, 14, and 23 mA. (e) Small-signal responses for the CSHR at the bias currents of 6, 8, 14, and 21 mA. Source: Xiao et al. [18]. © 2017, Optical Society of America.

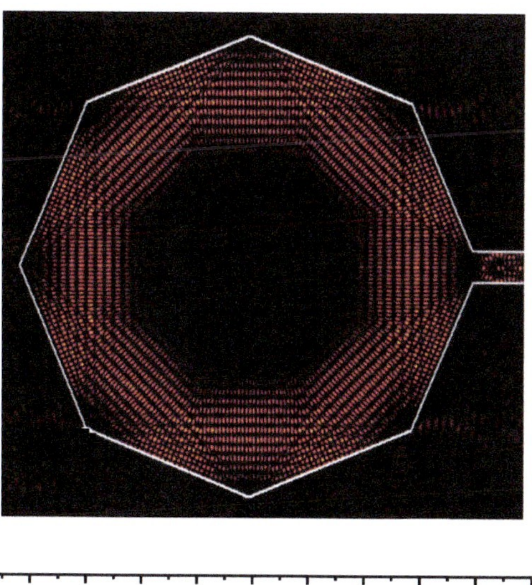

(a)

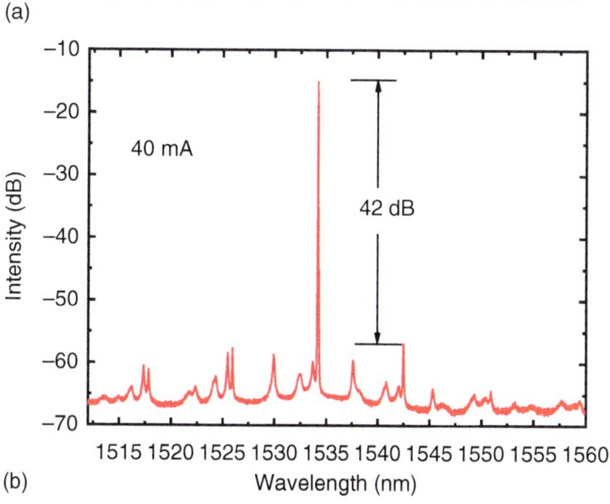

(b)

Figure 8.18 (a) The simulated intensity distribution of $|H_y|^2$ for the fundamental mode at 1512 nm. (b) Lasing spectra at 40 mA and TEC temperature of 287 K, for an octagonal microlaser with a side length of 10.8 and a 2-μm-width vertex output waveguide. Source: Zou et al. [26]. © 2014, Springer Nature Switzerland AG.

The lasing spectra from 1523 to 1543 nm are plotted in Figure 8.19a at injection currents ranging from 8 to 22 mA. The resonant peak can be fitted by a Lorentz function to determine the mode wavelength and the full-width at half-maximum (FWHM). The corresponding mode wavelengths and FWHMs vs. the injection currents are plotted in Figure 8.19b as the solid and open circles, respectively, for the modes around 1532 nm. The mode wavelength decreases from 1532.21 to 1531.97 nm with the injection current increases from 8 to 12 mA, and then nearly linearly increases with the injection current above 12 mA. The initial blue shift of mode wavelength is caused by free carrier dispersion before reaching the threshold current, while the red shift is mainly caused by the heating effect of the injection current, which becomes the prime cause due to the saturation of the carrier density

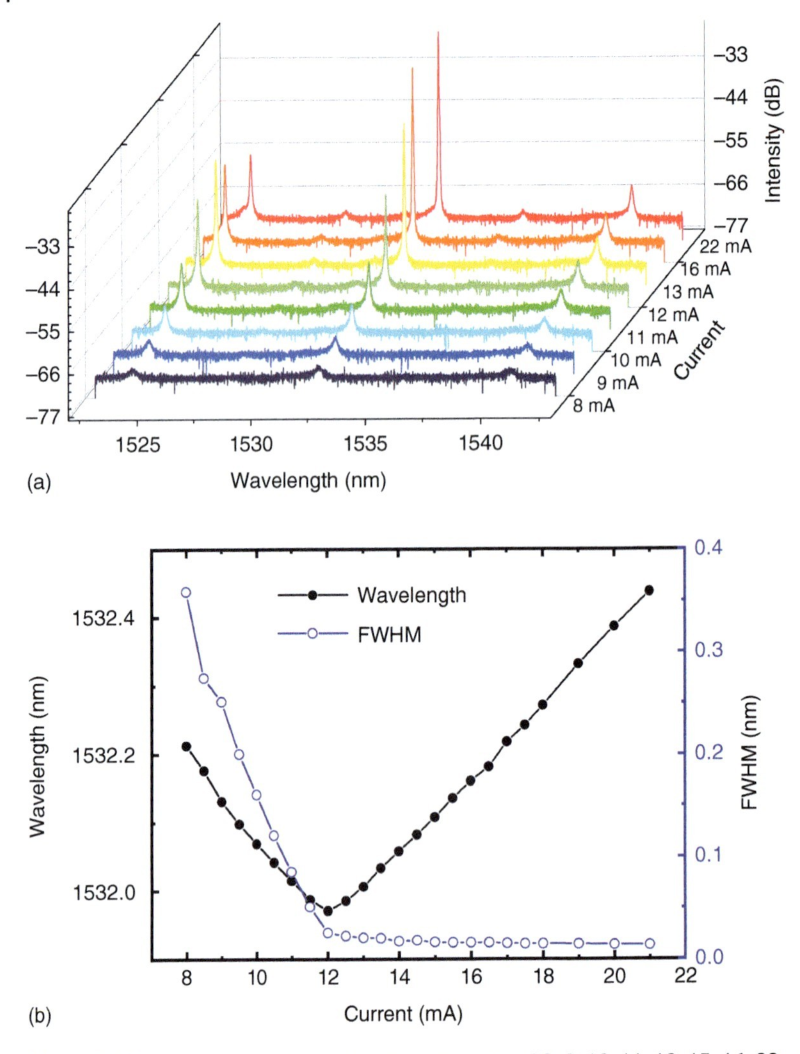

Figure 8.19 (a) Lasing spectra at injection currents of 8, 9, 10, 11, 12, 13, 16, 22 mA, and (b) the mode wavelength and FWHM vs. the injection current for the mode around 1532 nm at the TEC temperature of 287 K. Source: Zou et al. [26]. © 2014, Springer Nature Switzerland AG.

above the threshold. The FWHM of the main mode firstly quickly decreases from 0.359 to 0.0506 and 0.0255 nm from 8 to 11.5 and 12 mA, and then gradually decreases with the increase of injection current. At 13.5 mA, we get the FWHM of 0.02 nm, the same as the resolution of the optical spectrum analyzer.

8.6 Summary

In this chapter, the mode characteristics in polygonal microcavities, particularly in the hexagonal microcavities, and the waveguide-coupled hexagonal microcavity lasers have been studied. The symmetry properties of the polygonal cavities are described by the point group C_{Nv}, which reveal the degeneracy of the confined modes. The WGMs in polygonal microcavities have multiple angular components that form the same representations of the point group C_{Nv}. In the hexagonal microcavities, both triangular and hexagonal WGMs are predicted by the ray analyses and proved by the numerical simulation. Mode coupling between the degenerate triangular WGMs belonging to different triangular POs occurs and results in removal of the mode degeneracy. The mode Q factors of the hexagonal and triangular WGMs show distinct dependences on the refractive index ratio for the hexagonal microcavities, which can be well explained by the ray model. The WGMs cannot be simply explained by the ray model in the wavelength-scale hexagonal microcavities, because the transverse distributions of the light rays cannot be neglected. The field distributions of the WGMs in the wavelength-scale hexagonal microcavities are quite similar to those in the circular microcavities, and the corresponding mode Q factors are not necessarily positively related to the angular mode number. Waveguide-coupled unidirectional emission hexagonal microcavity lasers have been demonstrated with excellent single transverse mode lasing. By replacing the flat sides with circular arcs, the Q factors of the WGMs can be enhanced greatly for low-threshold lasing. The deformed hexagonal microcavity semiconductor lasers were also demonstrated for high-speed direct modulation, dual-transverse mode lasing, and random bit generation. The lasing characteristics of an octagonal microlaser were also presented to emphasize lasing mode linewidth vs. injection current.

Compared with circular cavities, the polygonal microcavities exhibit different mode properties for the WGMs, such as field distributions, mode structures, symmetries, and degeneracies. The unique properties of the WGMs in the polygonal microcavities also allow further regulation of lasing modes with cavity deformation, spatially selective current injection, and output waveguide coupling, for the practical application of the microcavity lasers. The corresponding research works on the polygonal microcavities and hexagonal microcavity lasers play a significant role in both fundamental physics studies and potential device applications.

References

1 Vahala, K.J. (2003). Optical microcavities. *Nature* 424: 839–846.

2 Ward, J. and Benson, O. (2011). WGM microresonators: sensing, lasing and fundamental optics with microspheres. *Laser Photonics Rev.* 5: 553–570.

3 He, L.N., Ozdemir, S.K., and Yang, L. (2013). Whispering gallery microcavity lasers. *Laser Photonics Rev.* 7: 60–82.

4 McCall, S.L., Levi, A.F.J., Slusher, R.E. et al. (1992). Whispering-gallery mode microdisk lasers. *Appl. Phys. Lett.* 60: 289–291.

5 Levi, A.F.J., McCall, S.L., Pearton, S.J., and Logan, R.A. (1993). Room temperature operation of submicrometer radius disk laser. *Electron. Lett.* 29: 1666–1667.

6 Huang, Y.Z., Guo, W.H., and Wang, Q.M. (2000). Influence of output waveguide on mode quality factor in semiconductor microlasers with an equilateral triangle resonator. *Appl. Phys. Lett.* 77: 3511–3513.

7 Poon, A.W., Courvoisier, F., and Chang, R.K. (2001). Multimode resonances in square-shaped optical microcavities. *Opt. Lett.* 26: 632–634.

8 Wiersig, J. (2003). Hexagonal dielectric resonators and microcrystal lasers. *Phys. Rev. A* 67: 023807.

9 Song, Q.H., Ge, L., Wiersig, J., and Cao, H. (2013). Formation of long-lived resonances in hexagonal cavities by strong coupling of superscar modes. *Phys. Rev. A* 88: 023834.

10 Xu, C.X., Dai, J., Zhu, G.P. et al. (2014). Whispering-gallery mode lasing in ZnO microcavities. *Laser Photonics Rev.* 8: 469–494.

11 Yang, Y.D. and Huang, Y.Z. (2007). Symmetry analysis and numerical simulation of mode characteristics for equilateral-polygonal optical microresonators. *Phys. Rev. A* 76: 023822.

12 Richens, P.J. and Berry, M.V. (1981). Pseudointegrable systems in classical and quantum-mechanics. *Physica D* 2: 495–512.

13 Lebental, M., Djellali, N., Arnaud, C. et al. (2007). Inferring periodic orbits from spectra of simply shaped microlasers. *Phys. Rev. A* 76: 023830.

14 Wherrett, S. (1986). *Group Theory for Atoms, Molecules and Solids*. London: Prentice-Hall.

15 Boriskina, S.V., Benson, T.M., Sewell, P., and Nosich, A.I. (2005). Optical modes in 2-D imperfect square and triangular microcavities. *IEEE J. Quantum Electron.* 41: 857–862.

16 Tang, M., Yang, Y.D., Weng, H.Z. et al. (2019). Ray dynamics and wave chaos in circular polygonal microcavities. *Phys. Rev. A* 99: 033814.

17 Huang, M.H., Mao, S., Feick, H. et al. (2001). Room-temperature ultraviolet nanowire nanolasers. *Science* 292: 1897–1899.

18 Xiao, Z.X., Huang, Y.Z., Yang, Y.D. et al. (2017). Single-mode unidirectional-emission circular-side hexagonal resonator microlasers. *Opt. Lett.* 42: 1309–1312.

19 Wang, F.L., Yang, Y.D., Huang, Y.Z. et al. (2018). Single-transverse-mode waveguide-coupled deformed hexagonal resonator microlasers. *Appl. Opt.* 57: 7242–7248.

20 Yang, Y.D., Tang, M., Wang, F.L. et al. (2019). Whispering-gallery mode hexagonal micro/nano-cavity lasers. *Photonics Res.* 7: 594–607.

21 Yang, Y.D., Huang, Y.Z., and Wang, S.J. (2009). Mode analysis for equilateral-triangle-resonator microlasers with metal confinement layers. *IEEE J. Quantum Electron.* 45: 1529–1536.

22 Kouno, T., Suzuki, S., Kishino, K. et al. (2015). Optical properties of arrays of hexagonal GaN microdisks acting as whispering-gallery-mode-type optical microcavities. *Phys. Status Solidi* 212: 1017–1020.

23 Fang, H.H., Ding, R., Lu, S.Y. et al. (2013). Whispering-gallery mode lasing from patterned molecular single-crystalline microcavity array. *Laser Photonics Rev.* 7: 281–288.

24 Lin, J.D., Huang, Y.Z., Yang, Y.D. et al. (2011). Single transverse whispering-gallery mode AlGaInAs/InP hexagonal resonator microlasers. *IEEE Photonics J.* 3: 756–764.

25 Ma, C.G., Xiao, J.L., Xiao, Z.X. et al. (2020) Spontaneous chaotic microlasers for random number generation. The paper is not published. The arXiv link of the preprint is: arXiv: 2005. 09470.

26 Zou, L.X., Huang, Y.Z., Lv, X.M. et al. (2014). Dynamic characteristics of AlGaInAs/InP octagonal resonator microlaser. *Appl. Phys. B* 117: 453–458.

9

Vertical Loss for 3D Microcavities

9.1 Introduction

High quality (Q) optical microcavities have become a research hot topic in recent years due to their wide applications in strong-coupling cavity quantum electrodynamics, enhancement and suppression of spontaneous emission, novel light sources, and dynamic filters in optical communication [1]. Whispering-gallery mode (WGM) microdisk lasers were demonstrated in 1992 [2], and the microdisk lasers have been extensively studied in the following decades. The microdisks were supported by a pedestal forming strong vertical optical confinement for realizing high Q factors with a compact size. An electric injection microdisk laser at room temperature was demonstrated by using a double-layered disk structure, where the upper disk was used for electrical contact [3]. However, the microdisk on a pedestal prevents the heat dissipation and limits the current injection efficiency.

Microcavities with the vertical confinement of semiconductor materials will have better thermal conductivity and current injection efficiency than the microdisk on a pedestal, but with a low refractive index contrast for the vertical optical confinement. High efficiency microcavity lasers can be fabricated if WGMs still have a high Q-factor in such microcavities. InAs/GaAs quantum-dot microcylinder lasers and InGaN multiple-quantum-well spiral-shaped micropillars were demonstrated with vertical weak confinements [4, 5]. The quantum-dot microcylinder laser with a radius of 5 μm reached lasing, however, the quantum-well microcylinder lasers realized stimulated emission with the radius $R \geq 7$ μm [4]. But the microdisk laser with sub-micrometer radius was realized at room temperature [3]. The results indicate that the optical loss, especially vertical radiation loss, plays an important role in the microcylinder lasers, in addition to the increase of the surface recombination with decreasing radius.

In this chapter, a systematic research on the vertical radiation loss is presented for three-dimensional (3D) semiconductor microcavities with weak vertical confinement. In Section 9.2, numerical methods used in the simulation of 3D microcavities are presented. In Section 9.3, the mode coupling and vertical loss characteristics in microcylinder cavities are presented. In Section 9.4, the mode characteristics of 3D polygonal microcavities are presented. Finally, a summary is given in Section 9.5.

Microcavity Semiconductor Lasers: Principles, Design, and Applications, First Edition.
Yong-zhen Huang and Yue-de Yang.

9.2 Numerical Method for the Simulation of 3D Microcavities

Figure 9.1 schematically shows the cross-sectional view of a 3D microcavity, where n_1, n_2, n_3, and n_4 are the refractive indexes of the active layer, the upper cladding layer, the lower cladding layer, and the surrounding medium, respectively, and d is the thickness of the active layer. The entire structure is divided into two regions, where Ω_1 is the center region of the microcavity including the active layer of the microcavity and the upper and lower cladding layers, and Ω_2 is the surrounding region. Generally, Ω_2 is air in this chapter.

This section will introduce several numerical simulation methods for 3D microcavities. The finite-difference time-domain (FDTD) method has been described in detail in Chapter 3. Therefore, the effective index method (EIM) and the scattering matrix method are introduced in this section.

9.2.1 Effective Index Method

Since there is no analytical solution for the WGMs in 3D optical microcavities, and a real 3D numerical simulation requires great computing power, an efficient and accurate method is needed. As a very common method, EIM is used to transform 3D problem into two-dimensional (2D) problem by separating variable. 2D problems are greatly simplified in both theoretical analysis and numerical simulation than 3D problems. For ordinary microcavity lasers, the active layer is generally a thin slab, and EIM is very effective for analysis the thin slab structure.

As shown in Figure 9.1, separating variables is used in the Ω_1 region, and the non-uniform refractive index in the Ω_1 region is replaced by the effective refractive index [6]. This converts the problem into a uniform distribution of the refractive index in the z direction. Since the lateral dimension of the structure is much larger than the vertical dimension, the 2D waveguide structure in the plane is considered when obtaining the effective refractive index n_{eff}. The eigenvalue equation can be easily obtained similar to that introduced in Chapter 2:

$$\kappa d - m\pi = \tan^{-1}(\gamma_2/\kappa) + \tan^{-1}(\gamma_3/\kappa) \tag{9.1}$$

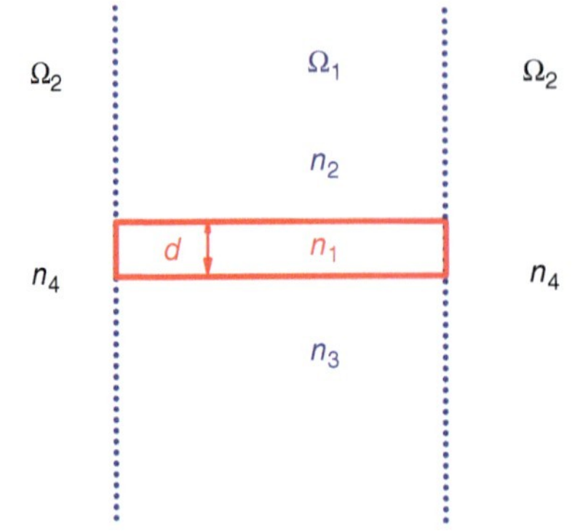

Figure 9.1 Schematic cross-section of a 3D microcavity.

Table 9.1 Variation of effective refractive index n_{eff} with the cladding refractive index n_2 for a thin slab waveguide with a thickness of 0.2 μm.

n_2	1	1.5	2	2.5	3	3.17
TE	2.657	2.710	2.791	2.916	3.115	3.215
TM	1.326	1.914	2.372	2.748	3.086	3.207

$$\kappa d - m\pi = \tan^{-1}\left(\frac{n_1^2}{n_2^2}\gamma_2/\kappa\right) + \tan^{-1}\left(\frac{n_1^2}{n_3^2}\gamma_3/\kappa\right) \tag{9.2}$$

where $\kappa = k_0\sqrt{n_1^2 - n_{eff}^2}$, $\gamma_2 = k_0\sqrt{n_{eff}^2 - n_2^2}$, $\gamma_3 = k_0\sqrt{n_{eff}^2 - n_3^2}$, and k_0 is wavenumber in vacuum. Eqs. (9.1) and (9.2) correspond to transverse-electric (TE) and transverse-magnetic (TM) modes of the slab waveguide, respectively. By solving Eqs. (9.1) and (9.2), the effective refractive index n_{eff} can be obtained, and then the non-uniform refractive index in the Ω_1 region is replaced by n_{eff}.

Here a 3D structure with the following structure is considered with the reference wavelength of 1.55 μm: the active layer thickness $d = 0.2$ μm, the refractive index of the active layer $n_1 = 3.4$ corresponding to the refractive index of the InGaAsP material with an emission wavelength around 1.55 μm, and both the upper and lower cladding layer are infinitely thick with the same refractive index n_2. The effective refractive indices are obtained by solving Eqs. (9.1) and (9.2) with different n_2, and listed in Table 9.1.

When $n_2 = 1$, the effective refractive index of TM mode is very low, and the TM mode in the microcavity vertically confined by air has a large lateral loss resulting in a low Q factor. So the TM modes are generally not considered in the microdisk with strong confinement in vertical direction. As the refractive index n_2 increases, the effective refractive index n_{eff} of the TM mode increases faster than that of the TE mode. When $n_2 = 3.17$ (corresponding to the refractive index of the InP substrate), the effective refractive indices of the TM and TE modes are 3.207 and 3.215, respectively, and the difference is very small. Therefore, in a microcavity confined by a semiconductor material in the vertical direction, the lateral loss of the TM mode is equivalent to that of the TE mode, and both the TE and TM modes should be considered.

The electromagnetic fields of the modes in an infinite slab waveguide only have three components, and all the six components are not zero in a 3D microcavity, so EIM is based on the principal component approximation. The TE-like fundamental mode in a vertically symmetric thin slab is taken as an example. Three electromagnetic field components (H_z, $E_{\|}$) are symmetric relative to the mirror plane of the thin slab, and the other three components (E_z, $H_{\|}$) are anti-symmetric, where $\|$ indicates the field components in the plane of the thin slab. The symmetric component is about 1 order of magnitude greater than the anti-symmetric component, so its main component is (H_z, $E_{\|}$), which is basically the same as the TE mode of the slab waveguide. Therefore, the mode refractive index of the slab waveguide TE mode can be used as the effective refractive index for the TE-like mode. Similar to the TE-like mode, the effective refractive index of the TM-like mode can be replaced by the mode refractive index of the TM mode of the slab waveguide.

In a vertically asymmetric 3D structure, the effective refractive index approximation can also be used based on the dominant electromagnetic field components, although there is no mirror symmetry of a symmetric structure. Since the semiconductor microcavity laser generally adopts a thin slab active layer structure, in which only the fundamental transverse mode can be supported, the EIM is suitable for the investigation of the WGMs in the microcavities.

The 2D circular microcavity can be solved by solving the eigen equation to obtain the mode frequencies and Q factors [7], which is introduced in Chapter 4. The eigen equation of the WGMs in a circular microcavity with an effective index of n_{eff} and a radius of R surrounded by the medium with a refractive index of n_4 can be written as

$$J_\nu(kn_{\text{eff}}R)H_\nu^{(1)\prime}(kn_4R) - \eta J_\nu'(kn_{\text{eff}}R)H_\nu^{(1)}(kn_4R) = 0 \tag{9.3}$$

where k is the wavenumber, ν is the angular mode number, $\eta = n_4/n_{\text{eff}}$ and $\eta = n_{\text{eff}}/n_4$ for TE and TM modes, J_ν and $H_\nu^{(1)}$ represent the Bessel function and the first kind Hankel function, respectively, and a time dependence of $\exp(-i\omega t)$ is used.

Because n_{eff} is wavelength-dependent, an initial value is assumed for n_{eff} in the calculation, then the wavenumber k is obtained from Eq. (9.3), and finally n_{eff} is obtained by EIM using the calculated WGM wavelength. By repeating the above-mentioned cycle several times, the self-consistent mode wavelength and effective refractive index can be obtained. Then the mode lifetime τ and quality factor Q in the microcavity can be obtained from the imaginary part of k as

$$\tau = -\frac{1}{2cIm(k)} \tag{9.4}$$

$$Q = -\frac{Re(k)}{2Im(k)} \tag{9.5}$$

9.2.2 S-Matrix Method

The scattering matrix (S-matrix) techniques were used to analyze the mode characteristics [8] and the effects of boundary roughness on the Q factors [9] for 2D microdisk resonators. Here a 3D S-matrix method is presented to analyze WGMs in the circular microcavities with different vertical confinements [10].

As shown in Figure 9.2, R is the radius of the circular microcavity, d is the thickness of the active layer, and n_1, n_2, n_3, and n_4 are the refractive indices of the active layer, the upper confined layer, the lower confined layer, and the external region Ω_2, respectively. The microdisk with the confined layers forms the inner region Ω_1, in which additional boundaries with the distance of L and perpendicular to the z axis are introduced. Periodic conditions are applied to these boundaries to simulate the infinite situation in the z direction. The slab waveguide is assumed to be a single-mode waveguide, and only guided TE and TM modes are considered. In region Ω_1, the electromagnetic fields can be expressed as the summation of TE and

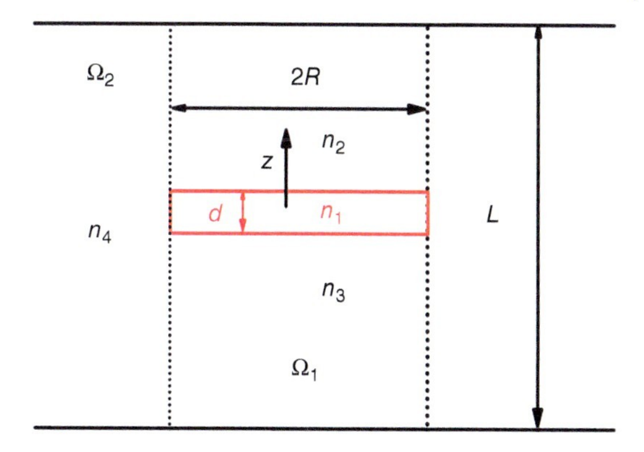

Figure 9.2 Schematic cross-section of a 3D circular microcavity.

TM modes. Omitting the common item of $\exp[i(-\omega t + \nu\varphi)]$, the field components in the region Ω_1 tangential to the circumference plane between Ω_1 and Ω_2 as [11]

$$E_z = Af(z)J_\nu(\kappa r)/J_\nu(\kappa R) \tag{9.6}$$

$$H_z = Bg(z)J_\nu(\gamma r)/J_\nu(\gamma R) \tag{9.7}$$

$$E_\varphi = A\frac{-\nu f'(z)J_\nu(\kappa r)}{\kappa^2 rJ_\nu(\kappa R)} + B\frac{ikg(z)J'_\nu(\gamma r)}{\gamma J_\nu(\gamma R)} \tag{9.8}$$

$$H_\varphi = A\frac{-ik\varepsilon_r(z)f(z)J'_\nu(\kappa r)}{\kappa J_\nu(\kappa R)} + B\frac{\nu g'(z)J_\nu(\gamma r)}{\gamma^2 rJ_\nu(\gamma R)} \tag{9.9}$$

where κ and γ are the transverse wavenumbers of E_z and H_z components, respectively, and the relative permittivity in the z direction ε_r is expressed as a function of z, J_ν is the Bessel function of order ν. The wave functions $f(z)$ and $g(z)$ in the z direction satisfy the TM and TE mode equations of the slab waveguide:

$$\frac{d}{dz}\frac{1}{\varepsilon_r(z)}\frac{d\varepsilon_r(z)f(z)}{dz} + k^2\varepsilon_r(z)f(z) = \kappa^2 f(z) \tag{9.10}$$

$$\frac{d^2g(z)}{dz} + k^2\varepsilon_r(z)g(z) = \kappa^2 g(z) \tag{9.11}$$

Considering the periodic boundary conditions, complete orthogonal and normalized basis functions are introduced:

$$|m\rangle = \frac{\exp(-ip_{mz})}{\sqrt{L}}, \qquad p_m = 2m\pi/L \tag{9.12}$$

where m is an integer number. Expanding $g(z)$ and $f(z)$ by using the basis functions, the column vectors g_m and f_m can be obtained:

$$f(z) = \sum_{m=-\infty}^{+\infty} f_m|m\rangle, \qquad f_m = \langle m|\, f(z)\rangle \tag{9.13}$$

$$g(z) = \sum_{m=-\infty}^{+\infty} g_m|m\rangle, \qquad g_m = \langle m|\, g(z)\rangle \tag{9.14}$$

In practical calculation, m is truncated from $-$M to M. Then, through the normalization relation $g_m^T g_m = 1$, $f_m^T f_m = 1$, where the superscript T means conjugate transpose. The electromagnetic fields $E_z, H_z, E_\varphi, H_\varphi$ of Eqs. (9.6)–(9.9) are expressed as the following column vectors:

$$E_{z,m} = Af_m J_\nu(\kappa r)/J_\nu(\kappa R) \tag{9.15}$$

$$H_{z,m} = Bg_m J_\nu(\gamma r)/J_\nu(\gamma R) \tag{9.16}$$

$$E_{\varphi,m} = A\frac{-\nu(-ip_m g_m)J_\nu(\kappa r)}{\kappa^2 r J_\nu(\kappa R)} + B\frac{ikg_m J_\nu'(\gamma r)}{\gamma J_\nu(\gamma R)} \tag{9.17}$$

$$H_{\varphi,m} = A\frac{-ik\varepsilon_r(z)f_m J_\nu'(\kappa r)}{\kappa J_\nu(\kappa R)} + B\frac{\nu(-ip_m g_m)J_\nu(\gamma r)}{\gamma^2 r J_\nu(\gamma R)} \tag{9.18}$$

$$s_m = \sum_{n=-M}^{M} \langle m|\varepsilon_r(z)|n\rangle f_n \tag{9.19}$$

In the Ω_2 region, the field can also be expressed as the summation of TE and TM modes including the incident and outward waves, and they can be expanded as column vectors by use of the basis functions [11]

$$E_{z,m} = C_m^{\text{in}}\frac{H_\nu^{(2)}(q_m r)}{H_\nu^{(2)}(q_m R)} + C_m^{\text{out}}\frac{H_\nu^{(1)}(q_m r)}{H_\nu^{(1)}(q_m R)} \tag{9.20}$$

$$H_{z,m} = D_m^{\text{in}}\frac{H_\nu^{(2)}(q_m r)}{H_\nu^{(2)}(q_m R)} + D_m^{\text{out}}\frac{H_\nu^{(1)}(q_m r)}{H_\nu^{(1)}(q_m R)} \tag{9.21}$$

$$\begin{aligned} E_{\varphi,m} = {} & \frac{ip_m\nu}{rq_m^2}\left[C_m^{\text{in}}\frac{H_\nu^{(2)}(q_m r)}{H_\nu^{(2)}(q_m R)} + C_m^{\text{out}}\frac{H_\nu^{(1)}(q_m r)}{H_\nu^{(1)}(q_m R)}\right] \\ & + \frac{ik}{q_m}\left[D_m^{\text{in}}\frac{H_\nu^{(2)\prime}(q_m r)}{H_\nu^{(2)}(q_m R)} + D_m^{\text{out}}\frac{H_\nu^{(1)\prime}(q_m r)}{H_\nu^{(1)}(q_m R)}\right] \end{aligned} \tag{9.22}$$

$$\begin{aligned} H_{\varphi,m} = {} & \frac{-ikn_4^2}{q_m}\left[C_m^{\text{in}}\frac{H_\nu^{(2)\prime}(q_m r)}{H_\nu^{(2)}(q_m R)} + C_m^{\text{out}}\frac{H_\nu^{(1)\prime}(q_m r)}{H_\nu^{(1)}(q_m R)}\right] \\ & - \frac{ip_m\nu}{rq_m^2}\left[D_m^{\text{in}}\frac{H_\nu^{(2)}(q_m r)}{H_\nu^{(2)}(q_m R)} + D_m^{\text{out}}\frac{H_\nu^{(1)}(q_m r)}{H_\nu^{(1)}(q_m R)}\right] \end{aligned} \tag{9.23}$$

where $H_\nu^{(1)}$ and $H_\nu^{(2)}$ are the first- and second-kind Hankel functions of order ν corresponding to the outward and incident waves with the time dependence of $\exp(-i\omega t)$, respectively, and $q_m = (k^2 - p_m^2)^{1/2}$. In the circular microcavity circumference plane, the tangential electric and magnetic field components should be continuous. From the boundary condition, we can obtain equations between coefficients A, B, C_m^{in}, C_m^{out}, D_m^{in}, and D_m^{out}. By eliminating the coefficients A and B, the relation between the incident and the outward wave coefficients can be established as

$$\begin{pmatrix} C_m^{\text{out}} \\ D_m^{\text{out}} \end{pmatrix} = \begin{bmatrix} O_{11} & O_{12} \\ O_{21} & O_{22} \end{bmatrix} \begin{pmatrix} C_m^{\text{in}} \\ D_m^{\text{in}} \end{pmatrix} \tag{9.24}$$

where $O_{\text{i,j}}$ $(i, j = 1, 2)$ are $(2M+1)\times(2M+1)$ matrices. Considering that practical incident waves have q_m as real positive values, then m is restricted in the range from $-N$

to N, where N is the largest integer number smaller than $kn_4L/2$. So the corresponding scattering matrix can be established as

$$S(k) = \begin{bmatrix} O_{11}(i,j) & O_{12}(i,j) \\ O_{21}(i,j) & O_{22}(i,j) \end{bmatrix} \tag{9.25}$$

where i and j are from $-N$ to N. The scattering matrix is unitary, and its determinant can be expressed as $\det(S) = \exp(-i\theta)$. Mode resonances can be identified from the Wigner delay time, which can be calculated from as [7]

$$\tau(k) = 2\pi \frac{d\theta}{dk} \tag{9.26}$$

The total phase of $\det(S)$ would have the variation of 2π around the resonance, and isolated Lorentzian peaks would appear in $\tau(k)$. The mode wavelengths and Q factors can be calculated from the central wavenumber k_c and the full width at half-maximum (FWHM) Δk of the Lorentzian peaks as

$$\lambda = \frac{2\pi}{k_c}, \qquad Q = \frac{k_c}{\Delta k} \tag{9.27}$$

It should be noted that because periodic boundary conditions are used in the vertical direction, the 3D S-matrix method cannot reflect the vertical radiation loss. In the case of vertical loss, the S-matrix method is not applicable. However, it still provides a new way for the study of 3D microcavity.

9.3 Control of Vertical Radiation Loss for Circular Microcavities

The numerical methods, such as 3D FDTD, EIM, and S-matrix, are introduced in Chapter 3 and Section 9.2 for 3D microcavity simulation. The 3D FDTD method can give accurate results since it is based on the Maxwell's equation without any approximations. However, 3D FDTD simulation for microcavities requires a great amount of calculation memory and very long calculation time. The EIM is often used to simplify the 3D problem to 2D because it gives accurate mode wavelengths although the Q factors are not reliable. The S-matrix method can give the mode wavelength and quality factor accurately for the circular microcavities without vertical radiation loss. If the condition of prohibition of the vertical radiation loss is obtained, then the scattering matrix method can be used to avoid the large amount of calculation power required by the 3D FDTD method. In this section, the vertical radiation losses and the control method for 3D circular microcavities are studied systematically.

9.3.1 Mode Coupling and Vertical Radiation Loss

In a circular microcylinder with a weak vertical waveguiding, the vertical radiation will play an important role in determining the Q-factors of the WGMs. The TE and TM WGMs in the microcylinder are denoted as $TE_{\nu,m}$ and $TM_{\nu,m}$, respectively, where the vertical mode number is ignored and ν and m are the angular and radial mode numbers. The WGMs will couple with the vertical propagating $EH_{\nu,m}$ and $HE_{\nu,m}$ modes in the corresponding circular waveguide of the cladding layers. The cut-off wavelengths of the guided and radiated $EH_{\nu,m}$ and $HE_{\nu,m}$ modes can be obtained as their vertical propagation constant $\beta = 0$. The definition is different from the cut-off condition of the EH and HE modes in the optical fiber, as the radiation modes are neglected in the fiber system but should be considered here [12]. At $\beta = 0$, the EH and HE modes become polarized modes, and correspond to TE and TM modes, respectively.

For a circular microcylinder with the radius $R = 1$ μm, the center layer thickness $d = 0.2$ μm, and the vertical refractive index distribution of $n_2/n_1 = n_2/3.4$, the mode wavelengths of the $TE_{7,1}$ and $TM_{7,1}$ WGMs obtained from the 2D eigenvalue Eq. (9.3) under the effective index approximation are plotted as the dash and solid lines in Figure 9.3, respectively, and the cut-off wavelengths of the radiated $EH_{7,1}$ and $HE_{7,1}$ are plotted as dash-dot and dot lines. Two crossover points exist between $TE_{7,1}$ and $TM_{7,1}$ WGMs near $n_2 = 2.3$ and between $TE_{7,1}$ and $HE_{7,1}$ near $n_2 = 2.65$. It can expect that the mode coupling between $TE_{7,1}$ and $HE_{7,1}$ will result in a large vertical radiation loss as the mode wavelength of $TE_{7,1}$ is smaller than the cut-off wavelength of $HE_{7,1}$ mode. The mode wavelength of $TM_{7,1}$ is larger than that of $TE_{7,1}$ as $n_2 > 2.3$, and does not cross with those of the $HE_{7,1}$ and $EH_{7,1}$ modes. So the mode coupling with the vertically propagating modes does not exist for the TM WGMs, and the TM WGMs will have high Q-factors. It should be noted that the above phenomena exist for the TE and TM WGMs with different mode numbers and are mainly affected by the refractive index difference $n_1 - n_2$. For a microcylinder with $R = 1$ μm, $d = 0.2$ μm, and $n_1 = 3.5$, the crossover points are observed at $n_2 = 2.45$ and 2.85 as the mode

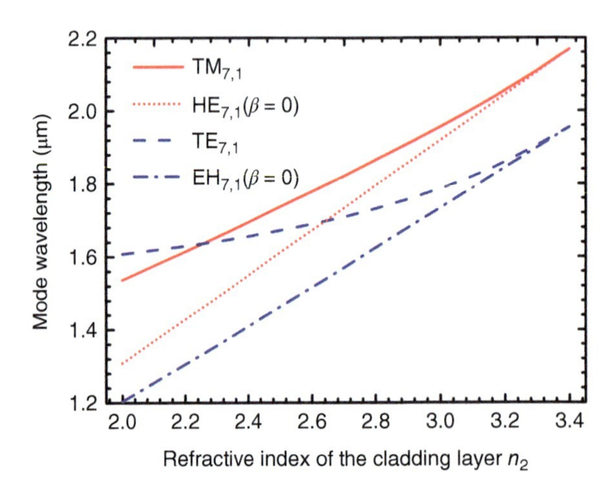

Figure 9.3 Mode wavelengths of $TE_{7,1}$ and $TM_{7,1}$ WGMs obtained from 2D eigenvalue equation under the effective index approximation vs. n_2 are plotted as solid and dash lines, respectively, for the symmetric microcylinder with radius $R = 1$ μm, $d = 0.2$ μm, and $n_1 = 3.4$. The cut-off wavelengths of the radiated $EH_{7,1}$ and $HE_{7,1}$ modes in the corresponding microwire with the refractive index n_2 are plotted as the dash-dot and dot lines. Source: Yang et al. [12]. © 2007, American Physical Society.

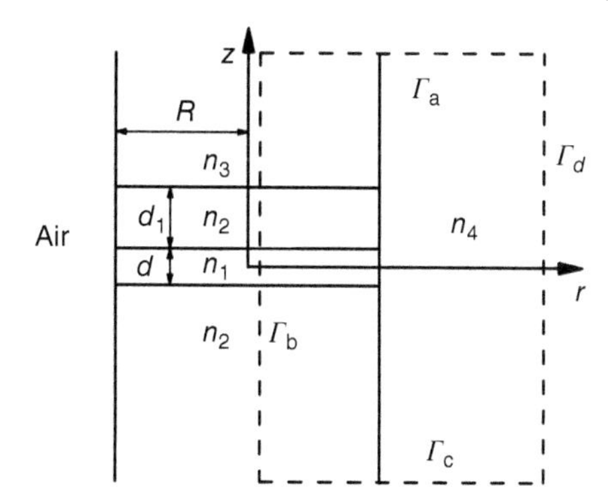

Figure 9.4 Cross section of a microcylinder with the vertical refractive index distribution $n(z)$ of $n_2/n_1/n_2/n_3$ surrounded by the medium with a refractive index of n_4, and the thicknesses of center layer and upper cladding layer of d and d_1. The FDTD simulation region is bounded by Γ_a, Γ_b, Γ_c, and Γ_d. Source: Yang et al. [12]. © 2007, American Physical Society.

index $m = 9$. When $n_2 = 3.4$, the wavelength of $TM_{7,1}$ ($TE_{7,1}$) and the cut-off wavelength of $HE_{7,1}(EH_{7,1})$ are the same, because both of them correspond to the $TM_{7,1}$ ($TE_{7,1}$) with the refractive index of 3.4.

To verify the above argument, the mode characteristics for circular microcylinders as shown in Figure 9.4 are investigated by 3D FDTD method in cylindrical coordinate, with the FDTD calculating window bounded by Γ_a, Γ_b, Γ_c, and Γ_d, where R, d, and d_1 are the radius, the thicknesses of the center layer and the upper cladding layer, respectively. The refractive index distribution in the z-direction is $n(z)$ of $n_2/n_1/n_2/n_3$, where n_1 is the refractive index of the center layer and n_2 is that of the cladding layers and the substrate, the refractive index of the surrounded medium is n_4. The perfect matched layer (PML) absorbing boundary condition in circular cylindrical coordinates is used on Γ_a, Γ_c, and Γ_d, with the boundaries Γ_a, Γ_c, and Γ_d are placed 5, 5, and 4 μm away from the center layer's upper, lower and lateral boundaries, respectively. The spatial steps Δz and Δr are set to be 10 and 20 nm, respectively, and the time step Δt is chosen to satisfy the Courant condition. At the inner boundary Γ_b at $r = 4\Delta r$, the condition $\psi_\nu \propto r^\nu$ is used for the E_z and H_z field components based on the asymptotic behavior of the Bessel function. In the simulation, an exciting source with a cosine impulse modulated by a Gaussian function $P(x_0, y_0, t) = \exp[-(t-t_0)^2/t_w^2]\cos(2\pi ft)$ is added to one component of the electromagnetic fields at a point (x_0, y_0) inside the microcylinder, where t_0 and t_w are the times of the pulse center and the pulse half width, respectively, and f is the center frequency of the pulse. The time variation of a selected field component at some points inside the microcylinder is recorded as a FDTD output, next the Padé approximation with Baker's algorithm is used to transform the FDTD output from the time-domain to the frequency-domain, and then the mode frequencies and Q-factors are calculated from the obtained intensity spectrum.

Firstly, a vertically symmetric microcylinder with $n_3 = n_2$, $n_4 = 1$, and the $n(z)$ of $n_1/n_2 = 3.4/n_2$, $d = 0.2$ μm, and $R = 1$ μm is considered. For the vertical fundamental mode, the electric field and magnetic field of TE WGMs are symmetric and anti-symmetric relative to $z = 0$ plane in the symmetric microcylinder, respectively, and TM WGMs are reversed. So symmetric sources relative to $z = 0$ plane applied on electric field and magnetic field can be used to excite the TE and TM WGMs, respectively.

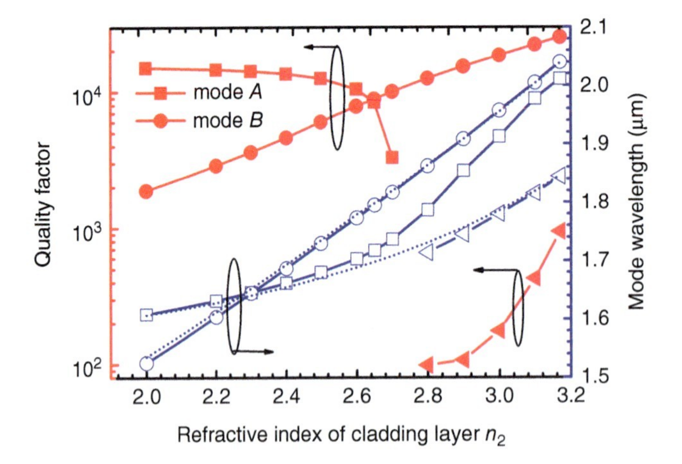

Figure 9.5 Mode wavelengths (Q-factors) of $TE_{7,1}$ (Mode A) and $TM_{7,1}$ (Mode B) WGMs vs. n_2 are plotted as the open (solid) squares and circles, respectively, for the symmetrical microcylinder with the radius $R = 1$ μm, $d = 0.2$ μm, and the $n(z)$ of $n_1/n_2 = 3.4/n_2$. The dotted lines are mode wavelengths of $TE_{7,1}$ and $TM_{7,1}$ WGMs obtained from 2D eigenvalue equations under EIM. The open and solid triangles are the mode wavelength and Q-factor of $TE_{7,1}$ WGM after the mode coupling. Source: Yang et al. [12]. © 2007, American Physical Society.

The mode wavelengths and Q-factors of the $TE_{7,1}$ (Mode A) and $TM_{7,1}$ (Mode B) WGMs are plotted vs. n_2 in Figure 9.5 as the open and solid squares and circles, respectively. The wavelengths of $TE_{7,1}$ and $TM_{7,1}$ WGMs have a crossover point at $n_2 = 2.3$, which indicates that mode coupling between the two WGMs does not exist in the symmetric microcylinder. The Q-factor of the $TE_{7,1}$ WGM gradually decreases from 1.5×10^4 to 1.0×10^4 as n_2 increases from 2 to 2.6, and then rapidly decreases as $n_2 > 2.6$. As $n_2 > 2.8$, two peaks appear in the TE mode field spectra, which are correspondence to the anti-crossing mode coupling between $TE_{7,1}$ and $HE_{7,1}$ modes. After the anti-crossing mode coupling, $TE_{7,1}$ WGM becomes the vertical propagation $HE_{7,1}$ mode with a much smaller Q-factor, which is not presented in Figure 9.5 because the corresponding mode peak obtained from different time series of FDTD output is unstable. Furthermore, the $HE_{7,1}$ mode transfers to $TE_{7,1}$ WGM with the mode wavelength and Q-factor marked as open and solid triangles in Figure 9.5, respectively, the corresponding Q-factor increases from 98 to 941 as n_2 increases from 2.8 to 3.17. The mode wavelengths of $TE_{7,1}$ and $TM_{7,1}$ WGMs obtained from the 2D eigenvalue equation [7] under the effective index approximation are also plotted as the dot lines in Figure 9.5, which agrees very well with the results of 3D FDTD simulation for $TM_{7,1}$ WGM. But the results of the eigenvalue equation for $TE_{7,1}$ WGM agree with the wavelength of Mode A as $n_2 < 2.6$ and the wavelength of the mode marked by triangles as $n_2 > 2.8$, respectively. The Q-factor of the $TM_{7,1}$ WGM increases from 1.9×10^3 to 2.5×10^4 as n_2 increases from 2 to 3.17, which is much larger than that of the $TE_{7,1}$ WGM as $n_2 > 2.7$. Because the horizontal radiation loss decreases with the increase of n_2 and the vertical radiation is absent, the Q-factor of the $TM_{7,1}$ WGM can increase with the increase of n_2 from 2.0 to 3.17.

Table 9.2 Mode wavelengths and Q-factors obtained by 3D FDTD simulation and the scattering-matrix for $TM_{\nu,1}$ WGMs in the symmetric microcylinder with $R = 1$ μm, $d = 0.2$ μm, $n_1 = 3.4$, and $n_2 = 3.17$.

	3D-FDTD		S-matrix	
ν	λ (μm)	Q	λ (μm)	Q
7	2.0398	2.52×10^4	2.0386	2.60×10^4
8	1.8281	1.23×10^5	1.8270	1.30×10^5
9	1.6586	6.54×10^5	1.6576	6.71×10^5
10	1.5195	3.50×10^6	1.5186	3.66×10^6

The mode wavelengths and Q-factors obtained by the 3D FDTD simulation and the S-matrix method are shown in Table 9.2 for the $TM_{\nu,1}$ WGMs in the symmetric microcylinder with the same structure parameters as shown in Figure 9.5 and $n_2 = 3.17$. The two methods yield almost the same mode wavelengths and Q-factors as the angular mode number $\nu = 7, 8, 9, 10$. The TM WGMs have Q-factors larger than 2.5×10^4, which indicates that the vertical radiation loss is very small for the TM WGMs. In contrast, 3D FDTD simulation gives Q-factor of 941, 878, 678, and 541 for $TE_{\nu,1}$ WGMs as $\nu = 7, 8, 9$, and 10, respectively, due to the strong vertical radiation loss. Accounting that the vertical waveguiding is mainly related to the refractive index difference and the difference between 3.4 and 3.17 is a medial value in GaAs and InP material systems, it can be expected that high Q-factor TM WGMs are easily to realize in real semiconductor microcylinders.

The microcylinders with $d = 0.2$ μm, $n_2 = 3.17$, $n_1 = 3.4$, and the radius $R = 2$, 3, 4, 5, and 6 μm are considered. The wavelengths of TE and TM WGMs obtained by 3D FDTD simulation and the cutoff wavelengths of the corresponding HE and EH modes obtained by eigenvalue equation are listed in Table 9.3 [13], where the mode index ν is chosen to keep the mode wavelength around 1.55 μm. The cutoff wavelengths of HE and EH modes are obtained from Eq. (9.3) for TM and TE modes with the refractive index of 3.17, respectively. With the increase of the

Table 9.3 Mode wavelengths of WGMs and cut-off wavelength of radiation modes for microcylinders with vertical refractive indices 3.17/3.4/3.17.

R (μm)	1	2	3	4	5	6
ν	9	21	33	45	58	70
λ-TM WGM (μm)	1.658	1.589	1.584	1.586	1.565	1.574
λ-TE WGM (μm)	1.534	1.535	1.548	1.562	1.549	1.560
λ (HE_{off}) (μm)	1.641	1.572	1.566	1.569	1.548	1.556
λ (EH_{off}) (μm)	1.511	1.511	1.526	1.539	1.527	1.537
Q (TE)	680	470	580	870	$\sim 10^4$	$\sim 10^6$

The Q factors obtained by 3D FDTD method are given for the TE WGMs.
Source: Huang and Yang [13]. © 2008, Optical Society of America.

radius, the wavelengths of the TE WGMs are larger than the cutoff wavelengths of the corresponding HE and EH modes at the radius between 5 and 6 μm. So the TE WGMs can also have high Q-factor in the microcylinder as $R > 5$ μm. Because the Q factors of TM WGMs are very high and difficult to obtain from the intensity spectrum exactly, only the Q factors of the TE WGMs obtained from 3D FDTD simulation are listed in Table 9.3. The results indicate that a critical lateral size exists for obtaining high Q-factor TE WGMs. For the high radial-order modes, the situation will be different. As the mode coupling between the modes with different radial-order is not forbidden, the vertical radiation loss typically exists for the high radial-order WGMs with a few number of loss channels, unless the refractive index contrast or the radius of the microcylinder is very large. Hoverer, the mode coupling between different radial-order modes is usually weaker than that between the same radial-order modes, so the dominant loss channel results from the coupling between the same radial-order modes [13].

9.3.2 Semiconductor Microcylinder Lasers with the Sizes Limited by Vertical Radiation Loss

The microcylinder lasers with radii R ranging from 1.5 to 5.5 μm are fabricated using an AlGaInAs/InP laser wafer with seven pairs of compressively-strained multi-quantum-wells (MQWs) [14]. Figure 9.6a schematically shows the fabrication processes of the microcylinder lasers. An 800 nm SiO_2 layer is firstly deposited on the laser wafer, and resonator patterns are transferred onto the SiO_2 layer using standard photolithography and inductively coupled-plasma (ICP) etching techniques. The patterned SiO_2 is used as a hard mask for ICP etching laser wafer with a depth of about 4 μm. Figure 9.6b shows the scanning electron microscope (SEM) image of a microcylinder laser with a radius of 5 μm after ICP etching. A 200 nm SiN_x layer is deposited on the wafer, and spin-coated bisbenzocyclobutene (BCB) layer is used to create a planar cladding layer. The BCB film is etched to expose the top of microcylinder resonators. After that, a 450 nm SiO_2 layer is deposited on the whole wafer and a contact window is opened by photolithography and ICP etching for current injection on the top of each resonator. Finally, a Ti/Pt/Au p-electrode is deposited by e-beam evaporation and lift-off process, forming a small size pattern. The laser wafer is mechanically lapped down to a thickness of about 120 μm, and an Au–Ge–Ni metallization layer is deposited as n-electrode. Figure 9.6c schematically shows the cross-sectional-view of the fabricated device.

After cleaving the laser sample, the lasers are characterized by putting them with the p-side up on a heat-sink mounted on a thermoelectric cooler (TEC). A 50 μm-core-diameter multimode fiber butt-coupled to the cleaving facet, which is about 20 μm distance from the rim of microcylinder, is used to collect the emission power. For a 3.75 μm-radius AlGaInAs/InP microcylinder laser, the applied voltage and coupled power vs. the continuous injection current are shown in Figure 9.7 at a TEC temperature of 288 K. The threshold current is about 0.56 mA corresponding to a current density of 1.3 kA cm^{-2}. The black dotted lines indicate the different slopes below and above threshold, and the corresponding slope efficiencies are 0.5

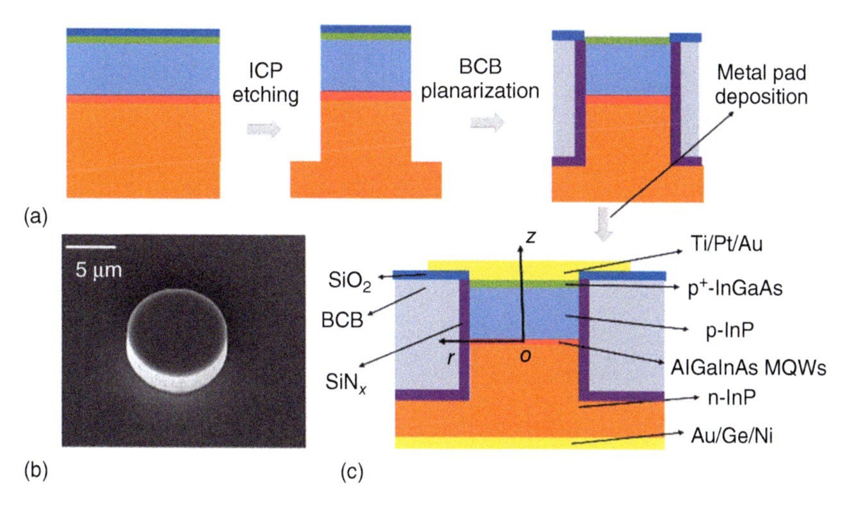

Figure 9.6 (a) Schematic diagram for the fabrication processes of microcylinder lasers, (b) SEM image of a 5 μm-radius microcylinder resonator after ICP etching, (c) cross-sectional schematic diagram of the microcylinder laser. Source: Yang et al. [14]. © 2015, Optical Society of America.

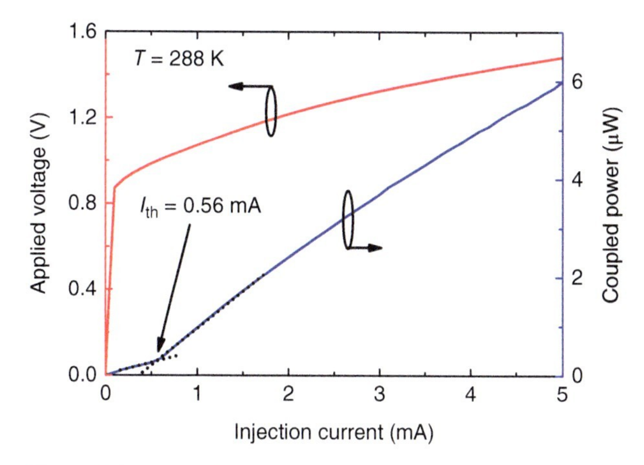

Figure 9.7 Applied voltage and fiber coupled power vs. the continuous injection current for a 3.75 μm-radius microlaser. Source: Yang et al. [14]. © 2015, Optical Society of America.

and 1.5 mW/A, which correspond to the spontaneous and stimulated emissions, respectively. The spontaneous emission factor β is estimated as $\sim 10^{-3}$ by fitting the mode power vs. injection current around the threshold. A series resistance of $\sim 100\,\Omega$ is estimated from the voltage–current curve in Figure 9.7 around the threshold current.

The lasing spectra are measured using an optical spectra analyzer (OSA) with a resolution of 0.05 nm. The measurement of lasing mode linewidth is then limited by the OSA resolution. Figures 9.8a–c show the measured spectra for the microcylinder lasers with radii of 3.75, 3.5, and 3.25 μm, under injection currents I of 1,

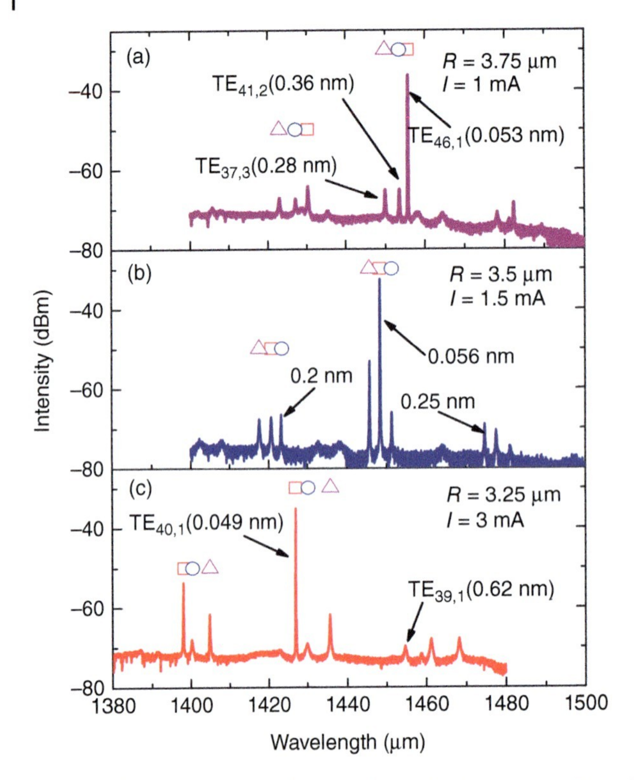

Figure 9.8 Lasing spectra for the microcylinder lasers with (a) $R = 3.75\,\mu\text{m}$ and $I = 1\,\text{mA}$, (b) $R = 3.5\,\mu\text{m}$ and $I = 1.5\,\text{mA}$, and (c) $R = 3.25\,\mu\text{m}$ and $I = 3\,\text{mA}$. Source: Yang et al. [14]. © 2015, Optical Society of America

1.5, and 3 mA, respectively. The threshold currents of the three devices are 0.56, 0.77, and 1.84 mA, which correspond to threshold current densities of 1.3, 2.0, and 5.5 kA cm^{-2}, respectively.

In Figure 9.8a, the main lasing modes are marked by squares, and the other two transverse modes are marked by circles and triangles for the 3.75 μm-radius microcylinder laser. The mode intervals of square, circle, and triangle modes are 25.4, 26.4, and 27.0 nm, respectively. The effective round trip optical path length L can be calculated from $n_g L = \lambda^2/\delta\lambda$ by taking $n_g = 3.5$ for the AlGaInAs/InP material. The optical path length ratios $\eta = L/(2\pi R)$ are obtained as 0.994, 0.953, and 0.927 for the square, circle and triangle modes, respectively. In the microcylinder lasers with compressively-strained QWs, the dominant modes are TE modes. The TE WG modes are marked as $\text{TE}_{\nu,m}$, where ν and m are the angular (longitude) and radial (transverse) mode numbers, respectively. Lower radial-order mode typically has mode field distribution closer to the rim and a relatively longer effective round trip length. Accordingly, we can attribute the square, circle and triangle modes to the first, second and third radial-order modes, respectively. The mode wavelength in the microcylinder resonator can be estimated under EIM, by comparing the experimental lasing wavelengths with the analytical mode wavelengths, the main lasing peak could be $\text{TE}_{46,1}$ with the linewidth of 0.053 nm limited by the OSA resolution and the experimentally used multimode fiber, and the two adjacent minor peaks at the short-wavelength side of the main peak could be $TE_{41,2}$ and $TE_{37,3}$, with the corresponding linewidth of 0.36 and 0.28 nm, respectively. It should be noticed that, because of the deviation of actual device size and refractive index, there may be an error of 1 in the angular mode number.

For the lasing spectra of the 3.25 μm-radius microcylinder laser shown in Figure 9.8c, the main lasing modes and two side modes are also marked as squares, circles, and triangles, respectively. The mode intervals of square, circle and triangle mode are 28.7, 29.8, 31.0 nm, and the corresponding optical path length ratios η are obtained as 0.973, 0.940, and 0.911, respectively. Thus, the square, circle, and triangle modes can be attributed to the first, the second, and the third radial-order modes, respectively. Compared with the analytical mode wavelengths, the main lasing peak could be $TE_{40,1}$, and the two adjacent minor peaks at the long-wavelength side of the main peak could be $TE_{35,2}$ and $TE_{31,3}$. The microcylinder lasers with $R > 3.75$ μm typically lase in the wavelengths ranging from 1445 to 1465 nm, which are around the gain peak of the AlGaInAs MQWs. However, the 3.25 μm-radius microcylinder laser lases at 1427 nm and exhibit a relatively high threshold current of 5.5 kA cm^{-2}. The reason is that a heavy injection is required to obtain enough gain at short wavelength. The peak at the 1456 nm could be $TE_{39,1}$, which has a linewidth of 0.62 nm. The results indicate the mode $TE_{39,1}$ at 1456 nm has much lower Q factor than the mode $TE_{40,1}$ at 1427 nm with a linewidth of 0.049 nm even they are the same transverse mode with an angular mode number difference of 1.

For the lasing spectra of the 3.5 μm-radius microcylinder laser shown in Figure 9.8b, the main lasing modes and two side modes are also marked as squares, circles, and triangles, respectively. The mode intervals of square, circle and triangle modes are 27.7, 28.0, 28.2 nm, and the optical path length ratios η are obtained as 0.965, 0.959, and 0.944, respectively. The optical path lengths are quite similar, and it is difficult to determine the radial-order of each peaks based on mode intervals. The results indicate that the mode wavelengths should be shifted from the original positions of WG modes. For the three groups of peaks at the wavelength around 1420, 1450, and 1480 nm, the highest Q mode shifts from the long-wavelength side to the short-wavelength side, with the linewidths of 0.20, 0.056, and 0.25 nm, as shown in Figure 9.8b.

The threshold current density vs. the radius of microcylinder lasers is plotted as circles in Figure 9.9 at the TEC temperature of 288 K. The solid line connects the lowest threshold current density of each radius laser. With the decrease of the radius, a dramatic increase of the threshold current is observed as $R < 3.5$ μm. The threshold current density increases from 1.3 to 5.5 kA cm^{-2} as the radius decreases from 3.75 to 3.25 μm. A relatively large variation in threshold current is observed for the 3.5 μm-radius microcylinder lasers. The lasing spectrum for a 3.5 μm-radius microcylinder laser with a relatively high threshold current density of 5.8 kA cm^{-2} is shown in the inset of Figure 9.9. The main lasing mode is at 1406 nm far away from the lasing mode in Figure 9.8b, so the threshold current is high.

The experimental results are briefly summarized as follows. The first radial-order mode lasing are realized in the microcylinder lasers with the radii of 3.75 and

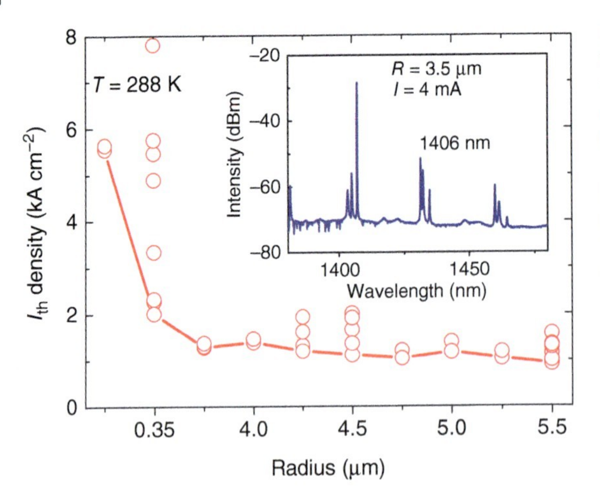

Figure 9.9 Measured threshold current and density vs. the radius of microcylinder lasers. Inset: lasing spectrum for a 3.5 μm-radius microcylinder laser with a relatively high threshold current. Source: Yang et al. [14]. © 2015, Optical Society of America.

3.25 μm, but the 3.5 μm-radius microcylinder laser show distinct lasing characteristics, including unexpected mode intervals and large variation of threshold currents. A dramatic increase of the threshold current density is found as $R < 3.5\,\mu m$, which indicates a drop of mode Q factors. The physical mechanisms of the above phenomena have not been understood clearly yet. In order to figure out the lasing mode characteristics, 3D FDTD simulations are performed to simulate the modes in the microcylinder resonators.

The FDTD simulated mode Q factors of the first radial-order modes in the microcylinder resonators with radii from 3 to 3.75 μm are plotted in Figure 9.10 as the functions of mode wavelengths. The Q factors of the modes around 1450 nm decrease from 7.5×10^4 to 1.2×10^3 as the microcylinder radius decrease from 3.75 to 3 μm, which agrees well with the increase of the experimental threshold current density as $R < 3.5\,\mu m$ For the 3.25 μm-radius microcylinder resonator, the Q factors are larger than 6×10^4 as wavelength are shorter than 1400 nm, which are mainly limited by the grid size of 25 nm used in the FDTD simulation. The mode Q

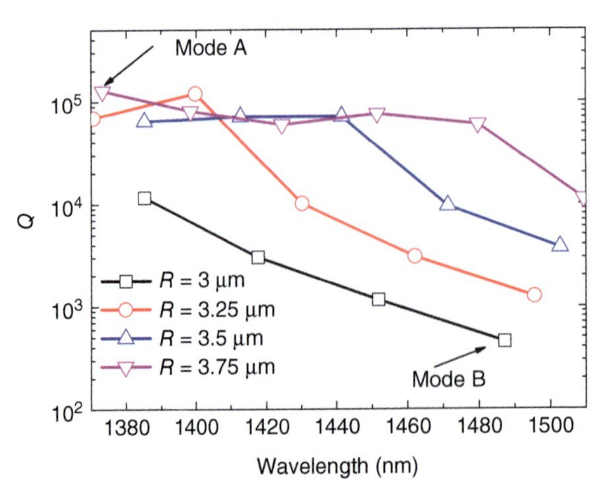

Figure 9.10 3D FDTD simulated mode Q factors of the first radial-order modes in the microcylinder resonators with radii ranging from 3 to 3.75 μm. Source: Yang et al. [14]. © 2015, Optical Society of America.

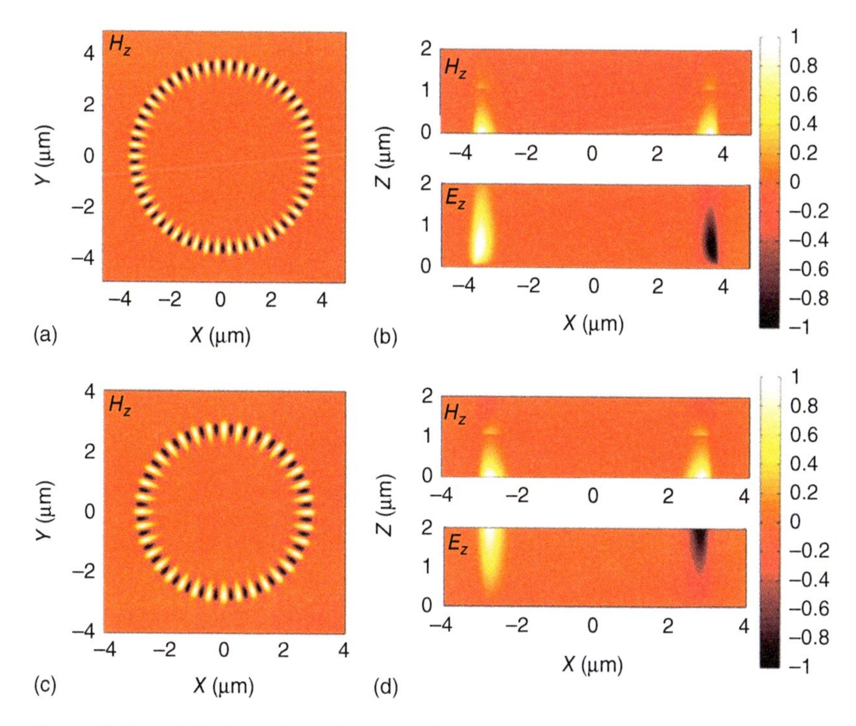

Figure 9.11 (a) Horizontal and (b) vertical plane field distributions of Mode A at $z = 0$, and (c) horizontal and (d) vertical plane field distributions of Mode B. The intensity in the region of $z > 1$ μm is magnified four times for the H_z distribution. Source: Yang et al. [14]. © 2015, Optical Society of America.

factors greatly decrease at the wavelength around 1430 nm, and decrease to ~10^3 at the wavelength of 1496 nm. Thus the 3.25 μm-radius microcylinder resonator can only have high-Q mode with the wavelength shorter than 1430 nm, and exhibit a relatively high threshold current density in the experiment. For the 3 μm-radius microcylinder resonator, the high Q modes are far away from the gain peak, and room-temperature lasing is not achieved for these microcylinder lasers.

Two representatively high Q and low Q modes in Figure 9.10 are marked as Mode A and Mode B, with the Q factors of 1.2×10^5 and 4.5×10^2, respectively. Figure 9.11a,b show the horizontal and vertical plane field distributions of the mode at 1373 nm in 3.75 μm-radius microcylinder resonator (Mode A). The intensity in the region of $z > 1$ μm is magnified four times for the H_z distribution. The field distribution shows an angular mode number of 50, and a strong confinement in vertical direction for both H_z and E_z field components. Figure 9.11c,d show the horizontal and vertical plane field distributions of the mode at 1487 nm in 3 μm-radius microcylinder resonator (Mode B). The field distribution shows an angular mode number of 36, and a leakage in vertical direction for both H_z and E_z field components. Thus, the high Q and low Q modes have distinct vertical radiation loss properties.

The degrade of mode Q factors in the microcylinder resonator is because of the vertical radiation loss induced by the mode coupling between the TE WG modes

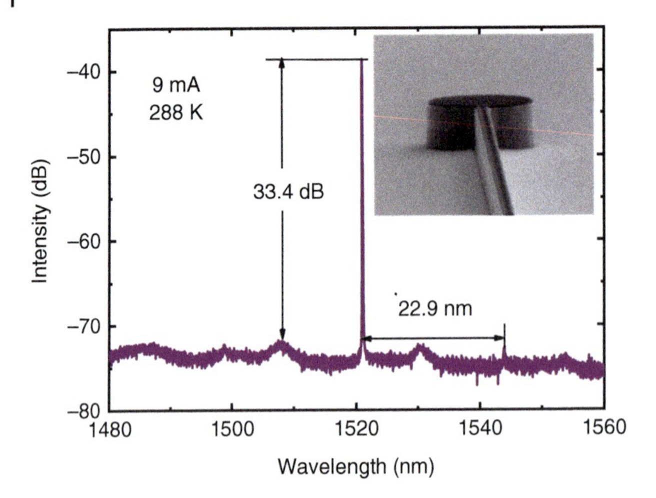

Figure 9.12 Lasing spectra of a 5-μm-radius microdisk laser at 288 K. The inset is a SEM image of a microdisk resonator after dry etching.

and the vertical propagation mode, as the wavelength of TE WG modes are shorter than the corresponding propagation mode. Therefore, the vertical radiation loss limits the sizes of the microcavity lasers vertically confined by semiconductor material. The radii of the experimentally demonstrated microcylinder lasers are smaller than that shown in Table 9.3, because the laser wafer has a gain peak around 1450 nm. To further reduce the laser size, it should find an effective way to reduce vertical radiation loss.

Finally, lasing spectra of an InAlGaAs/InP microdisk laser with a radius of 5 μm is presented in Figure 9.12, where the inset is a SEM image of a microdisk connected with a 1-μm-wide output waveguide. Continuous wave lasing operation is realized with a threshold current of 4 mA at 298 K. Single mode operation with a side mode suppression ratio of 33.4 dB is realized with dominant mode wavelength of 1520.9 nm and a longitudinal mode wavelength interval of 22.9 nm as shown in Figure 9.12. Between the adjacent longitudinal modes, non-lasing high order modes with relatively large linewidths could also be recognized around 1508.1 and 1530.5 nm. For the microdisk connected with a 1-μm-wide output waveguide, the threshold current decreases with the decrease of the radius of the microdisk from 10 to 7 μm, but increases as the radius down to 5 μm.

9.3.3 Cancelation of Vertical Radiation Loss by Destructive Interference

A microcylinder with the asymmetric vertical refractive index distribution of $n_3/n_2/n_1/n_2 = 1/3.17/3.4/3.17$, $d = 0.2$ μm, and $R = 1$ μm is considered as shown in Figure 9.4. Here the refractive indices n_3 and n_4 are set to 1. The mode Q factor of the $TE_{9,1}$ mode vs. the thickness of the upper cladding layer d_2 is calculated and plotted in Figure 9.13. The dotted line at $Q = 550$ is the mode Q factor of $TE_{9,1}$ mode in the microcylinder with an infinite upper cladding layer. A strong oscillation of

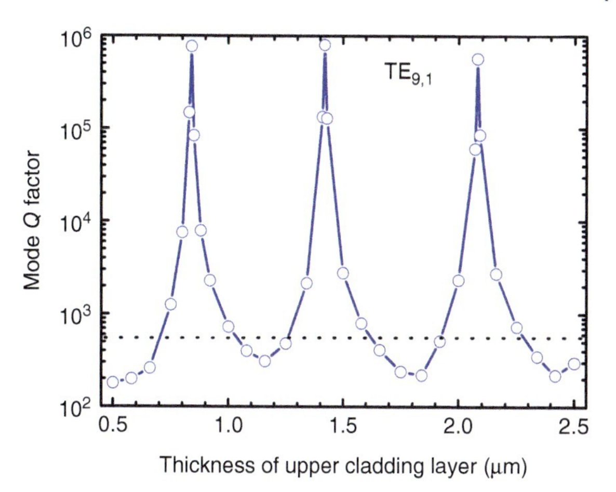

Figure 9.13 Mode Q factor of $TE_{9,1}$ mode vs. the thickness of upper cladding layer d_1 in the microcylinder with $R = 1$ obtained by FDTD simulation. The dashed line is the mode Q factor for the microcylinder resonator with an infinite upper cladding layer. Source: Yang et al. [15]. © 2010, Optical Society of America.

the Q factor is found with the increase of the thickness of upper cladding layer, and the peak values of 7.69×10^5, 8.02×10^5, and 5.70×10^5 are obtained at $d_1 = 0.84$, 1.42, and 2.08 μm, respectively. The peak values of the Q factor are close to the Q factor of 9.87×10^5 obtained by the S-matrix method [10], which neglects the vertical radiation loss. The results indicate that the vertical radiation loss almost vanishes for the $TE_{9,1}$ mode at $d_2 = 0.84$, 1.42, and 2.08 μm.

Using a long optical pulse with a narrow bandwidth to excite only one mode, the field distribution can be obtained with an impulse at $t_w = 10^4\Delta t$, $t_0 = 3t_w$, and $f = 195$ THz. Figure 9.14a,b depict the z-directional electric field E_z, and Figure 9.14c,d depicts the z-directional magnetic field H_z, for $TE_{9,1}$ mode at $d_1 = 1.42$ and 1.50 μm, respectively. The component E_z is confined well in the core and upper cladding layer as $d_1 = 1.42$ μm, but oscillates in the lower cladding layer as $d_1 = 1.50$ μm corresponding to a vertical loss.

Because the Q factor of $TE_{9,1}$ mode in the microcylinder with an infinite upper cladding layer is only 550, it can be expected that the enhancement of mode Q factor is caused by the reflection from the upper boundary of the upper cladding layer. In the microcylinder resonator, the TE WGM can couple to the HE vertical propagation mode of the upper and lower cladding layers when the mode wavelength of the TE WGM is smaller than the cut-off wavelength of the HE mode, which results in a vertical radiation loss. The energy couples from the TE WGM to the HE modes in both the upper and lower cladding layers, and propagates in the z and the $-z$ directions. The leaked HE mode in the upper cladding layer will be reflected by the upper boundary of the upper cladding layer, and then propagates through the upper cladding layer and the core layer to the lower cladding layer. Then two leaked HE modes will interfere with each other destructively or constructively, so as to decrease or increase the vertical radiation loss through the lower cladding layer. The propagation constant of the HE mode can be obtained by solving the eigen-equation of the circular waveguide [11] at a wavelength of 1.5315 μm, which is the WG mode wavelength of $TE_{9,1}$. Because E_z and H_z have a difference of π in the reflection phase shift, we focus on one component of the electromagnetic fields. If all the phase shifts are

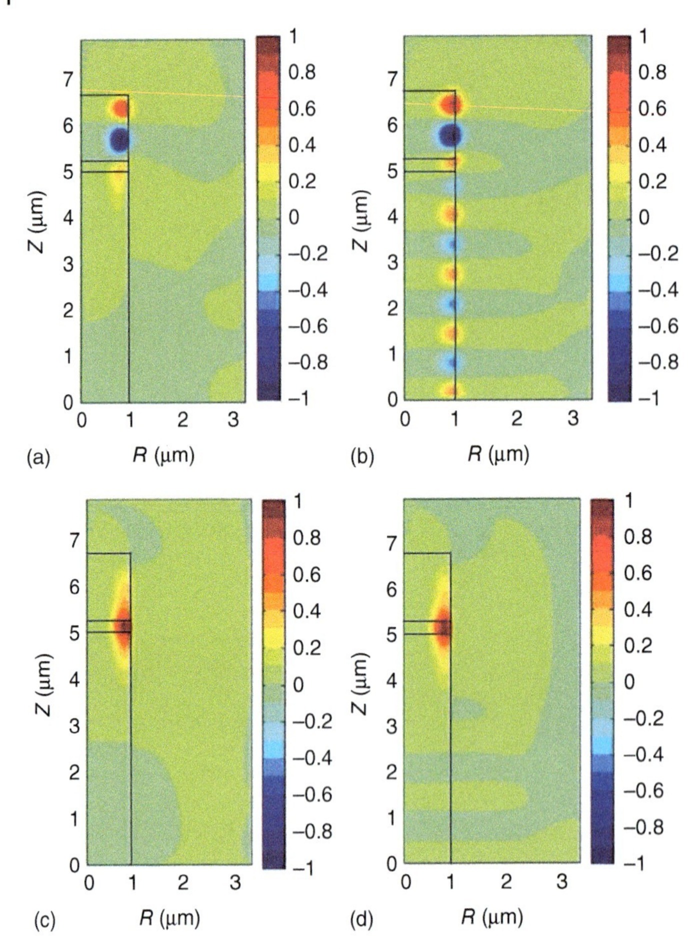

Figure 9.14 Field distributions of E_z for $\mathrm{TE}_{9,1}$ mode at the upper cladding layer thickness d_1 of (a) 1.42 μm and (b) 1.50 μm, and H_z at d_1 of (c) 1.42 μm and (d) 1.50 μm. Source: Yang et al. [15]. © 2010, Optical Society of America.

for the E_z component, the phase difference between the two HE modes in the lower cladding layer can be calculated by

$$\Phi = \phi_1 + 2\beta_2 d_1 + \beta_1 d + \phi_2 \tag{9.28}$$

where β_2 and β_1 are the propagation constants of HE modes in the cladding layer and the core layer, ϕ_1 is the phase difference between the excited HE modes at the upper and lower boundaries of the core layer from the TE WGM, and ϕ_2 is the phase shift of the HE mode reflected on the upper boundary of upper cladding layer. In the microcylinder with an infinite upper cladding layer, the component E_z of the TE WGM is anti-symmetric relative to $z = 0$ plane, which means $\phi_1 = \pi$. In fact, this is also a good approximation for the microcylinder with the thick upper cladding layer, and is suitable for our simulation when $d_2 > 0.5$ μm. Because the HE mode has the dominant component (E_z, H_r, H_φ) when the vertical propagation constant is small,

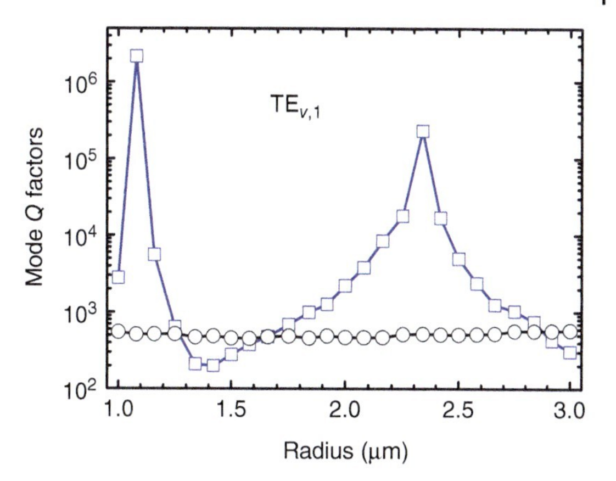

Figure 9.15 The Q factors of modes $TE_{m,1}$ with the wavelengths near 1.55 μm vs. the radius of microcylinder resonator obtained by FDTD simulation. The squares and circles are for the resonator with $d_1 = 1.5$ μm and an infinite upper cladding layer. Source: Yang et al. [15]. © 2010, Optical Society of America.

the TM approximation is used to calculate the phase shift ϕ_2 at the incident angle of $\arccos(\beta_2/n_2k_0)$ similar to a slab waveguide, where k_0 is the wavenumber in vacuum. Then the destructive interference condition can be obtained from $\Phi = (2n+1)\pi$ as $d_1 = (0.64n + 0.17)$ μm. When $n = 1, 2, 3$, it gives $d_1 = 0.81$, 1.45, and 2.09 μm, which are in good agreement with the FDTD results in Figure 9.13.

In the microcylinder resonator with the constant thickness $d_1 = 1.5$ μm, the mode Q factors of $TE_{v,1}$ modes vs. the radius R are calculated and plotted in Figure 9.15 as open squares. The Q factors of the same modes in the microcylinder resonator with infinite upper cladding layer are plotted as open circles. The mode wavelengths are chosen near 1.55 μm with the corresponding angular mode number v increases from 9 to 33 with a step of 1, as R increases from 1 to 3 μm. Two peak values appear at $R = 1.08$ and 2.34 μm, the corresponding angular mode numbers are 10 and 25, and the Q factors are 2.16×10^6 and 2.33×10^5, respectively. The enhancement of Q factors is caused by the destructive interference similar to the microcylinder resonator with varying thickness of the upper cladding layer. The phase difference Φ in Eq. (9.28) is found to be 4.90π and 2.85π at $R = 1.08$ and 2.34 μm, respectively, which correspond to $n = 2$ and 1. However, they are slightly smaller than $(2n+1)\pi$ because some approximations are used in the calculation of the reflection phase shift.

For the first radial order mode $TE_{v,1}$, the vertical leakage can be totally canceled because only one propagation leaky mode exists, but high radial order modes will have different characteristics. For a microcylinder resonator with $R = 1.5$ μm, the mode Q factor of $TE_{11,2}$ vs. the thickness of upper cladding layer d_1 is calculated and plotted in Figure 9.16. Oscillation of the Q factor is also found with an increase of the upper cladding layer thickness, two peak values of 3.70×10^3 and 2.37×10^3 are found at $d_1 = 1.06$ and 1.88 μm, respectively. The magnitudes of the Q factors are two orders smaller than the value of 2.96×10^5 obtained by the S-matrix method which neglects the vertical radiation loss. The result indicates that the vertical radiation loss does not vanish. The reason is that the $TE_{11,2}$ mode can couple to three propagation modes $HE_{11,2}$, $HE_{11,1}$, and $EH_{11,1}$. The destructive interference condition for the $HE_{11,2}$ can be obtained as $d_1 = 0.81n + 0.24$ at the mode wavelength 1.5609 μm of $TE_{11,2}$. When $n = 1$ and 2, it gives $d_1 = 1.05$ and 1.86 μm, which agree well with the FDTD results. However, the destructive interference cannot be realized at the same

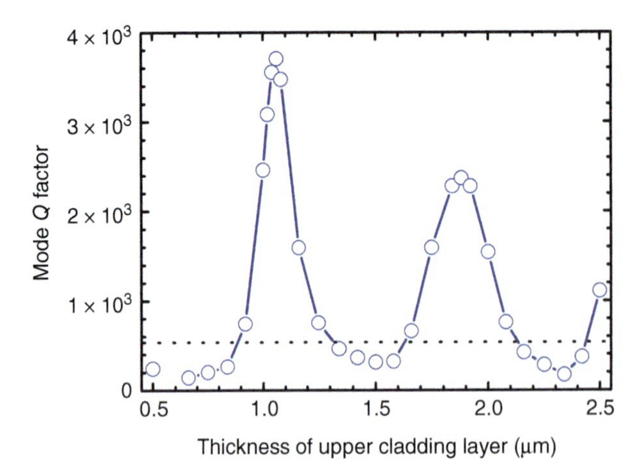

Figure 9.16 The mode Q factor of $TE_{11,2}$ mode vs. the thickness of upper cladding layer in the microcylinder with $R = 1.5\ \mu m$ obtained by the FDTD simulation. The dashed line is the mode Q factor for the microcylinder resonator with an infinite upper cladding layer. Source: Yang et al. [15]. © 2010, Optical Society of America.

value of d_1 for the three propagation modes, and the vertical radiation loss does not vanish totally. Because the coupling between the modes with the same radial mode number is 1 order in magnitude larger than that between the modes with different radial mode numbers, the Q factor is mainly determined by the coupling with the $HE_{11,2}$ mode.

The above discussions mainly focus on the influence of the thickness of the upper cladding layer for vertical radiation loss cancelation. In fact, changing the thickness of the center layer can also results in the cancelation of vertical radiation loss. In a uniform microcylinder waveguide, the HE and EH modes transform to pure TM and TE WGMs with three electromagnetic components (E_z, H_r, H_φ) and (H_z, E_r, E_φ) when $\beta = 0$. Although the HE and EH modes with nonzero β usually have all six electromagnetic components, the dominant components are (E_z, H_r, H_φ) and (H_z, E_r, E_φ) for the HE and EH modes, respectively, as β is small. The incident angle θ of the HE or EH mode light rays on the boundaries between the core and cladding layers equals to arccos $(\beta_1 c/\omega n_1)$, where β_1 is the vertical propagating constant of the HE or EH modes in the core layer. When the incident angle θ is larger than the critical angle of the totally internal reflection $\arcsin(n_2/n_1)$, the mode light ray can be totally reflected. The phase shifts for HE and EH modes are defined as that of the electromagnetic components E_z and H_z, respectively. The propagating modes can be confined in the core layer as the WGMs when the following resonant condition is satisfied

$$2\beta_1 d + \phi_{\text{up}} + \phi_{\text{low}} = 2(l + 2)\pi \tag{9.29}$$

where l is the vertical mode number, and ϕ_{up} and ϕ_{low} are reflection phase shifts on the upper and lower boundaries of the core layer, respectively. The reflection phase shifts are defined in the range from 0 to 2π, and can be calculated under TE (TM) approximation for the EH (HE) modes. The TE and TM WGMs in the microcylinder with thick core layer are marked as $TE_{l,v,m}$ and $TM_{l,v,m}$, where l, v and m are the vertical, angular and radial mode numbers, respectively. In the following simulation, the mode number is omitted for the fundamental vertical modes as $l = 0$.

Based on the dominant electromagnetic field components, the WGMs can be marked as TM and TE modes corresponding to the HE and EH propagating modes in the core layer as shown in Figure 9.17. Carefully viewing the field behavior at the boundaries between the core and cladding layers, it can be found that a EH or HE propagating mode incident on the boundary will produce not only a reflected and a transmitted mode of its own type, but also many transformed modes. As a result of the coupling between different modes at the discontinuity boundary, all the modes in the core and cladding layers will be generated, including propagating and decaying mode away from the boundary. It should be emphasized that the mode coupling at the boundary is not caused by any surface or sidewall roughness. Because the HE/EH modes in a microcylinder waveguide are much more complicated than the TE/TM modes in a ridge waveguide and the WGMs have both horizontal and vertical losses, FDTD method is used to simulate the mode characteristics for the WGMs in the microcylinder resonators.

At the boundaries between the core and cladding layers, many modes will be generated, but only the propagating modes in the cladding layer are considered as the vertical leakage loss in the analysis. At first, only the mode coupling between the same-radial-order modes is considered, which is usually significantly larger than that between different-radial-order modes. As shown in Figure 9.17, the light ray is plotted along the angular direction for the propagating modes in the core and cladding layers. In the case of the TE WGM, the EH generated field will be evanescent in the cladding layer due to the total internal reflection, but the transformed HE mode at the boundary can propagate in the cladding layer, and results in vertical leakage loss. The leakage process is illustrated in Figure 9.17. Starting at the lower boundary, a EH mode is guided by the microylinder waveguide as the solid ray incident on the lower boundary, where the angle of incidence is larger than the total internal reflection angle, thus total internal reflection occurs for the EH mode. However, due to the step discontinuity in the z direction, mode coupling between EH and HE modes generates additional small transmitted and reflected HE propagating modes as the dashed rays 1 and 2. The reflected HE mode light ray 2 traverses across the core layer and propagates in the upper cladding layer.

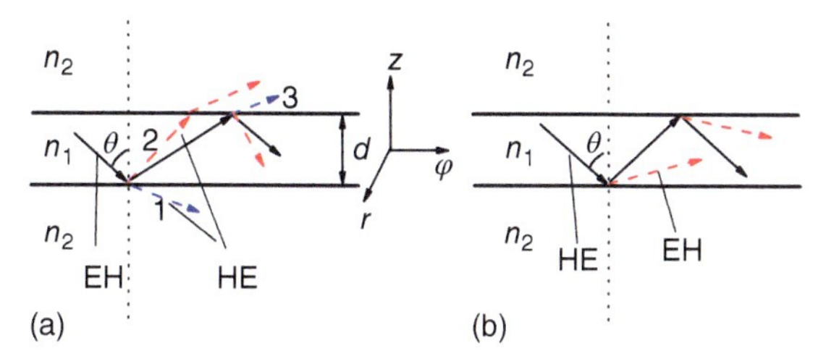

Figure 9.17 The schematic diagram of EH and HE mode light rays along the angular direction for (a) TE and (b) TM WGMs. The coupling between EH and HE modes at the discontinuity boundaries is shown. Source: Yang and Huang [16]. © 2011, IEEE.

At the upper boundary, the EH mode generates additional small transmitted and reflected HE propagating modes. Then the interference between light rays 2 and 3 can happen in the upper cladding layer. The phase difference between two mode light rays 2 and 3 can be written as

$$\Phi = \beta_{1\mathrm{HE}} d + \phi_1 + \phi_2 \tag{9.30}$$

where $\beta_{1\mathrm{HE}}$ is the propagating constant of HE mode (light ray 2) in the core layer, ϕ_1 is the phase difference between the reflected and transmitted HE mode light rays 2 and 1 at the lower boundary, and ϕ_2 is the phase difference between two transmitted HE mode light rays 1 and 3 at the boundaries and equals to $(l+1)\pi$ considering the different dominant fields of HE and EH modes. The interference between the two HE leakage modes can enhance or reduce the leakage loss, and the maximum and minimum leakage losses correspond to Φ equals to $2n\pi$ and $(2n+1)\pi$, respectively. In addition, the first-radial-order TE WGMs have no vertical leakage loss in the microcylinder resonator as radius larger than 5 μm, because the transmitted HE mode becomes decaying mode in the cladding layer.

For the vertically symmetric microcylinder resonator as shown in Figure 9.4 with $n_1 = 3.4$ and $n_3 = n_2 = 3.17$, the mode wavelength is calculated from Eq. (9.29) under the TE approximation, and the phase difference Φ can obtained from Eq. (9.30) by calculating the propagating constant of $\mathrm{HE}_{9,1}$ mode in the core layer and assuming $\phi_1 = 0$. The analytical mode wavelength and phase difference vs. the thickness of core layer are plotted in Figure 9.18. The destructive interference can be obtained as $d = 1.08$ and 2.27 μm at $\Phi = (2n+1)\pi$ with $n = 1$ and 2, and the constructive interference can be obtained as $d = 0.5$ and 1.68 μm at $\Phi = 2n\pi$ with $n = 1$ and 2, respectively.

For the case of TM WGM as shown in the right side of Figure 9.17, the HE mode field will be evanescent in the cladding layer due to the total internal reflection. The transformed EH mode usually has smaller vertical propagating constant than HE mode, and then also cannot propagate in the cladding layer. Thus the mode coupling at the boundaries does not lead to a leakage loss since both the generated HE and

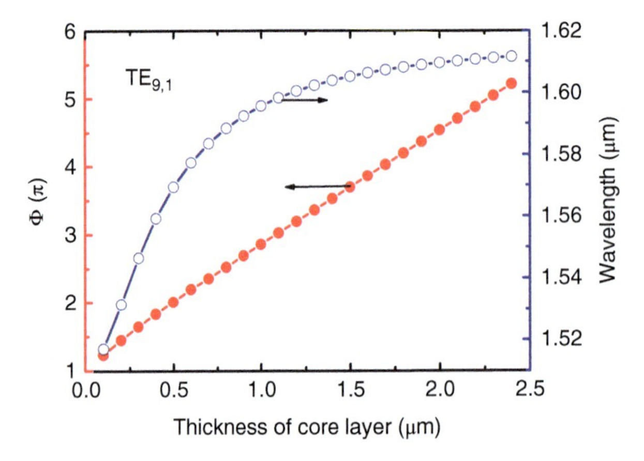

Figure 9.18 The analytical phase difference and mode wavelength vs. the thickness of the core layer. Source: Yang and Huang [16]. © 2011, IEEE.

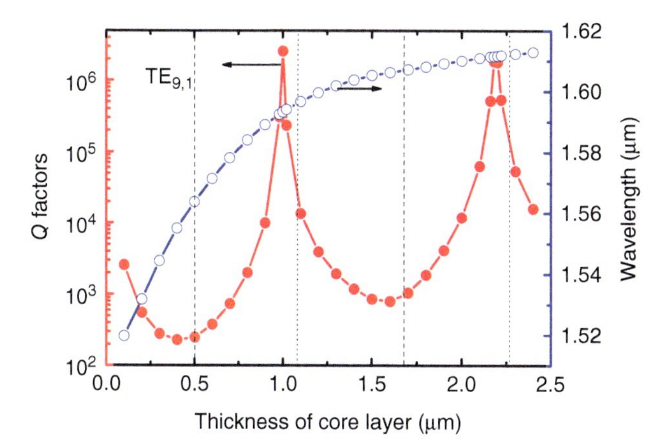

Figure 9.19 Mode Q factors and wavelengths of $TE_{9,1}$ mode vs. the thickness of the core layer in the microcylinder with $R = 1$ μm. Source: Yang and Huang [16]. © 2011, IEEE.

EH modes decaying away from the boundaries in the cladding layers. And both the HE and EH mode are confined in the core layer.

For the high-radial-order WGMs, the WGMs can couple to the lower-radial-order propagating modes in the cladding layer, and results in vertical leakage loss for both TE and TM WGMs. Because both the lower-radial-order HE and EH modes can propagate in the cladding layer, the vertical leakage loss usually induced by multiple leakage modes.

The mode Q factor and wavelength vs. the thickness of the core layer d is calculated and plotted in Figure 9.19 for the $TE_{9,1}$ mode by the FDTD method and the Padé approximation. The mode wavelength increases with the increase of the thickness of core layer, according to the increase of the efficient refractive index. A strong oscillation of the Q factor is found with the increase of the thickness of the core layer, and the peak values of 2.60×10^6 and 1.85×10^6 are obtained at $d = 1$ and 2.18 μm, respectively. The maximum value of Q factor is still limited by the grid size in the vertical direction, which is 10 nm in the simulation, and the destructive interference cannot be exactly fulfilled. However, more than 3 orders of magnitude increase in Q factors indicates that the leakage loss is almost totally canceled at the destructive interference condition, and the generated reflected and transmitted HE modes have similar amplitude. The valley values of Q factors are 230 and 800 at $d = 0.4$ and 1.6 μm, respectively. One possible reason for the increase of the valley Q values is that the vertical propagating constant decreases as the mode wavelength increases, and then the HE/EH mode coupling become weak. The analytical constructive and destructive interference thicknesses obtained from Figure 9.18 are marked by vertically dashed and dotted lines in Figure 9.19, which are about 0.08–0.1 μm larger than that obtained by the FDTD simulation. $\phi_1 = 0$ is used in Figure 9.18 under TE approximation, which may cause the difference, because the transmitted and reflected HE modes have a small phase difference.

Figure 9.20a,c depict the z-directional simulated magnetic field H_z, and Figure 9.20b,d depict the z-directional simulated electric field E_z, for $TE_{9,1}$ mode at

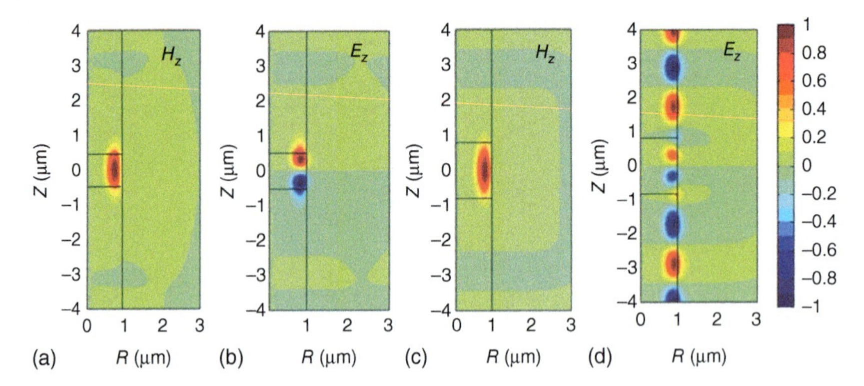

Figure 9.20 Field distributions of H_z for $TE_{9,1}$ mode at core layer thickness of (a) 1 μm and (c) 1.60 μm, and E_z at core layer thickness of (b) 1 μm and (d) 1.60 μm in the microcylinder with $R = 1$ μm. Source: Yang and Huang [16]. © 2011, IEEE.

$d = 1$ and 1.60 μm, respectively. The TE WGMs with even vertical mode number l have symmetric field component H_z and antisymmetric component E_z relative to the middle plane at $z = 0$. The component E_z is confined well in the core layer at $d = 1$ μm corresponding to the high Q factor in Figure 9.18, but oscillates in the cladding layers at $d = 1.60$ μm corresponding to the low Q factor with vertical leakage loss.

The vertical first order modes can be supported as the thickness of core layer d larger than the cut-off thickness $\lambda/2\sqrt{n_1^2 - n_2^2}$, which is about 0.63 μm for the mode wavelength at 1.55 μm. The mode Q factor of the first-vertical-order mode $TE_{1,9,1}$ oscillates with the increase of the thickness of core layer, similar to the fundamental vertical mode. For the high radial-order modes, the mode coupling between different-radial-order modes should be considered, which leads a vertical leakage loss for both TE and TM modes. The Q factors of high radial-order WGMs also oscillate with the increase of the thickness of core layer due to the interference of the vertical radiation modes.

The cancelation of vertical radiation loss predicted by numerical simulation is not observed in the experiment, the reason is the absorption of the top InGaAs and metal layers and fabrication induced additional losses. To further decrease the device size with low threshold current density, a carful design on the vertical structure to suppress the vertical radiation loss is required for the AlGaInAs/InP microcylinder lasers.

9.4 Verical Radiation Loss for Polygonal Microcavities

In Section 9.3, the vertical radiation properties of the WGMs in 3D circular microcavities are studied. The coupling between the WGMs with different angular mode numbers in a circular microcavity is forbidden owing to its circular rotational symmetry. Differently, the WGMs in a polygonal microcavity will have multiple angular components, and mode coupling can happen between the WGMs with the same

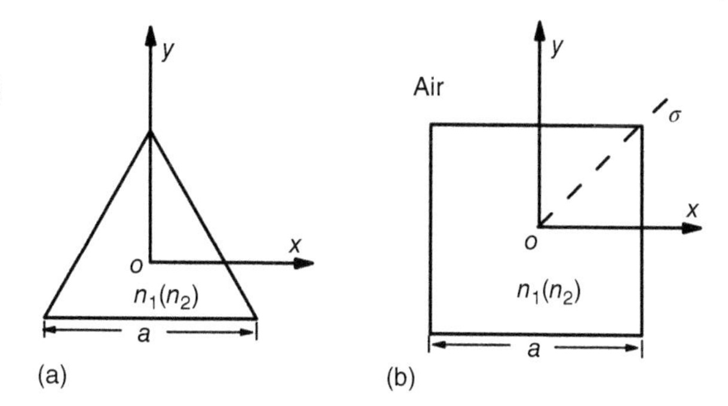

Figure 9.21 Schematic diagrams of (a) equilateral-triangular and (b) square microcavities with the side length a laterally confined by air. Source: Huang and Yang [13]. © 2008, IEEE.

symmetry [17, 18]. In addition, the 3D FDTD simulation in cylindrical coordinate can be performed as a 2D FDTD calculation through the transformation of the angular direction, so the calculation time can be greatly reduced. However, the WGMs in a polygonal microcavity should be simulated by a real 3D FDTD method. The required computing power will be greatly increased compared to the circular microcavity. In this section, the vertical radiation loss of WGMs in 3D semiconductor confined equilateral-triangular and square microcavities are studied.

The schematic diagrams of equilateral-triangular and square microcavities are shown in Figure 9.21, where the vertical waveguide structures are similar to the microcylinder shown in Figure 9.4 with $n_1 = 3.4$, $n_3 = n_2 = 3.17$, and $d = 0.2\,\mu\text{m}$, which only support the fundamental mode in the vertical direction. The WGMs in the 3D equilateral-triangular and square microcavities vertically confined by semiconductor material are simulated by full 3D FDTD method.

9.4.1 3D Equilateral-Triangular Microcavity with Weak Vertical Waveguiding

Since the vertical waveguide only support the fundamental mode, the confined TE and TM WGMs in 3D equilateral-triangular microcavities are marked as $TE_{m,l}$ and $TM_{m,l}$ ignoring the vertical mode index with the same definition as in Chapter 6, where m and l are the transverse and longitudinal mode numbers, respectively. The field distributions of TE (TM) modes can be expressed as symmetric and anti-symmetric magnetic (electric) field distributions relative to $x = 0$ plane. The symmetric and anti-symmetric modes marked by superscripts "e" and "o" are accidentally degenerate modes with different Q-factors as l is a multiple of 3. The intensity spectra obtained by 3D FDTD simulation and Padé approximation under different symmetric condition relative to $x = 0$ plane are shown in Figure 9.22a,b for TE and TM modes, respectively.

The first-order transverse modes are observed between two fundamental modes for the TM modes in Figure 9.22b. However, the peak of the first-order TE-like mode does not appear in Figure 9.22a because the Q-factors of $TE_{1,23}$ and $TE_{1,21}$ are too small, which are less than 100 obtained by 2D FDTD under the EIM. In Table 9.4, the mode Q-factors and wavelengths obtained by the 3D FDTD simulation are compared with those of 2D FDTD simulation under effective index approximation for

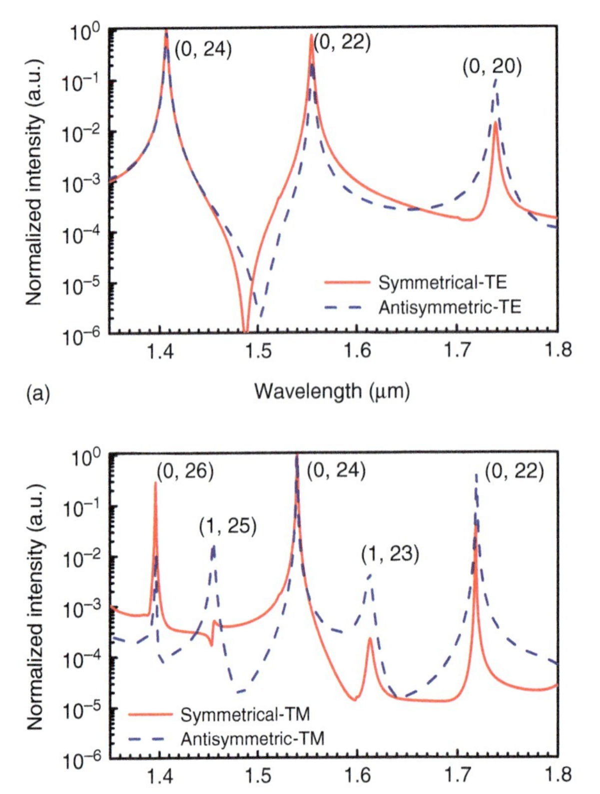

Figure 9.22 Intensity spectra obtained by 3D FDTD simulation and Padé approximation for the 3D equilateral-triangular microcavity with $a = 3\,\mu\text{m}$ under symmetric and anti-symmetric conditions relative to the $x = 0$ plane for (a) TE and (b) TM modes. Source: Huang and Yang [13]. © 2008, IEEE.

the confined TE and TM modes. For the corresponding slab waveguide with the thickness of 0.2 μm, the effective index of 3.215 and 3.207 obtained at the wavelength 1.55 μm are used in the 2D FDTD simulation for TE and TM modes, respectively. The Q-factors of 3D FDTD simulation are smaller than those of the 2D FDTD simulation, because the vertical radiation loss is not considered in the 2D FDTD simulation. Assuming $1/Q_{3D} = 1/Q_{2D} + 1/Q_l$, where Q_{3D} and Q_{2D} are Q-factors obtained by the 3D and 2D FDTD simulations, Q_l determined by the vertical radiation loss is also listed in Table 9.4. Q_l of TM modes is much larger than that of TE modes, because the TM modes have smaller vertical loss than TE-like modes in the equilateral triangular microcavity with the vertical optical confinement of semiconductors.

9.4.2 3D Square Microcavity with Weak Vertical Waveguiding

A 3D square microcavity with the side length $a = 2\,\mu\text{m}$ and vertically confined by semiconductor material is also considered. Ignoring the vertical mode index, the TE and TM modes are marked as $\text{TE}_{p,q}$ and $\text{TM}_{p,q}$ similar to 2D square resonators in Chapter 7, where the mode numbers p and q denote the number of wave nodes in the x and y directions, respectively. The high Q-factor modes in the microsquare are with an even number of $p - q$, which have antisymmetric magnetic (electric) field about the diagonal mirror plane σ of the square for TE (TM) WG-like modes.

Table 9.4 Mode Q-factors and wavelengths for the equilateral-triangular microcavity with $a = 3\,\mu m$ obtained by 3D FDTD simulation and 2D FDTD simulation under EIM.

		$TE_{0,22}$	$TE^o_{0,24}$	$TE^e_{0,24}$	$TM_{0,22}$	$TM^o_{0,24}$	$TM^e_{0,24}$
3D FDTD	Q_{3D}	6.2×10^2	5.9×10^2	5.6×10^2	1.7×10^3	2.1×10^3	1.4×10^3
	λ (μm)	1.556	1.408	1.408	1.719	1.540	1.541
2D FDTD	Q_{2D}	1.6×10^3	5.7×10^3	3.1×10^3	2.8×10^3	5.5×10^3	2.5×10^3
	λ (μm)	1.549	1.399	1.399	1.718	1.539	1.539
	Q_l	1.0×10^3	6.6×10^2	6.8×10^2	4.3×10^3	3.4×10^3	3.2×10^3

Table 9.5 Mode Q-factors and wavelengths for the square microcavity with $a = 2\mu m$ obtained by 3D FDTD simulation and 2D FDTD simulation under the EI approximation.

		$TE^o_{4,6}$	$TE^o_{5,7}$	$TM^o_{4,6}$	$TM^o_{5,7}$
3D FDTD	Q_{3D}	4.8×10^2	3.8×10^2	4.1×10^3	6.5×10^3
	λ (μm)	1.529	1.318	1.657	1.410
2D FDTD	Q_{2D}	3.1×10^3	1.6×10^4	6.0×10^3	2.3×10^4
	λ (μm)	1.513	1.304	1.636	1.392
	Q_l	5.7×10^2	3.9×10^2	1.3×10^4	8.9×10^3

As most of application only care about the high Q modes, only the modes with the same parity of p and q are considered in the following simulation, that is, the modes with the same symmetry about the $x = 0$ and $y = 0$ planes. The intensity spectra obtained by 3D FDTD simulation and Padé approximation with different symmetry conditions are shown in Figure 9.23a,b for TE and TM modes, respectively. The solid and dashed lines correspond to the modes with the mode numbers p and q are both odd and even, respectively. It can be found that the FWHMs of the TM modes are significantly smaller than that of the TE modes, that is, the Q factor of the TM modes are much larger than that of the TE modes in 3D square microcavity with weak vertical confinement.

Mode Q-factors and wavelengths obtained by 3D FDTD simulation and 2D FDTD simulation under the effective index approximation are listed in Table 9.5 for WG modes in the square microcavity. 3D FDTD simulation also gives smaller Q-factors than 2D FDTD simulation. Similar to equilateral triangular microcavity, Q_l of TM modes is much larger than that of TE modes in the square microcavity too. The results show that TM modes have less vertical radiation loss than TE modes in the square microcavity.

9.5 Summary

In this chapter, the effective index method, S-matrix and FDTD method are introduced for the investigation of 3D microcavities. The effective index method can

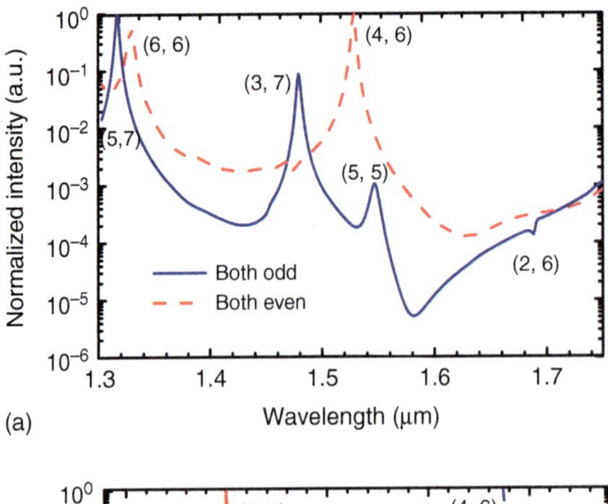

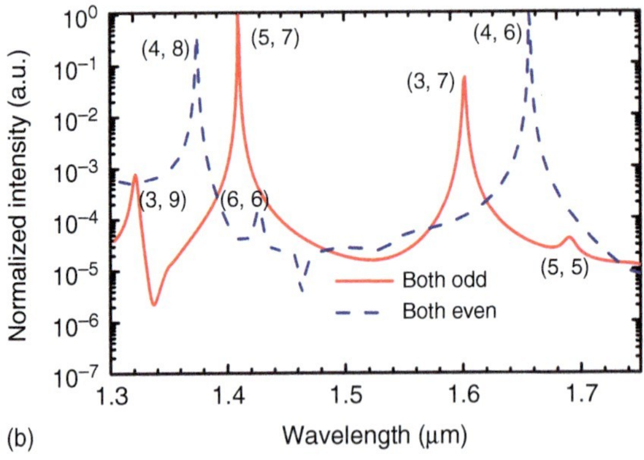

Figure 9.23 Intensity spectra of (a) TE modes and (b) TM modes obtained by the 3-D FDTD simulation and Padé approximation under different symmetric conditions relative to $x = 0$ and $y = 0$ planes. Source: Huang and Yang [13]. © 2008, IEEE.

give the mode wavelength accurately and quickly, and the 3D S-matrix method can determine both the mode wavelength and quality factor under the conditions of the absent of vertical radiation loss.

The 3D FDTD method was used to simulate the change of mode characteristics in a 3D cylindrical microcavity. It was found that in a cylindrical microcavity with a weakly constrained structure, the TM mode has no longitudinal coupling loss at all, so it can maintain a high quality factor in a weakly confined structure. Therefore, TM mode can be used to develop simple microcavity filters, electrical injection cylindrical semiconductor lasers, and even single photon radiation sources. The TE mode requires a strong vertical confinement structure or a larger size to ensure that its wavelength is greater than the cut-off wavelength of the guided wave mode in the substrate to obtain a high quality factor. However, the vertical radiation loss of TE modes can be canceled by design the cavity structures, such as thicknesses of the upper cladding layer and the core layer, the radius.

The 3D FDTD method was used to simulate the triangular and square microcavities. It was found that TM modes also have vertical radiation loss in the triangular and square microcavities, because modes in triangles and squares can couple with other modes with the same angular components. However, in triangles and squares, the vertical radiation loss of the TM mode is still much smaller than that of the TE mode.

References

1 Vahala, K.J. (2003). Optical microcavities. *Nature* 424: 839–846.

2 McCall, S.L., Levi, A.F.J., Slusher, R.E. et al. (1992). Whispering-gallery mode microdisk lasers. *Appl. Phys. Lett.* 60: 289–291.

3 Levi, A.F.J., McCall, S.L., Pearton, S.J., and Logan, R.A. (1993). Room temperature operation of submicrometer radius disk laser. *Electron. Lett.* 29: 1666–1667.

4 Arzberger, M., Böhm, G., Amann, M.-C., and Abstreiter, G. (2001). Continuous room-temperature operation of electrically pumped quantum dot microcylinder lasers. *Appl. Phys. Lett.* 79: 1766–1768.

5 Chern, G.D., Tureci, H.E., Stone, A.D. et al. (2003). Unidirectional lasing from InGaN multiple-quantum-well spiral-shaped micropillars. *Appl. Phys. Lett.* 83: 1710–1712.

6 Frateschi, N.C. and Levi, A.F.J. (1996). The spectrum of microdisk lasers. *J. Appl. Phys.* 80: 644–653.

7 Hentschel, M. and Richter, K. (2002). Quantum chaos in optical systems: the annular billiard. *Phys. Rev. E* 66: 056207.

8 Laeri, F. and Nöckel, J.U. (2001). Nanoporous compound materials for optical applications – microlasers and microresonators. In: *Handbook of Advanced Electronic and Photonic Materials*, vol. 6 (ed. H.S. Nalwa), 58–90. Academic Press.

9 Rahachou, A.I. and Zozoulenko, I.V. (2004). Scattering matrix approach to the resonant states and *Q* values of microdisk lasing cavities. *Appl. Opt.* 43: 1761–1772.

10 Luo, X.S., Huang, Y.Z., Guo, W.H. et al. (2006). Investigation of mode characteristics for microdisk resonators by S-matrix and three-dimensional finite-difference time-domain technique. *J. Opt. Soc. Am. B: Opt. Phys.* 23: 1068–1073.

11 Marcuse, D. (1982). *Light Transmission Optics*, 2e, 290–294. Van Nostrand Reinhold.

12 Yang, Y.D., Huang, Y.Z., and Chen, Q. (2007). High-Q TM whispering-gallery modes in three-dimensional microcylinders. *Phys. Rev. A* 75: 013817.

13 Huang, Y.Z. and Yang, Y.D. (2008). Mode coupling and vertical radiation loss for whispering-gallery modes in 3-D microcavities. *J. Lightwave Technol.* 26: 1411–1416.

14 Yang, Y.D., Xiao, J.L., Liu, B.W., and Huang, Y.Z. (2015). Mode characteristics and vertical radiation loss for AlGaInAs/InP microcylinder lasers. *J. Opt. Soc. Am. B: Opt. Phys.* 32: 439–444.

15 Yang, Y.D., Huang, Y.Z., Guo, W.H. et al. (2010). Enhancement of quality factor for TE whispering-gallery modes in microcylinder resonators. *Opt. Express* 18: 13057–13062.

16 Yang, Y.D. and Huang, Y.Z. (2011). Investigation of vertical leakage loss for whispering-gallery modes in microcylinder resonators. *J. Lightwave Technol.* 29: 2754–2760.

17 Yang, Y.D. and Huang, Y.Z. (2007). Symmetry analysis and numerical simulation of mode characteristics for equilateral-polygonal optical microresonators. *Phys. Rev. A* 76: 023822.

18 Yang, Y.D., Huang, Y.Z., and Chen, Q. (2007). Comparison of Q-factors between TE and TM modes in 3D microsquares by FDTD simulation. *IEEE Photonics Technol. Lett.* 19: 1831–1833.

10

Nonlinear Dynamics for Microcavity Lasers

10.1 Introduction

Nonlinear dynamics of semiconductor lasers subject to optical injections or optical feedback have been widely investigated for potential applications, such as processing of microwave signals through photon technology and optical fiber transmission [1–6], secured communications [7, 8], and random bit generation [9–12], etc. Photonic microwave technology provides the processing of microwave signals through photon technology and optical fiber transmission, with unique advantages in low propagation loss and free of electromagnetic interference. Optical injection locking (OIL) [13], optical phase-lock loop [14], monolithic dual-wavelength lasers [15–17], and external or direct modulation on lasers [18, 19] were traditionally applied for photonic microwave generation. With the merit of wide tunable frequency range and not requiring electronic microwave sources, nonlinear dynamics in optically injected semiconductor lasers, especially period-one oscillation and four-wave mixing (FWM), have been investigated for photonic microwave generation in distributed feedback (DFB) lasers [20–24], vertical-cavity surface-emitting lasers [25], as well as quantum-dot DFB lasers [26]. An electro-absorption modulator integrated with a DFB laser subject to optical injection was used to generate microwave signal [27], monolithically integrated DFB lasers were experimentally demonstrated for tunable and narrow linewidth millimeter-wave generation [28], suppression of chaos was investigated in integrated twin-DFB lasers for millimeter-wave generation [29], continuous tuning radio frequency (RF) signals over seven octaves were realized based on a side-band-injection-locked laser [30], and 39-GHz millimeter-wave carrier was generated in dual-mode injection-locking of a directly encoded colorless laser diode [31]. In addition, similar as the improvement of the modulation bandwidth for semiconductor lasers by passive optical feedback [32], the modulation bandwidth can be greatly enhanced for semiconductor lasers under OIL state. The resonance-frequencies greater than 100 GHz was demonstrated for strong optical injection locked vertical-cavity surface-emitting lasers [33], enhanced modulation bandwidth was reported for a Fabry–Perot semiconductor laser subject to light injection from another laser [34], 40-Gb s^{-1} directly-modulated was realized for photonic crystal lasers under optical injection-locking [35], modulation characteristics enhancement was studied for monolithically integrated DFB lasers under mutual

Microcavity Semiconductor Lasers: Principles, Design, and Applications, First Edition.
Yong-zhen Huang and Yue-de Yang.

injection locking [36], and the circular polarization switching and polarization bistability were realized for optically injected spin-vertical cavity surface emitting laser [37]. Stable periodic dynamics in optically injected semiconductor lasers were investigated by mapping operating points, and limit-cycle oscillations insensitive to fluctuations were demonstrated [22, 38]. Nonlinear dynamic states of FWM, OIL, period-one and period-two oscillations were also demonstrated for AlGaInAs/InP microring and microdisk lasers under external optical injection, the enhancement of the 3 dB bandwidth due to OIL was realized for the microdisk lasers [39–41], and similar dynamic behaviors were observed for integrated twin-microdisk laser under mutually internal optical injection [42]. Furthermore, microwave signal was obtained from the electrode of square microlasers under optical injection, due to the oscillation of the carriers in the active region caused by the light beating between the lasing mode and the injection light [43].

The nonlinear dynamics of semiconductor lasers subject to optical injection have been extensively investigated by the numerical simulations of rate equations, using qualitative analyses such as perturbation, bifurcation, or multi-time scale analysis [44–47]. The simulated dynamical characteristics are in good agreement with experimental results related to the operating conditions, intrinsic laser parameters, and injection parameters [48, 49]. Perturbation analysis of the rate equations indicated that OIL can dramatically enhance the resonance frequency and modulation bandwidth [5, 50–52]. The complex dynamics for the optically injected laser were systematically analyzed by the bifurcation theory, including various routes and boundaries between stable locking to unlocking oscillations [53]. The important roles of the linewidth enhancement factor and the gain saturation factor on the nonlinear dynamics were proved for the optically injected semiconductor lasers [54, 55]. Analytical studies of the locking and unlocking boundary at negative detuning were shown a region of bistability resulting from competing attractors between locking and unlocking solutions for the coupled equations [56]. The nonlinear dynamical period-one oscillation was investigated analytically for microwave generation [57]. Rich nonlinear dynamics in integrated coupled lasers with short coupling delay were numerically simulated by solving coupled delay differential equations, showing excellent agreement with the experimental results [58].

With the merits of small footprint and high Q-factor, whispering-galley mode (WGM) microlasers are potential light sources for photonic integrated circuits and suitable for realizing optical injection inside the monolithically integrated devices. In this chapter, we summarize the achievement of nonlinear dynamics for microdisk lasers subject to optical injection. In Section 10.2, rate equations especially coupling rate coefficient are derived for microlasers under optical injection. In Section 10.3, dynamic behaviors are numerically investigated based on single mode rate equations with an optical injection term, and the modulation bandwidth improvement is studied based on small signal modulation approximation in Section 10.4. In Section 10.5, experimental results of the dynamical characteristics are reported for a microdisk laser subject to optical injection. In Section 10.6, optical injection behaviors are studied for a square microlaser under optical injection with microwave signal directly measured from the electrode of the microlaser. In

Section 10.7, dynamical characteristics are presented for a twin-microdisk laser with mutually optical injection via a connected waveguide, and discussion and conclusion are presented in Section 10.8.

10.2 Rate Equation Model with Optical Injection

In this section, a comprehensive rate equation is introduced to analyze the dynamical characteristics of an optically injected microlaser. The complex lasing mode field E with a frequency ω_0 and a phase ϕ is related to a mode photon density s [50]:

$$E(t) = E_0(t)e^{-i(\omega_0 t+\phi)} = \sqrt{\frac{\hbar\omega_0 s}{2\varepsilon_0 n_g^{\,2}}}e^{-i(\omega_0 t+\phi)}, \tag{10.1}$$

where $\hbar\omega_0$ is photon energy, ε_0 the permittivity in vacuum, and n_g the group index. With an injection field as $E_{\text{inj}}(t)e^{-i(\omega_{\text{inj}}t+\phi_{\text{inj}})}$, the complex lasing mode field amplitude satisfies following rate equation [44, 59, 60]:

$$\frac{dE_0e^{-i\phi}}{dt} = \frac{1}{2}(1+i\alpha)\left[\Gamma v_g g(n,s) - \alpha_i v_g - \frac{1}{\tau_{\text{pc}}}\right]E_0e^{-i\phi} + \kappa_c E_{\text{inj}}(t)e^{-i(\Delta\omega t+\phi_{\text{inj}})}, \tag{10.2}$$

where α is the linewidth enhancement factor related to the variation of complex refractive index with carrier density, $v_g = c/n_g$ is the group velocity of the lasing mode, Γ is the optical confinement factor indicating the spatial overlap between the active region volume and the mode volume, α_i is the internal loss except the output coupling loss, κ_c is the coupling rate coefficient, Δf is the frequency detuning with $\Delta\omega = \omega_{\text{inj}} - \omega_0 = 2\pi\Delta f$, and the output-coupling lifetime is determined by a passive mode quality factor Q:

$$\tau_{\text{pc}} = \frac{Q}{\omega_0}. \tag{10.3}$$

Dividing (10.2) into the real and the imaginary parts, we have

$$\frac{dE_0}{dt} = \frac{1}{2}\left[\Gamma v_g g(n,s) - \alpha_i v_g - \frac{1}{\tau_{\text{pc}}}\right]E_0 + \kappa_c E_{\text{inj}}\cos\psi, \tag{10.4}$$

$$\frac{d\phi}{dt} = -\frac{1}{2}\alpha\left[\Gamma v_g g(n,s) - \alpha_i v_g - \frac{1}{\tau_{\text{pc}}}\right] + \kappa_c\frac{E_{\text{inj}}}{E_0}\sin\psi, \tag{10.5}$$

$$\psi = \phi_{\text{inj}} - \phi + (\omega_{\text{inj}} - \omega_0)t = \Delta\phi + \Delta\omega t. \tag{10.6}$$

For a microcavity laser with coupling output waveguide as shown in Figure 10.1, the laser mode has a output field term $\sqrt{1-r^2}E_{0\text{w}}$ or $E_0/2\tau_{\text{pc}}$ based on Figure 10.1 or Eq. (10.4), respectively, and similar the injected field term of $\sqrt{1-r^2}E_{\text{injw}}$ or $\kappa_c E_{\text{inj}}$. Assuming the equivalence of the contributions for mode fields in the output waveguide to the terms in the rate equation between the output and injected field terms,

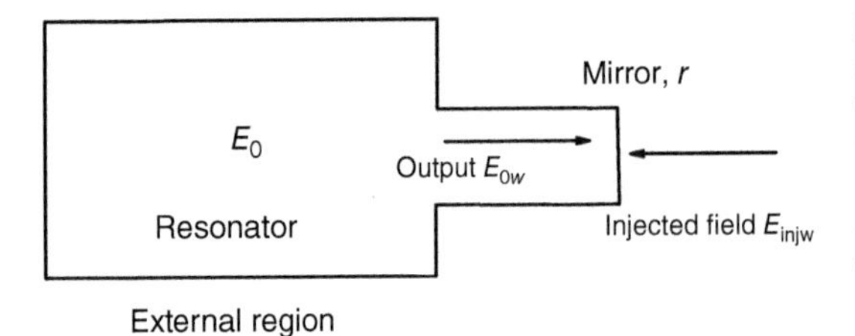

Figure 10.1 Schematic of a resonator with an output waveguide. E_0 is the lasing mode field in the resonator, E_{0w} is lasing mode field in the output waveguide, and E_{injw} is the injected mode field.

we can have

$$\frac{\kappa_c E_{\text{inj}}}{E_0/2\tau_{\text{pc}}} = \frac{E_{\text{injw}}}{E_{0w}}, \tag{10.7}$$

i.e. the output coupling term can be replaced by the injected term in the rate equation similar as the output field is feedback. The highest coupling rate coefficient is induced from (10.7) as

$$\kappa_c = \frac{1}{2\tau_{\text{pc}}} = \frac{\omega_0}{2Q}, \tag{10.8}$$

where an approximation of $E_{\text{injw}}/E_{\text{inj}} = E_{0w}/E_0$ is taken, i.e. the injected field in the output waveguide coupling into the resonator mode as high as the ratio of the mode field in the output waveguide to that in the resonator. For a Fabry–Perot cavity laser with a cavity length L, we have $\tau_{\text{pc}} = -L/(v_g \ln r^2)$ and the coupling rate coefficient as

$$\kappa_c = \frac{v_g}{2L} \ln \frac{1}{r^2} = \frac{-\ln r^2}{\tau_{\text{rt}}}, \tag{10.9}$$

where $\tau_{\text{rt}} = 2L/v_g$ is the cavity round trip time. Compared with the optical injected term in the tradition rate equation, the factor $\sqrt{1-r^2}$ is replaced by $-\ln r^2$ for the coupling rate coefficient (10.9). Finally, the rate equations under the optical injection are written as:

$$\frac{ds}{dt} = \left[\Gamma v_g g(n,s) - \alpha_i v_g - \frac{1}{\tau_{\text{pc}}}\right] s + 2\kappa_c \sqrt{S_{\text{inj}} S} \cos\psi + \Gamma\beta B n^2, \tag{10.10}$$

$$\frac{d\psi}{dt} = \frac{\alpha}{2}\left[\Gamma v_g g(n,s) - \alpha_i v_g - \frac{1}{\tau_{\text{pc}}}\right] - \kappa_c \sqrt{\frac{S_{\text{inj}}}{S}} \sin\psi + \Delta\omega, \tag{10.11}$$

$$\frac{dn}{dt} = \frac{\eta I}{qV_a} - An - Bn^2 - Cn^3 - v_g g(n,s)s, \tag{10.12}$$

where the rate equation for photon density s is obtained by multiplying $2E_0$ in (10.4) and adding a spontaneous emission term, but phase shift caused by spontaneous emission is ignored in (10.11). n and s are the carrier density and the lasing mode photon density, I is the injection current with an injection efficiency η, q is the electron charge, V_a is the volume of the active region for the microlaser, β is the spontaneous emission rate, and A, B, and C are the defect, bimolecular, and Auger recombination coefficients, respectively. The differential $d\phi_{\text{inj}}/dt$ is added in the left side of (10.11), which can be approximately assumed to be zero for the injection field.

10.3 Dynamical States of Rate Equations with Optical Injection

In this section, we simulate dynamical states for a microdisk laser at $\lambda = 1550\,\text{nm}$ similar as tested microlasers with the device parameters as shown in Table 10.1. The gain coefficient $g(n, s)$ is assumed to be a three-parameter (N_{tr}, N_s, g_0) logarithmic function in order to ensure a more accurate gain fitting for the strained quantum well material, and the nonlinear gain effects are taken into account [61]:

$$g(n,s) = \frac{g_0}{1+\varepsilon s} \ln\left(\frac{n+N_s}{N_{tr}+N_s}\right), \tag{10.13}$$

where ε is the gain suppression factor, and N_{tr} and N_s are the transparency carrier density and logarithmic gain parameter. A second-order Runge–Kutta integration method is utilized to simulate the differential rate equations (10.10)–(10.12) under different injection strength defined as:

$$R_{inj} = \frac{s_{inj}}{s_{fr}}, \tag{10.14}$$

where s_{fr} is the free-running steady-state lasing mode photon density. The time series outputs of the carrier density as well as the photon density and phase are obtained with the integration time step and time span of 200 fs and 100 ns. Then the nonlinear dynamics and oscillation characteristics are investigated for the optically injected laser by Fourier transform or convergence rule for the resulting time series. Furthermore, various modulation properties like the small-signal modulation response or noise performance can also be investigated by adding the current modulation signal term [40, 41, 62] or the Langevin noise terms [63] to the rate equations.

By solving the rate equations, we identify the laser dynamics by the orbits of phase portraits in the [Re(E), Im(E)] plane, where the complex mode field E is related to the photon density s and the phase ψ as in (10.1). The optical spectra, the RF microwave spectra, i.e. beating signal of the optical spectra, the optical intensity time series, and the phase portraits are plotted in Figure 10.2a for OIL state, (b) period-one (P1) oscillation state, (c) period-two (P2) oscillation state, (d) period-four oscillation (P4) state, and (e) chaotic state, respectively, at $R_{inj} = -5\,\text{dB}$ and $\alpha = 3$:

(a) At $\Delta f = -10\,\text{GHz}$ in Figure 10.2a, the injection locking state has a constant optical power and a single point in the phase portrait plane, with the original lasing mode is totally suppressed and the optical spectrum features a single peak at the frequency of the injected light.
(b) At $\Delta f = 5\,\text{GHz}$ in Figure 10.2b, the P1 state has the optical spectrum with extra FWM peaks, in addition to the injected mode and lasing mode marked by arrow and square symbols, the RF spectrum shows peaks at the oscillating frequency $f_r = 11.09\,\text{GHz}$ and $2f_r$, the output power oscillates at the same frequency, and the phase portrait exhibits a near circular trajectory. The oscillating frequency f_r is larger than Δf because of the red shift of the lasing mode caused by the refractive index variation with the carrier density under the optical injection.

Table 10.1 Parameters used in rate equations for an AlGaInAs/InP microlaser at $\lambda = 1550$ nm.

Parameters	Definition	Value
N_{tr} (cm^{-3})	Transparency density	1.2×10^{18}
N_s	Logarithmic gain parameter	$0.92 N_{tr}$
n_g	Group refractive index	3.5
η	Current injection efficiency	0.7
β	Spontaneous emission factor	1×10^{-2}
Γ	Confinement factor	0.2
R (μm)	Radius of microdisk laser	6
d_a (nm)	Thickness of the active region	100
A (s^{-1})	Defect recombination coefficient	1×10^{8}
B (cm^3 s^{-1})	Bimolecular recombination coefficient	1×10^{-10}
C (cm^6 s^{-1})	Auger recombination coefficient	1×10^{-28}
Q	Mode quality factor	1×10^{4}
α_i (cm^{-1})	Internal loss	6
g_0 (cm^{-1})	Material gain parameter	1500
ε	Gain suppression factor	1.5×10^{-17}
I (mA)	Injection current	8
κ_c (ns^{-1})	Coupling rate coefficient	60.8
α	Linewidth enhancement factor	3

(c) At $\Delta f = 7$ GHz in Figure 10.2c, the P2 state appears with two oscillation frequencies $f_r = 11.77$ GHz and $f_r/2$ in the optical intensity time series and in optical spectrum peaks, and the phase portrait exhibits a periodic orbit comprising two sequential cycles.

(d) At $\Delta f = -16.2$ GHz in Figure 10.2d, a P4 state shows sequentially four cycles in the phase orbit and the corresponding modulation frequencies of $f_r = 18.56$ GHz, $f_r/4, f_r/2$, and $3f_r/4$, respectively.

(e) At $\Delta f = -14.5$ GHz in Figure 10.2e, the laser transforms from the periodic oscillation to the chaotic state routing. A complicated transient with an irregular and unpredictable output signal is observed in the chaotic state, with the broadening optical spectrum and the disorganized orbits in the phase portrait.

Based on the numerical results, we can obtain the bifurcation diagrams to summarize the dynamic characteristics. The bifurcation diagrams of photon density extrema normalized to the free-running photon density s_{fr} as $s/s_{fr} - 1$ are presented in Figure 10.3a,b with the injection strength R_{inj} and the detuning frequency Δf as the bifurcation parameters, respectively. The bifurcation diagrams at $\Delta f = 2$ GHz in Figure 10.3a show the routes from injection locking to period oscillations and further toward the chaos instabilities, and then period oscillation again, and finally the large range of injection locking state with the increase of injection strength. At

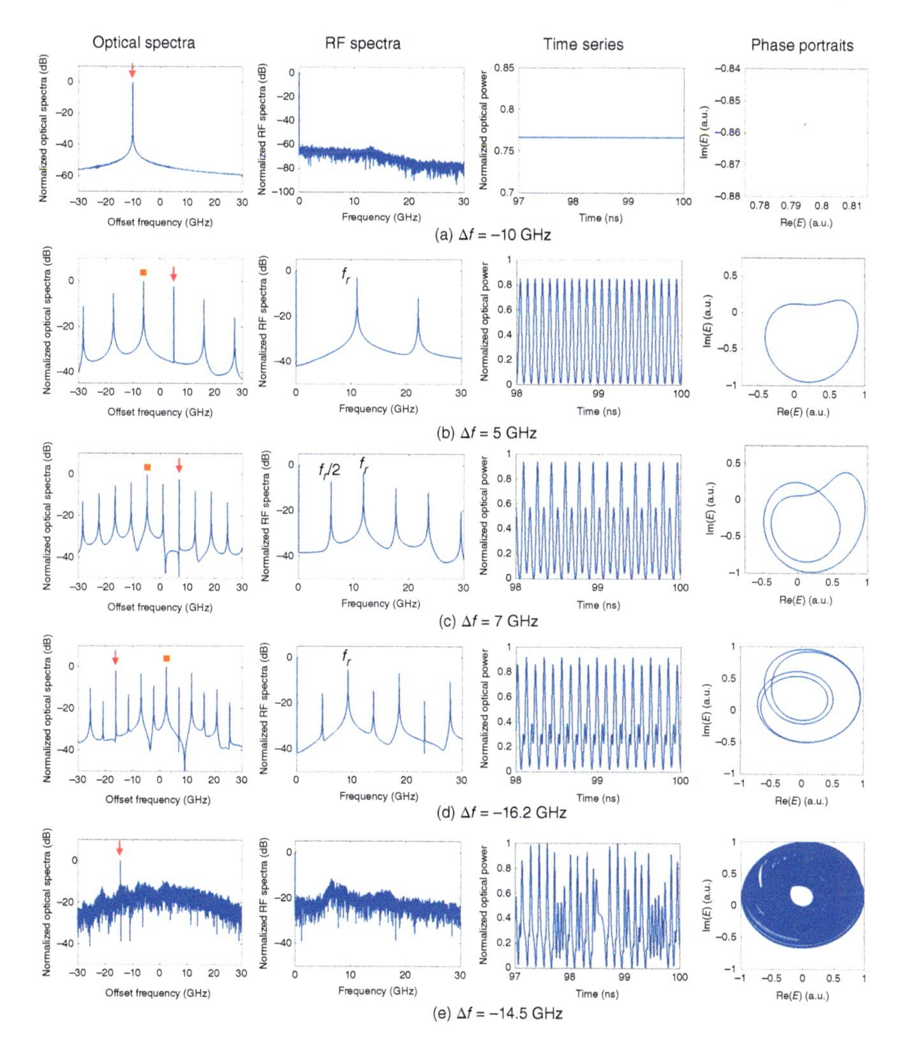

Figure 10.2 Calculated optical spectra, microwave spectra, optical intensity time series, and phase portraits for the microlaser at (a) OIL, (b) P1, (c) P2, (d) P4, and (e) chaotic, at $R_{\text{inj}} = -5$ dB. The symbols of the arrow and square in the optical spectra indicate the positions of the injection light and original lasing mode, respectively.

$R_{\text{inj}} = -5$ dB in Figure 10.3b, the locking state appears in the minus detuning range surrounded by P1, P2, and chaos states at two sides, with the boundaries as Hopf bifurcation and Saddle-Node bifurcation.

Furthermore, the dynamical states for the microlasers subject to optical injection and the transforming routes between different states are systematically investigated based on the stability diagrams over the injection parameter space (R_{inj}, Δf). Various dynamic regions characterized by different colors and clear boundaries are shown in Figure 10.4. The stable locking state marked by white is separated from other regions by a Hopf bifurcation line and a Saddle-Node bifurcation line. Above the Hopf bifurcation line, a large region of P1 dynamics is marked by transforming colors with the

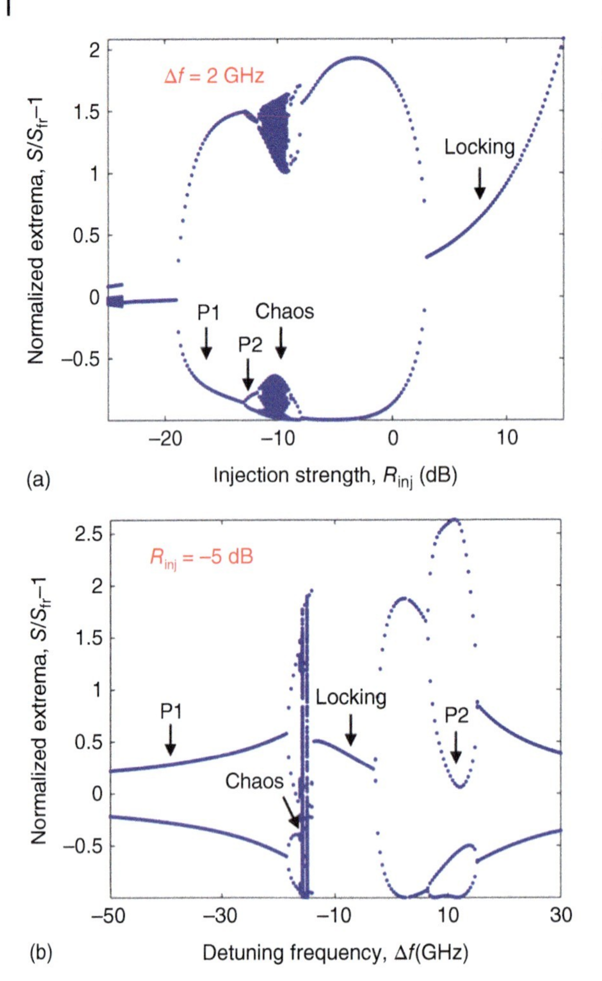

Figure 10.3 Bifurcation diagrams for the microlaser under optical injection at (a) $\Delta f = 2$ GHz and (b) $R_{\text{inj}} = -5$ dB.

oscillating frequencies in gigahertz marked by the contour curves. The P2 is marked by dark red in the stability map, which is observed at relatively low injection strength and small detuning frequency. Besides, some small separate regions of more complicated nonlinear dynamics marked in black can be observed including high-order periodic, quasi-periodic, and chaotic dynamics. As for the negative detuning side, the locking and unlocking transitions are separated by the Saddle-Node bifurcation, which shows complicated dynamical states around the boundary including P2 oscillations. This unlocked region can be explained by the crossing of the Saddle-Node bifurcation by unstable periodic orbit bifurcations [46]. When it gradually moves away from the Saddle-Node bifurcation, a large region of P1 oscillations is observed with the P1 frequency close to corresponding detuning frequency. In addition, the impact of the linewidth enhancement factor on the nonlinear dynamics of the optically injected microlaser was numerically investigated [64], and more asymmetry and larger regions of high-order periodic and chaotic behaviors in the stability map

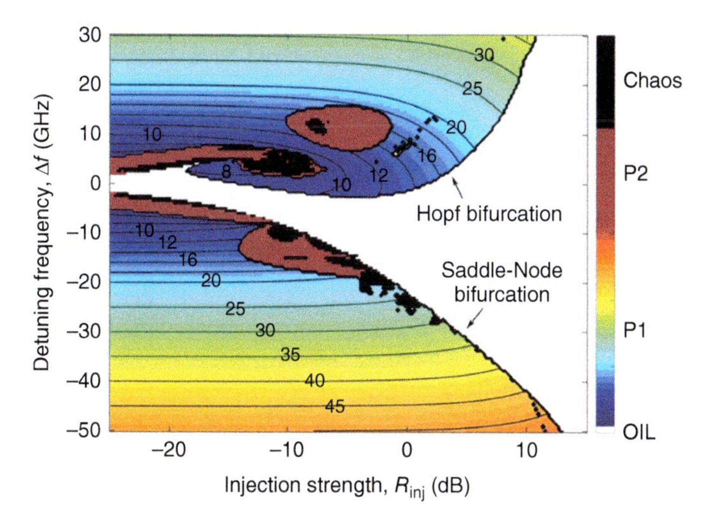

Figure 10.4 Calculated stability diagrams of the microlaser under the optical injection as a function of the injection strength and detuning frequency. The colored regions represent various dynamical states identified by the number of photon density extrema, with the OIL, P1, and P2 states marked in white, blue and yellow, red and the more complex dynamics like chaos and quasi-periodicity in black. The contour curves represent the P1 frequency in gigahertz. Source: Ma et al. [64]. © 2016, IEEE.

were observed at $\alpha = 5$. The results indicate that the linewidth enhancement factor plays an important role in the emergence and transition of the nonlinear behavior in the optical injected lasers.

Finally, the dependence of the nonlinear dynamics on the passive mode Q factor was numerically investigated for optically injected microlasers. The passive mode Q factor can be effectively modulated by varying the output waveguide width. The stability diagrams as functions of injection parameters (R_{inj}, Δf) are plotted in Figure 10.5 at $Q = 5\times10^3$, 1.5×10^4, 2×10^4, and 1×10^5, respectively. The practical mode Q_t factor is determined by the passive Q factor and a Q_i related to the material absorption loss α_i as $Q_i = n_g k/\alpha_i = 2.5\times10^4$ at $\alpha_i = 6\,\text{cm}^{-1}$. With the increase of the passive mode Q factor, the regions of P2 oscillation and chaotic states shrink significantly as shown from Figure 10.5a–d, especially on the positive detuning frequency side. In addition, the asymmetry between the positive and negative detuning sides gradually becomes unapparent, indicating weaker influence of the linewidth enhancement factor with a higher Q factor. Larger area of P1 dynamics can be observed by increasing the Q factor.

The results indicate that it is feasible to optimize the mode Q factor of microlasers for different applications of nonlinear dynamics. Small Q factors would be preferable for chaotic communication and random number generations, while larger Q factors would be preferred for photonic microwave generation. The microlaser with a large passive mode Q factor is less influenced by the optical injection because the coupling rate coefficient is inversely proportion to the passive mode Q factor from (10.3) to (10.8), which is required to be verified experimentally.

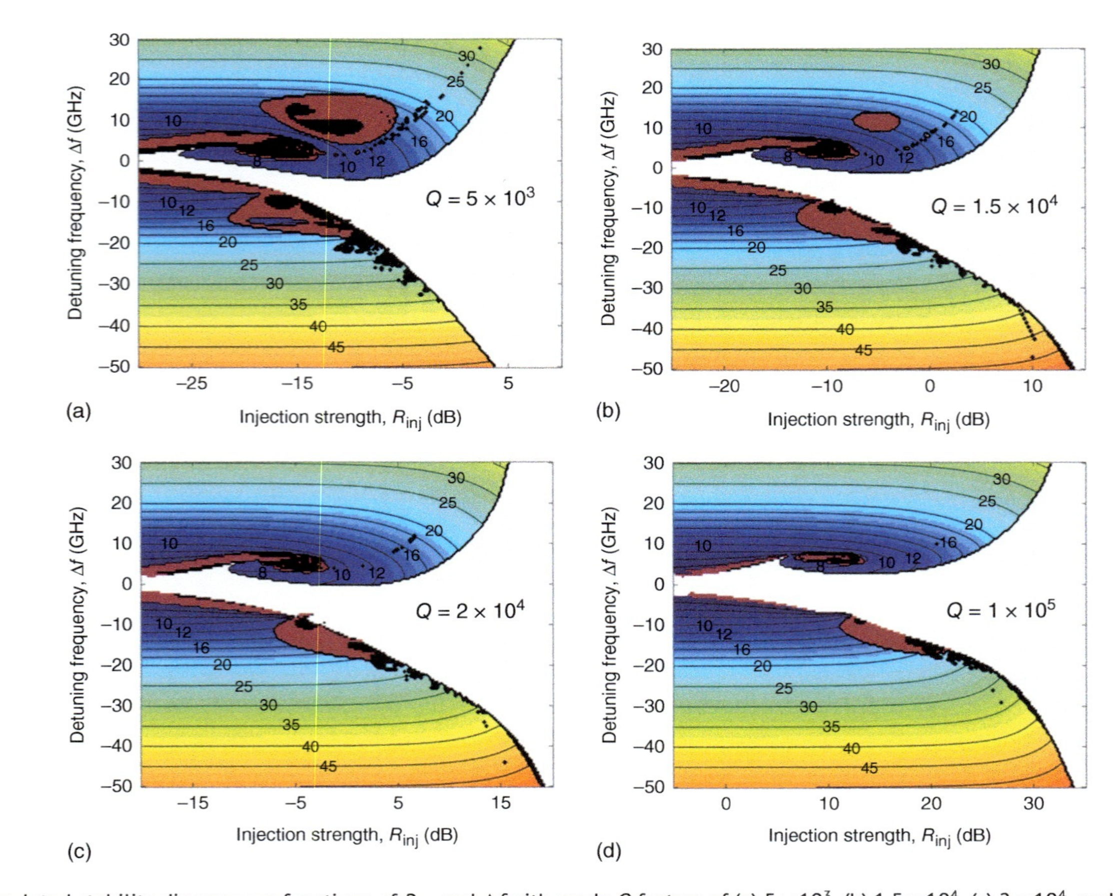

Figure 10.5 Calculated stability diagrams as functions of R_{inj} and Δf with mode Q factors of (a) 5×10^3, (b) 1.5×10^4, (c) 2×10^4, and (d) 1×10^5. Source: Ma et al. [64]. © 2016, IEEE.

10.4 Small Signal Analysis of Rate Equations

As one of the most important dynamic states, OIL provide a method for controlling and stabilizing the laser oscillations, such as improved side-mode suppression [65], enhanced relaxation oscillation frequency [5], reduced wavelength chirp [66], or reduced relative intensity noise [63]. Recently, nanolasers subject to optical injection was analyzed using rate equations accounting the influence of the Purcell cavity-enhanced factor and the spontaneous emission factor [67]. In this section, by analytically solving the single-mode rate equations as in [40, 68], we investigate the impact of OIL on the high-speed modulation performances for microdisk lasers. A small-signal time-harmonic perturbation is applied to the microlaser around its steady-state solutions as

$$\begin{aligned}
n(t) &= n_0 + \Delta n(t)\exp(i\omega t),\\
s(t) &= s_0 + \Delta s(t)\exp(i\omega t),\\
\psi(t) &= \psi_0 + \Delta\psi(t)\exp(i\omega t),\\
I(t) &= I_0 + \Delta I(t)\exp(i\omega t),
\end{aligned} \tag{10.15}$$

where n_0, s_0, and ψ_0 are the steady values of n, s, and ψ obtained by solving the steady state rate equations under the continuous-wave (CW) injection current I_0, and Δn, Δs, and $\Delta\psi$ are the small variables of n, s, and ψ corresponding to the small modulation current $\Delta I(t)\exp(i\omega t)$. Under the small signal approximation $\Delta n \ll n_0$, $\Delta s \ll s_0$, $\Delta\psi \ll \psi_0$, and $\Delta I \ll I_0$, the linearized equations for the small terms can be written in the matrix form:

$$\begin{bmatrix} \gamma_{nn}+i\omega & \gamma_{ns} & 0 \\ \gamma_{sn} & \gamma_{ss}+i\omega & \gamma_{s\psi} \\ \gamma_{\psi n} & \gamma_{\psi s} & \gamma_{\psi\psi}+i\omega \end{bmatrix} \cdot \begin{bmatrix} \Delta n \\ \Delta s \\ \Delta\psi \end{bmatrix} = \frac{\eta}{qV_a}\begin{bmatrix} \Delta I \\ 0 \\ 0 \end{bmatrix}, \tag{10.16}$$

where γ_{nn} and γ_{ns} are the derivatives of dn/dt relative to n and s; γ_{sn}, $\gamma_{ss,}$ and $\gamma_{s\psi}$ are the derivatives of ds/dt relative to n, s, and ψ; and $\gamma_{\psi n}$, $\gamma_{\psi s}$, and $\gamma_{\psi\psi}$ are the derivatives of $d\psi/dt$ relative to n, s, and ψ, respectively. The parameters derived from (10.10)–(10.12) are expressed by

$$\begin{aligned}
\gamma_{nn} &= A + 2Bn_0 + 3Cn_0^2 + v_g a_n s_0\\
\gamma_{ns} &= v_g g(n_0, s_0) - v_g a_s s_0\\
\gamma_{sn} &= -\Gamma v_g a_n s_0 - 2\Gamma B\beta n_0\\
\gamma_{ss} &= \frac{1}{\tau_{pc}} + \alpha_i v_g - \Gamma v_g g(n_0, s_0) + \Gamma v_g a_s s_0 - \kappa\sqrt{s_{inj}/s_0}\cos\psi_0\\
\gamma_{s\psi} &= 2\kappa\sqrt{s_{inj}\cdot s_0}\sin\psi_0\\
\gamma_{\psi n} &= -\alpha\Gamma v_g a_n/2\\
\gamma_{\psi s} &= \alpha\Gamma v_g a_s/2 - \kappa\sqrt{s_{inj}/s_0}\sin\psi_0/(2s_0)\\
\gamma_{\psi\psi} &= \kappa\sqrt{s_{inj}/s_0}\cos\psi_0
\end{aligned} \tag{10.17}$$

where the gain derivatives are $a_s = -\partial g(n,s)/\partial s|_{n=n_0,s=s_0}$ and $a_n = \partial g(n,s)/\partial n|_{n=n_0,s=s_0}$. Thus, the magnitude of the small-signal modulation response function can be obtained as:

$$H(\omega) = \frac{\Delta s}{\Delta I} = -\frac{\eta\gamma_{sn}}{qV_a}\cdot\frac{i\omega + Z}{(i\omega)^3 + A_0(i\omega)^2 + B_0 i\omega + C_0}, \tag{10.18}$$

where

$$\begin{aligned} A_0 &= \gamma_{ss} + \gamma_{nn} + \gamma_{\psi\psi}, \\ B_0 &= \gamma_{ss}\gamma_{nn} + \gamma_{\psi\psi}\gamma_{nn} + \gamma_{ss}\gamma_{\psi\psi} - \gamma_{s\psi}\gamma_{\psi s} - \gamma_{sn}\gamma_{ns}, \\ C_0 &= \gamma_{ss}\gamma_{\psi\psi}\gamma_{nn} + \gamma_{s\psi}\gamma_{\psi n}\gamma_{ns} - \gamma_{s\psi}\gamma_{\psi s}\gamma_{nn} - \gamma_{sn}\gamma_{ns}\gamma_{\psi\psi}, \\ Z &= \gamma_{\psi\psi} - \gamma_{s\psi}\gamma_{\psi n}/\gamma_{sn}, \end{aligned} \tag{10.19}$$

Using the parameters listed in Table 10.1 and referring to the locking range in the stability map in Figure 10.4, we simulate the small-signal modulation response $|H(\omega)|^2$ for the microlaser under OIL state. The small signal modulation responses are calculated at different injection conditions, and plotted in Figure 10.6a,b at $Q = 1\times10^4$ and Figure 10.6c,d at $Q = 2\times10^4$, respectively. Figure 10.6a shows the enhancement of modulation bandwidth and suppression of relaxation oscillation with the increase of injection strength at $\Delta f = -5$ GHz. The modulation frequency responses are also calculated with the variation of the frequency detuning Δf as

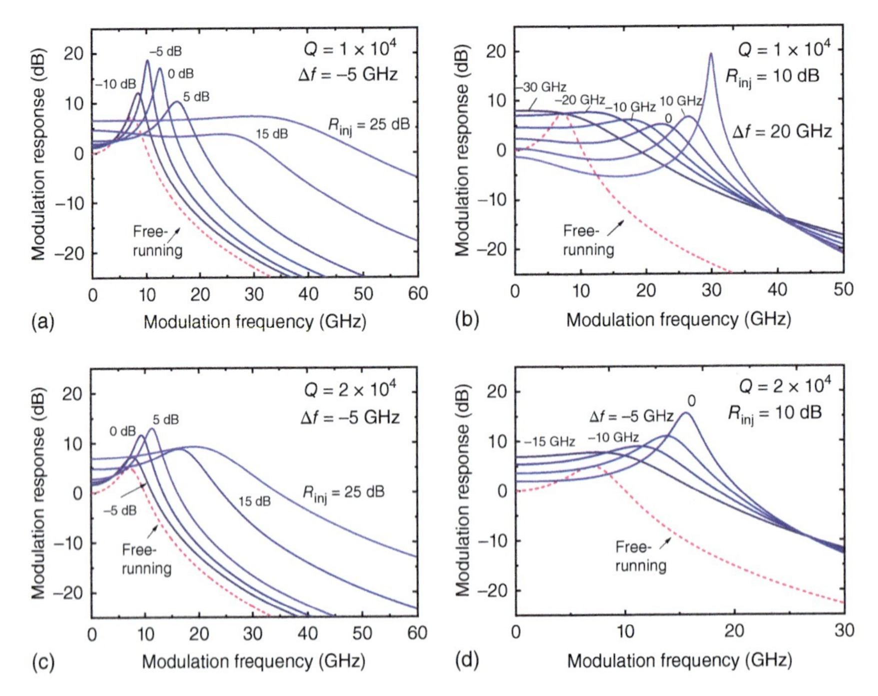

Figure 10.6 Calculated modulation response curves for the microdisk laser at free-running state and subject to external optical injection under different injection parameters at (a) $Q = 1\times10^4$ and $\Delta f = -5$ GHz, (b) $Q = 1\times10^4$ and $R_{inj} = 10$ dB, (c) $Q = 2\times10^4$ and $\Delta f = -5$ GHz, and (d) $Q = 2\times10^4$ and $R_{inj} = 10$ dB, respectively. Source: Modified from Huang et al. [68].

shown in Figure 10.6b. Strong optical injection ratio of $R_{inj} = 10$ dB is applied to the microlaser to generate a wide range of injection locking region. Greatly enhanced resonance frequency is observed with the increase of the frequency detuning. However, a large dip at the low frequency side appears under large frequency detuning. The asymmetry between the positive and negative detuning sides results from the influence of the non-zero linewidth enhancement factor. When the mode Q factor is 2×10^4, smaller resonance frequencies are observed in Figure 10.6c,d relative to Figure 10.6a,b, under the same injection ratio R_{inj}, due to the decrease of the injection rate coefficient for the case with a higher Q factor as indicated in Eq. (10.8). Thus, strong optical injection and the optimization of Q factor are effective methods for greatly enhancing the modulation bandwidth in microlasers.

10.5 Experiments of Optical Injection Microdisk Lasers

A microdisk laser with a radius of 6 μm directly connected with a 1.4-μm-wide output waveguide was used for optical injection test. After cleaving over the output waveguide to a length of about 20 μm, the microdisk laser was bonded p-side up on an AlN submount, and tested with the temperature controlled by a thermoelectric cooler (TEC). The output power and the applied voltage as functions of the CW injection current are plotted in Figure 10.7a, and lasing spectra at the injection current of 5, 14, and 28 mA are shown in Figure 10.7b. Single mode operation is observed at the injection current of 14 mA, with the lasing wavelength of 1543.51 nm, the side mode suppression ratio of 41 dB, and the single mode fiber (SMF) coupled power of 7.9 μW. The mode Q factor of the dominated mode is estimated to be 1.17×10^4 from the full width at half maximum (FWHM) just at the threshold current of 5 mA.

The experimental setup for investigating the optically injected microdisk laser is shown in Figure 10.8, where a tunable laser with a linewidth of 100 kHz is used for the optical injection into the microdisk laser through an optical circular. The optical power of the injected light is controlled by an erbium-doped fiber amplifier (EDFA) together with a tunable attenuator, and meanwhile 1% of the injected power is monitored by an optical power meter. The injection light polarization is carefully aligned with the lasing mode by adjusting a polarization controller to get the largest injection locking range at certain injection level. Once alignment, the polarization is maintained during the measurements. The laser output is amplified by an EDFA and filtered by a tunable band-pass filter (BPF), and split into two beams to measure the optical spectra by an optical spectrum analyzer (OSA) and detect by a photodetector (PD). A high-speed PD with a 3-dB bandwidth of 67 GHz is utilized to convert the optical light into electrical signal, which is subsequently measured by an electrical spectrum analyzer (ESA) or a vector network analyzer (VNA). The dynamic states under optical injection are examined by measuring the lasing spectra with an OSA at the resolution of 0.02 nm.

10.5.1 Nonlinear Dynamics Under Optical Injection

The injection optical power P_{inj} is corresponding to the optical power fed into the circulator. As the optical power injected into the microdisk is difficult to

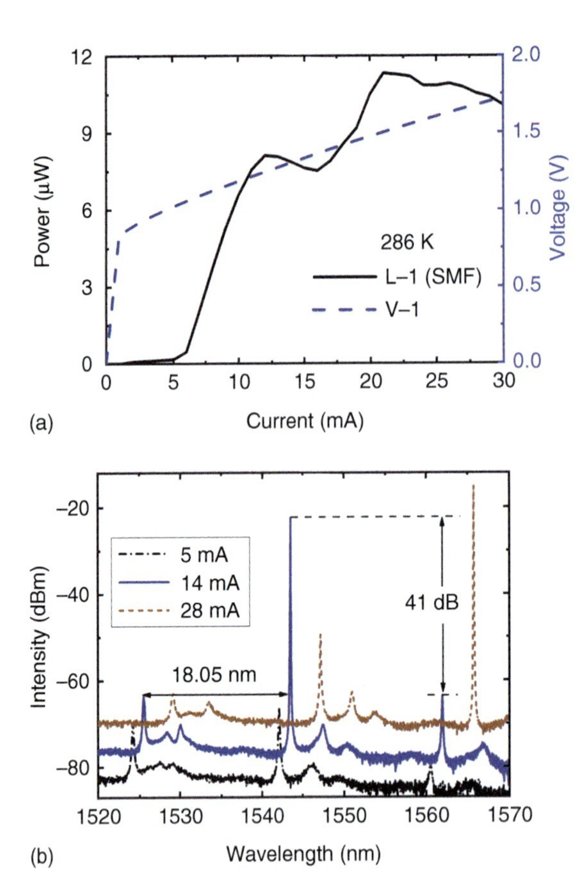

Figure 10.7 (a) Output power coupled into a single mode fiber and applied voltage vs. injection current, and (b) lasing spectra at the injection currents of 5, 14, and 28 mA, for a microdisk laser with a radius of 6 μm and a 1.4-μm-wide output waveguide at 286 K. Source: Ma et al. [64]. © 2016, IEEE.

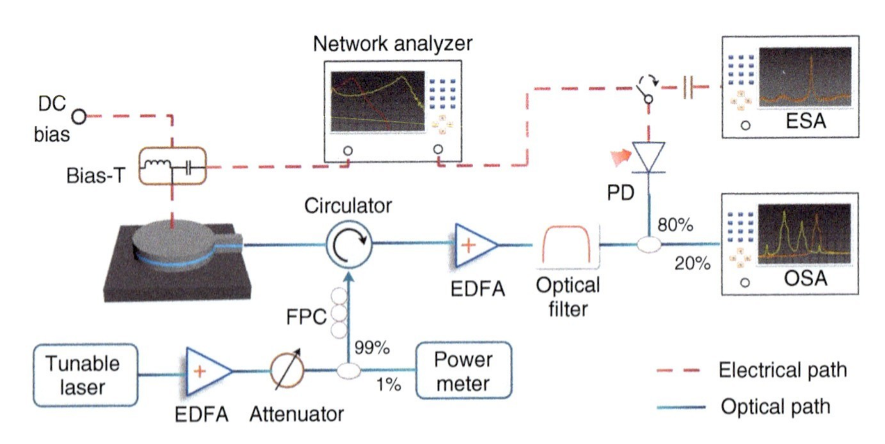

Figure 10.8 Experimental setup for studying nonlinear dynamics and small signal modulation characteristics. EDFA, erbium-doped fiber amplifier; PD, high-speed photodetector; FPC, fiber polarization controller; OSA, optical spectrum analyzer; ESA, electrical spectrum analyzer. Source: Ma et al. [64]. © 2016, IEEE.

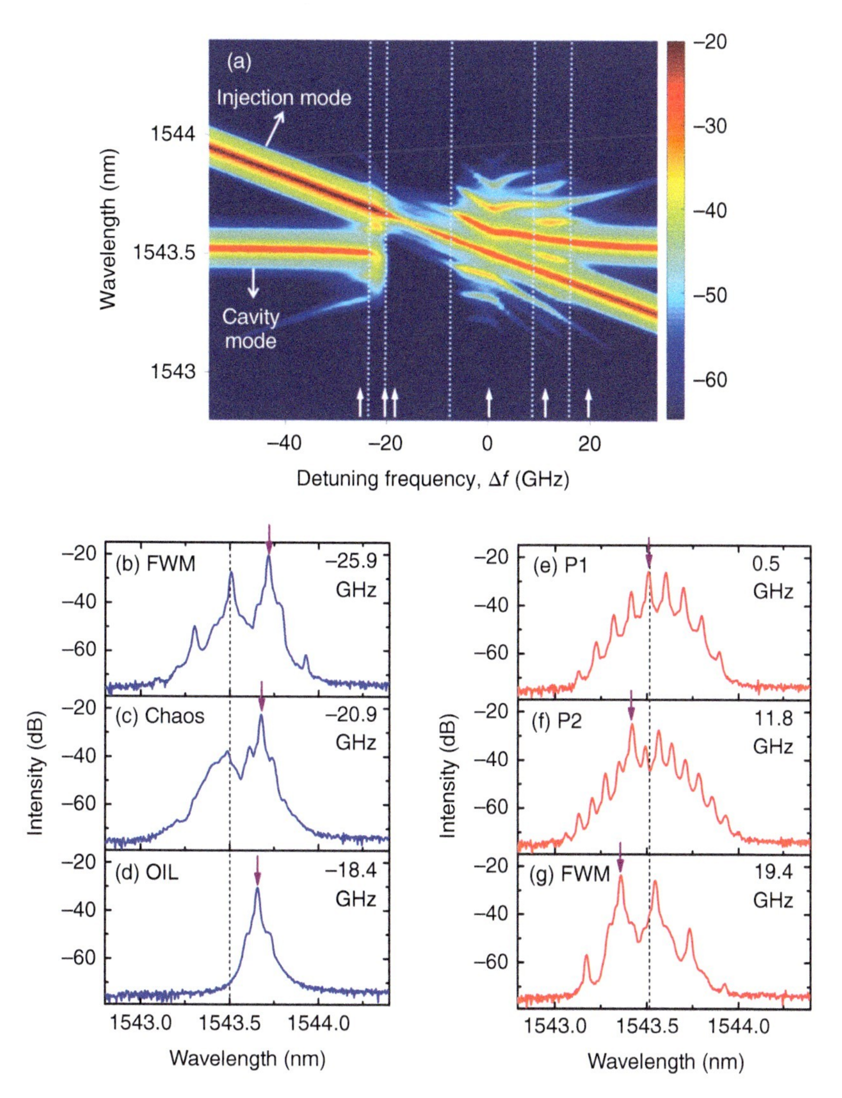

Figure 10.9 (a) Lasing spectra of the optically injected microdisk laser with the injection optical power of 1 mW at different frequency detuning Δf, at the biasing current of 14 mA. (b)–(g) Typical spectra measured at $\Delta f = -25.9, -20.9, -18.4, 0.5, 11.8$, and 19.4 GHz, corresponding to the arrows at the bottom of (a). The lasing mode is shown by a vertical dashed line for the free-running microdisk laser, and the injected mode is marked by the arrows in (b)–(g). Source: Huang et al. [68]. © 2016, IOP Publishing Ltd.

measure actually, we measured the photocurrent under certain injection power for the microdisk laser at zero bias voltage to estimate the injection efficiency. A photocurrent of 122 μA is detected at $P_{inj} = 3$ mW, which indicates an injection efficiency of 8.1% by assuming a photoelectric responsivity of 0.5 A W^{-1}. The optical spectra evolution vs. the detuning frequency Δf are plotted in Figure 10.9a at a fixed injection optical power $P_{inj} = 1$ mW, which indicates a transformation route between OIL, period oscillation and chaotic states, as illustrated in the bifurcation

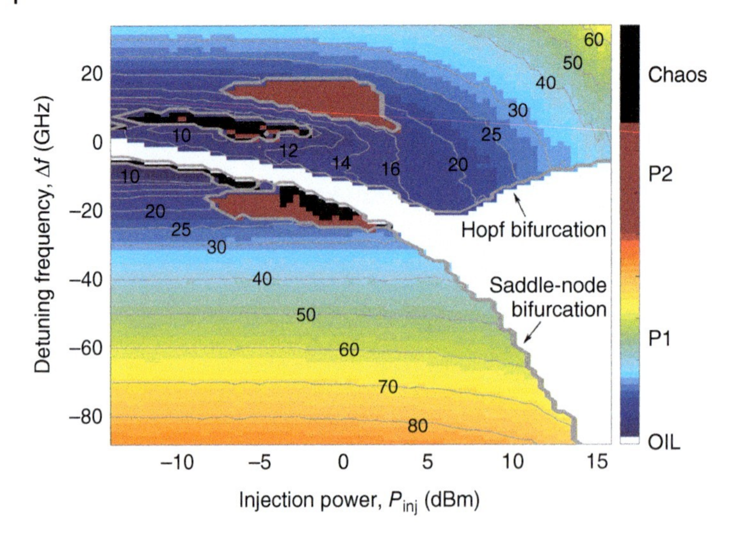

Figure 10.10 Measured stability diagrams of the microdisk laser at the biasing current of 14 mA under optical injection as functions of P_{inj} and Δf. The colored regions represent different dynamical regimes identified by the number of photon density extrema. The graduated color together with the contour curves represents the microwave frequency at P1 state in gigahertz. Source: Ma et al. [64]. © 2016, IEEE.

analysis in Figure 10.3. As Δf is swept from negative to positive detuning, the FWM exists at $\Delta f < -25$ GHz, P2 and chaos exist at -25 GHz $< \Delta f < -20.9$ GHz, OIL exists at -20.9 GHz $< \Delta f < -7.3$ GHz, P1 exists at -7.3 GHz $< \Delta f < 8.2$ GHz, P2 exists at 8.2 GHz $< \Delta f < 16.7$ GHz, and the subsequent FWM appears at $\Delta f > 16.7$ GHz again. Detailed lasing spectra of the FWM, chaos, OIL, P1, and P2 states are shown in Figure 10.9b–g at the frequency detuning $\Delta f = -25.9$, -20.9, -18.4, 0.5, 11.8, and 19.4 GHz, respectively, corresponding to the arrows located at the bottom of Figure 10.9a. For the P1 state at $\Delta f = 0.5$ GHz, multiple peaks are observed with a frequency difference of 12 GHz, due to the oscillation of the carrier density at the mode beating frequency of 12 GHz. For the FWM at $\Delta f = 19.4$ GHz, the oscillation of carrier density cannot follow the beating signal between the lasing mode and the injected mode.

The dynamical states for the microdisk laser are systematically investigated by measuring the nonlinear map over the injection parameter space (P_{inj}, Δf), as shown in Figure 10.10. The various dynamical states are identified from the measured optical spectra together with the photonic generated microwaves. The OIL, P1, P2, and chaotic states are marked in white, graduated colors from blue to red, dark red, and black, respectively. The stable OIL region is observed to be separated from other regions by a Hopf bifurcation line and a Saddle-Node bifurcation line, and the locking range is greatly increased with the injection strength. The upper and lower bounds of the OIL regime over the injection parameter space can be obtained as [69]:

$$-\frac{1}{2\pi}\kappa\sqrt{P_{inj}/P}\sqrt{1+\alpha^2} < \Delta f < \frac{1}{2\pi}\kappa\sqrt{P_{inj}/P}, \tag{10.20}$$

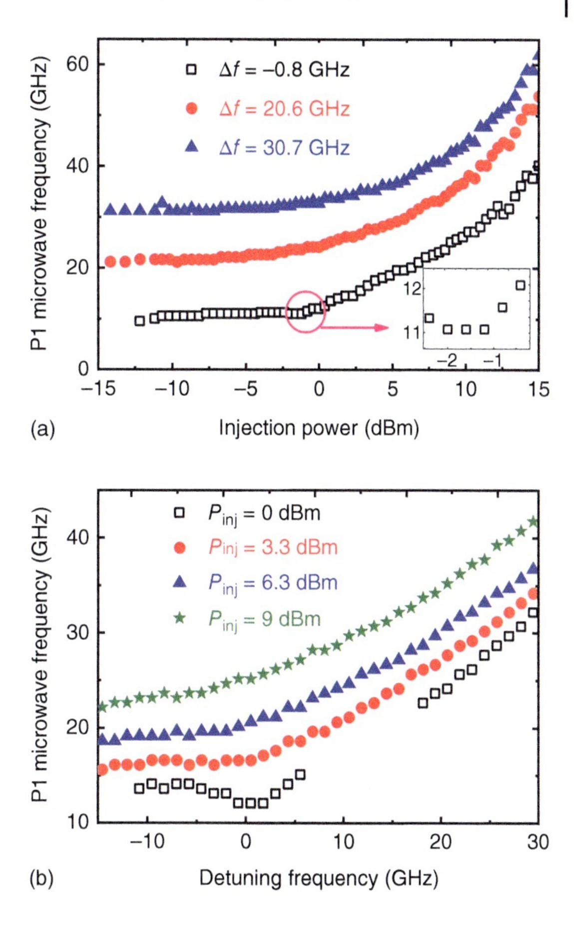

Figure 10.11 Measured P1 microwave frequency as functions of (a) the injection power P_{inj} and (b) the detuning frequency Δf, for the 6-μm-radius microdisk laser under optical injection. The curve for $P_{inj} = 0$ dBm is cut off by the P2 state. Inset in (a): Zoom-in view of P1 frequency. Source: Ma et al. [64]. © 2016, IEEE.

where κ is the injection coupling efficiency, and P is the lasing mode power. A factor of $\sqrt{1+\alpha^2}$ can be calculated from the ratio of the boundary slopes after transforming the horizontal axis in Figure 10.10 into the square root of the injection power $\sqrt{P_{inj}}$. Then, a linewidth enhancement factor of 3.0 is estimated by linearly fitting the slopes for the injection power from 5.5 to 13 dBm. Above the Hopf bifurcation and below the Saddle-Node bifurcation, large regions of P1 or FWM dynamics are observed, with the photonic microwave frequency marked by the contour curves in gigahertz. Besides, P2 and chaotic states can be identified as small separate islands in the stability diagrams.

The dependence of P1 microwave frequency on the injection power P_{inj} is illustrated in Figure 10.11a for the detuning frequencies Δf of –0.8, 20.6, and 30.7 GHz, respectively. When P_{inj} is relatively small, the generated P1 frequencies basically equal to the detuning frequencies for Δf of 20.6 and 30.7 GHz because the cavity mode is nearly undisturbed. With the increase of P_{inj}, the thermal effect and the decrease of carrier density cause the increase of P1 frequencies gradually, due to the cavity mode redshifts. However, for the case of $\Delta f = -0.8$ GHz, the P1 frequency is observed to be around 10 GHz at a low injection level, and subsequently shows a slight local minimum around $P_{inj} = -1.6$ dBm. The P1 frequency local minimum

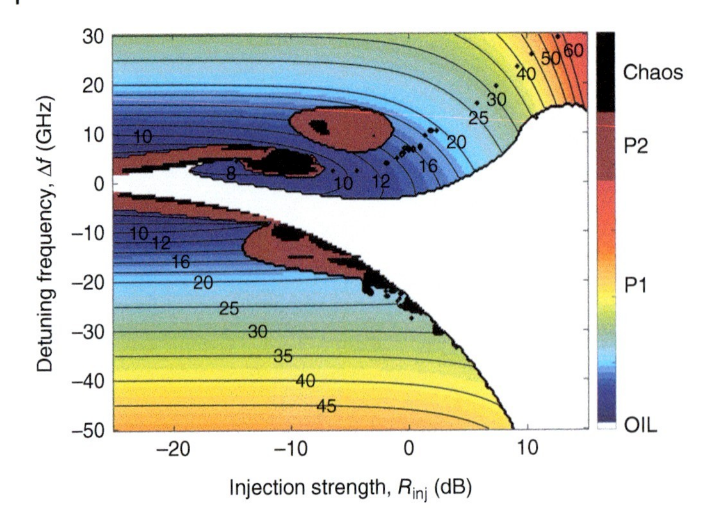

Figure 10.12 Calculated stability diagrams as functions of R_{inj} and Δf for the microlaser considering the heating effect due to the external optical injection. Source: Ma et al. [64]. © 2016, IEEE.

results from the competition between antiguidance effect related to the linewidth enhancement factor α and the injection pulling effect. The phase of the slave microlaser would be progressively locked when approaching the locking boundaries and the cavity mode frequency would be pulled toward that of the injected mode due to the pulling effect [4]. When the two effects balance out, a P1 frequency minimum would occur. In addition, the dependence of P1 frequency on the detuning frequency Δf is illustrated in Figure 10.11b for the injection powers P_{inj} of 0, 3.3, 6.3, and 9 dBm, respectively. Similarly, a local minimum of the P1 frequency with respect to Δf is observed around $\Delta f = 0$ for $P_{inj} = 0$ dBm.

10.5.2 Comparison Between Experiment and Simulated Results

Comparing the simulated and experiment results in Figures 10.4 and 10.10, an evident discrepancy is observed, especially the P1 frequency variation tendency along the Hopf and Saddle-Node bifurcations under high injection levels. The rise of temperature with the increase of optical injection power results in a redshift of the cavity mode wavelength and thus generates a larger frequency difference between the injection light and the practical cavity mode than the original value Δf determined at free-running state. By setting the injection condition far away from OIL region and then experimentally fitting the mode wavelength redshift $\delta\lambda$ vs. the optical injection power for a side mode, we obtain a linear relation of $\delta\lambda/P_{inj} = 0.00487$ nm mW^{-1}. Taking a revised frequency detuning term with such thermal effect into account in rate equations, we can obtain stability diagrams as shown in Figure 10.12, which is in agreement very well with Figure 10.10. In comparison with Figure 10.4, the Hopf and Saddle-Node bifurcations in Figure 10.12 exhibit gentler and steeper slopes respectively. Meanwhile, the P1 frequency around the Hopf bifurcation shows a larger region with contours perpendicular to the R_{inj} axis, indicating more area of P1 oscillations with low sensitivity to the fluctuations.

Finally, the injection strength is compared to confirm the definition of the coupling rate coefficient. The SMF coupled output power is 7.9 μW at the injection current of 14 mA as shown in Figure 10.7a. Assuming the output power has the same coupling efficiency as the optical injection between the microdisk laser to the SMF, which is estimated to be $\eta_{inj} = 8.1\%$ in Section 10.5.1. So the practical output power is $P_o = 0.097$ mW, and the injection optical power coupled into the microlaser is $\eta_{inj}P_{inj}$. Based on the equivalent assumption used in deriving the coupled rate coefficient, the injection strength can be calculated by

$$R_{inj} = \frac{\eta_{inj}P_{inj}}{P_o}, \tag{10.21}$$

which is −0.78 dB at the injected optical power $P_{inj} = 1$ mW and injection current 14 mA. Compared the results in Figures 10.10 and 10.12, the relation between theoretical injection strength and injected power is estimated to be $R_{inj} = P_{inj}(\mathrm{dBm}) - 3.5$ dB, which yields $R_{inj} = -3.5$ dB at $P_{inj} = 1$ mW. The obtained injection strengths of −0.78 and −3.5 dB are agreement rather well with a difference less than 3 dB. A much larger coupling rate coefficient of $\sqrt{\eta_{inj}(1-r)}/\tau_{rt}$ was used in [64], with a roundtrip time of the microdisk cavity $\tau_{rt} = 2\pi n_g R/c$ similar as in a Fabry–Perot cavity, and the injection strength was 20 dB smaller than that in Figure 10.12. During the proofreading process of [68], we thought that the coupling rate coefficient should be related to the passive mode lifetime for a microcavity, and gave an intuitive coupling rate coefficient of $\sqrt{\eta_{inj}}/\tau_{pc}$ near the result in (10.8). With the coupling rate coefficient of (10.8), we find that the numerical simulations are agreement very well with the experimental results.

10.5.3 Modulation Bandwidth Enhancement Under Optical Injection

Furthermore, the modulation responses of the microdisk laser under strong optical injection are measured at 286 K with the test setup as shown in Figure 10.8. The small signal modulation responses for the microlaser under different detuning frequencies are plotted in Figure 10.13a at the optical injection power of 13.8 dBm and the injection current of 14 mA. The measured small signal modulation curve is plotted as the dashed line for the free-running microdisk laser. The corresponding optical spectra are shown in Figure 10.13b, which indicates the transformation from OIL state into P1 state as Δf increases. The resonance frequency of free-running microdisk laser is 8.7 GHz obtained from the small signal modulation response. Significant enhancements in the frequency responses are observed with the strong optical injection. As Δf increases from −68.7 to 19.4 GHz, the resonance peak frequency and the 3-dB bandwidth enhances approximately from 22 to 47 GHz and 25 to 50 GHz, respectively, accompanied with an enhancement in the resonance peak gradually. The 3-dB bandwidth increases to about 37 GHz at $\Delta f = -12.1$ GHz. A 3-dB bandwidth of about 50 GHz is observed under P1 state at $\Delta f = 19.4$ GHz, with several dips up to 7 dB from 34 to 44 GHz and a resonance peak with a height

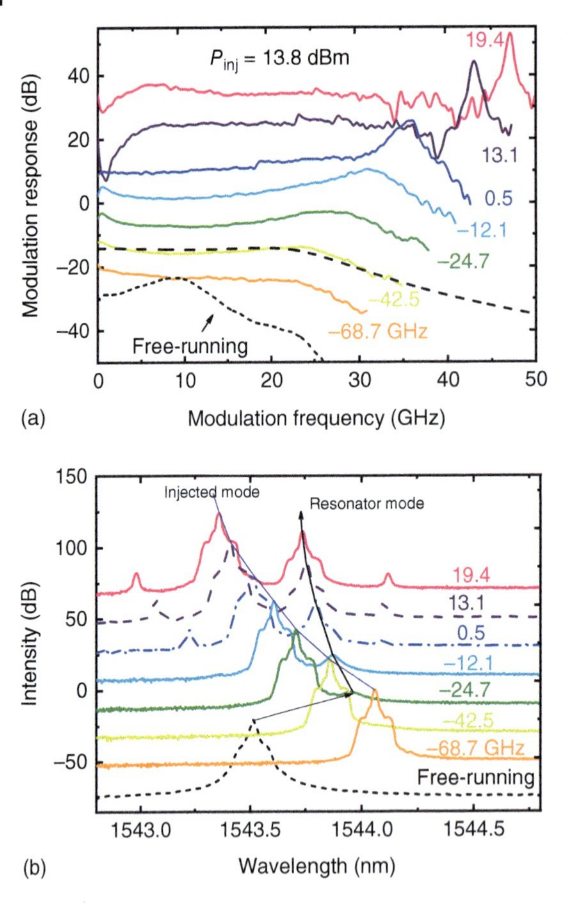

Figure 10.13 (a) Measured small signal modulation responses and (b) optical spectra for the microdisk laser under different detuning frequencies at P_{inj} = 13.8 dBm. The injected mode peaks are connected by a thin line and resonator mode peaks are marked by two arrows. The modulation response curves are offset by 20, 23, 35, 48, 60, 77, and 87 dB, and the optical spectra by 15, 35, 55, 75, 95, 115, and 135 dB at $\Delta f = -68.7$, −42.5, −24.7, −12.1, 0.5, 13.1, and 19.4 GHz, respectively. The dotted line in (a) is simulated modulation response curve at $\Delta f = -10$ GHz and R_{inj} = 13 dB. Source: Ma et al. [64]. © 2016, IEEE.

of 19 dB at 47 GHz. In addition, a low frequency dip centered at 1 GHz is observed at the P1 state with Δf = 13.1 and 19.4 GHz, compared to a small peak as Δf is less than 0.5 GHz. The small signal modulation curve calculated by Eq. (10.18) at $\Delta f = -10$ GHz and P_{inj} = 13.8 dBm is also plotted in Figure 10.13a as the dotted line, which is in agreement well with the experimental results at $\Delta f = -42.5$ GHz. The corresponding injection strength is R_{inj} = 13 dB calculated from Eq. (10.21) at P_{inj} = 13.8 dBm and P_o = 0.097 mW. Due to the resonant mode shift caused by the thermal effect with the injected optical power, the practical mode frequency detuning is not the value marked in Figure 10.13, which is calculated based on the free-running mode frequency. For a side mode far away from OIL region, we got mode wavelength red shift vs. the injection optical power at 0.004 87 nm mW^{-1} [64], which induces a wavelength red shift of 0.117 nm or mode frequency variation of −14.7 GHz at P_{inj} = 13.8 dBm. So the experimental frequency detuning $\Delta f = -42.5$ GHz should be replaced by a real value of −27.8 GHz. The difference of the theoretical and experimental frequency detuning is reduced to 17.8 GHz between the curves at −10 and −42.5 GHz in Figure 10.13a.

Finally, we detailly examine the wavelength shift of the lasing mode under the strong optical injection. At Δf = −24.7, −12.1, 0.5, 13.1, and 19.4 GHz, the

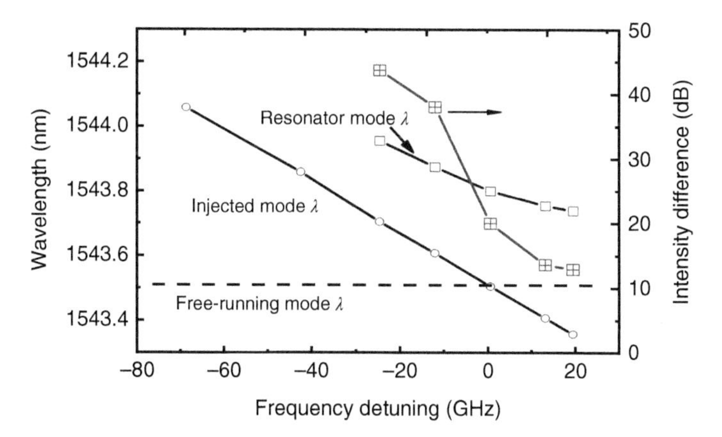

Figure 10.14 The injected and resonant mode wavelengths, and intensity difference between the injected mode and resonant mode vs. the detuning frequency obtained from the lasing spectra.

injected mode peaks and resonant mode peaks are clearly observed as marked in Figure 10.13b with the frequency difference of 32.1, 33.6, 37.1, 43.6, and 47.9 GHz, respectively, which are corresponding to the resonance peaks of the small signal modulation curves in Figure 10.13a due to the photon–photon resonance effects [36, 70, 71]. The injected mode wavelength, resonant mode wavelength, and intensity difference between the injected mode and the resonant mode are obtained from the lasing spectra in Figure 10.13b and plotted in Figure 10.14 as functions of the frequency detuning Δf, where the free-running mode wavelength is marked by the dashed line. At $\Delta f = 19.4$ GHz, the resonant mode wavelength red shift is 0.23 nm from the free-running lasing mode wavelength, the corresponding mode frequency shift is 28.9 GHz. Approximately taking 28.9 GHz as mode frequency shift due to the strong optical injection, we should reduce Δf from −42.5 to −13.6 GHz, which is nearly equal to the value of −10 GHz used in the simulation of small signal modulation curve. Furthermore, perfect OIL is realized at $\Delta f = -68.7$ and −42.5 GHz as the corresponding resonant mode peak cannot be observed from the lasing spectra. But near OIL is also classified at $\Delta f = -24.7$ and −12.1 GHz with the intensity difference between the injected mode and resonant mode is larger than 35 dB as shown in Figure 10.14. In addition to heating effect, we expect that the strong optical injection will cause spectral and spatial hole burning of carriers and result in different wavelength shifts for different modes.

10.6 Microwave Generation in Microlaser with Optical Injection

Under the optical injection, the light intensity inside the microlaser is modulated at the beat frequency between the injected light and the lasing mode, which leads to the oscillations of the carrier density and the corresponding quasi Fermi energies E_{Fn}

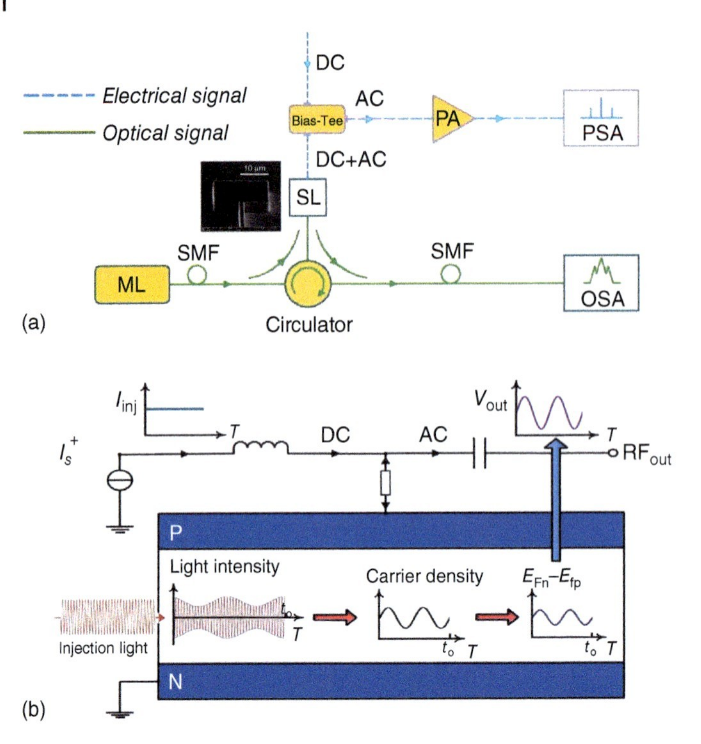

Figure 10.15 (a) Schematic of the experiment system setup. ML, master laser; SL, slave laser, i.e. microsquare laser; SMF, single mode fiber. Inset is SEM image for the microsquare laser after dry etching. (b) Principle schematic of photonic generated microwave inside the microcavity laser. T, time; $E_{Fn} - E_{Fp}$, the difference of the Fermi levels. Source: Liu et al. [43]. © 2015, IEEE.

and E_{Fp}. Hence, the microwave signal RF_{out} can be measured through the AC terminal of the bias-tee connected to the microlaser electrode, as shown in Figure 10.15a. An AlGaInAs/InP microsquare laser with a side length of 20 μm and a 1.5-μm-wide waveguide connected to the midpoint of one side was used as the slave laser (SL) for generating microwave under optical injection [43]. The microsquare laser is bonded p-side up on an AlN submount with a thin-film resistance about 40 Ω in series and mounted on a TEC to control the temperature. The microwave generation system was schematically shown in Figure 10.15a. A tunable laser is used as a master laser (ML), with its output light injected into the SL through a SMF and an optical circulator. The SL output was measured by an OSA through the output port of the circulator. The DC-bias current was fed to the SL through a 40-GHz-bandwidth bias-tee and a 40-GHz-bandwidth high-frequency probe. The microwave AC signal was extracted from the bias-tee and sent to a microwave power amplifier (PA) which has a 4 dB noise figure, a 30 dB power gain and a 15 GHz bandwidth from 3 to 18 GHz. A 26.5-GHz-bandwidth power spectrum analyzer (PSA), with a resolution bandwidth (RBW) of 100 kHz, was used to measure the amplified microwave signal. Under the optical injection, the light intensity inside the microsquare resonator is modulated at the beat frequency between the injected light and the lasing mode, which leads

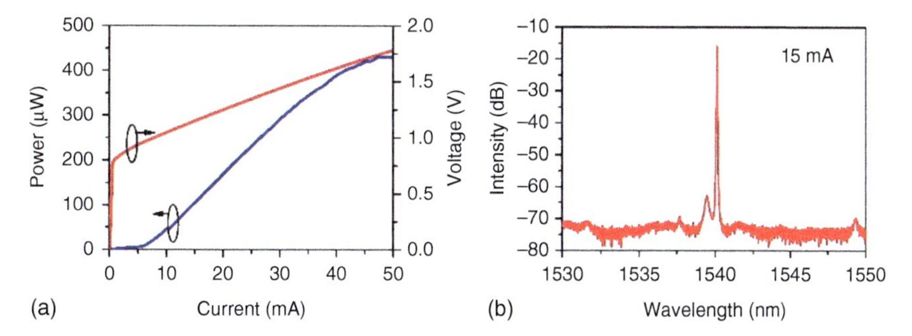

Figure 10.16 (a) Multiple mode fiber coupled optical power and applied voltage vs. DC current, and (b) lasing spectrum at the current of 15 mA for a microsquare laser with a side length of 20 μm and a 1.5 μm wide midpoint output waveguide. Source: Liu et al. [43]. © 2015, IEEE.

to the oscillations of the carrier density and the corresponding quasi Fermi energies E_{Fn} and E_{Fp}. Hence, the microwave signal $\mathrm{RF_{out}}$ can be measured through the AC terminal of the bias-tee connected to the electrode, as shown in Figure 10.15b.

At the free-running state, i.e. without optical injection, the multimode fiber coupled power and applied voltage vs. injection current were measured at TEC temperature of 288 K, and plotted in Figure 10.16a. The maximum output power coupled into a multimode optical fiber is 430 μW at a CW current of 45 mA, and the threshold current I_{th} is 5.5 mA. The lasing spectrum at the injection current of 15 mA is shown in Figure 10.16b, which shows a single mode operation at 1540.10 nm with a side-mode suppression ratio of 48 dB.

Taking the DC current of 15 mA and the TEC temperature of 288 K, we measured the lasing spectra and the corresponding microwave spectra for the microlaser subject to optical injection at the ML power of 1 mW. The lasing spectra of the microsquare laser under optical injection are plotted in Figure 10.17a at detuning frequency $\Delta f = 2.4$, 4.0, 5.0, 6.5, 9.0, 12.6, and 16.6 GHz, respectively, which is defined as the ML frequency minus the free-running microsquare laser frequency. The red vertical line indicates the lasing mode of the free-running microsquare laser at 194.7923 THz.

When $\Delta f = 16.6$ GHz, the microsquare laser is in the FWM state with the injected ML light, SL lasing mode and the FWM peak at 194.8089, 194.7917, and 194.7745 THz. The corresponding microwave has a peak 12 dB higher than the background at 17.21 GHz as shown in Figure 10.17b. The difference between the microwave frequency f_M and Δf originates from the red-shift of the SL lasing mode, due to the reduction of the free carrier intensity and thermal effect under the optical injection. The actual microwave frequency is frequency difference between the ML and the lasing mode of the SL under optical injection. At $\Delta f = 12.6$ GHz, the microsquare laser is still in the FWM state with a microwave peak of 18 dB at 13.24 GHz.

When $\Delta f = 9.0$ GHz, the lasing spectrum is dominated by the lasing mode and the injected light at 194.7912 and 194.8013 THz with two sidebands, separated by about 10.1 GHz. In addition to the FWM peaks, some small peaks appear at the

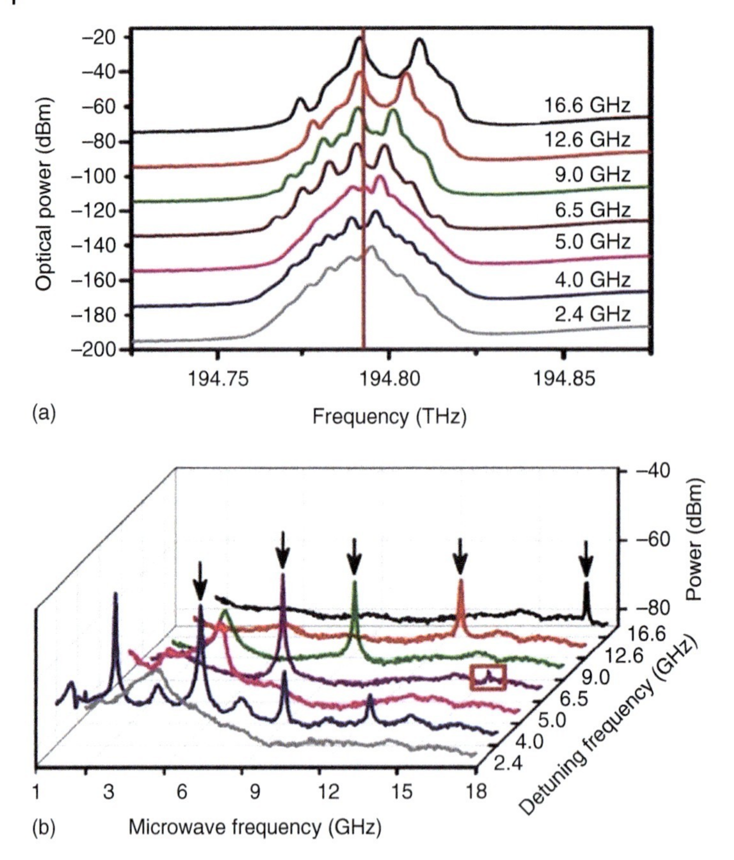

Figure 10.17 (a) Lasing spectra and (b) the corresponding microwave spectra for the microsquare SL at the injection current of 15 mA, subject to optical injection at a fixed ML power $P_i = 1$ mW and the detuning frequency $\Delta f = 2.4, 4.0, 5.0, 6.5, 9.0, 12.6$, and 16.6 GHz. The lasing spectra are offset 20 dB between adjacent spectra for clarity. The red vertical line indicates the free-running lasing peak at 194.7923 THz. Source: Liu et al. [43]. © 2015, IEEE.

halfway between the main peaks, due to the period-two oscillation. According to the bifurcation theory, the high order period oscillation appears as f_M is approximatively twice the relaxation oscillation frequency f_r [46]. Thus f_r, acting as a subharmonic, will dominate the period oscillation together with f_M. The corresponding microwave spectrum in Figure 10.17b has a main peak at 10.01 GHz with a FWHM of 25 MHz, and an additional broad peak at 5.0 GHz due to the P2 oscillation.

At $\Delta f = 6.5$ GHz, a period-one oscillation with five sidebands is observed besides the lasing mode and injected light at 194.7907 and 194.7988 THz, and the microwave spectrum has a peak at 8.12 GHz with a magnitude of 30 dB and a linewidth of 17.875 MHz. A small peak at 15.9 GHz is observed and marked by a small rectangular in Figure 10.17b, which is a harmonic signal from the high order sidebands.

At $\Delta f = 5.0$ GHz, the microsquare SL falls into chaotic state with a broadened lasing peak and a weak broadened microwave peak. Reducing the detuning frequency carefully, we observe a period-four (P4) oscillation at $\Delta f = 4.0$ GHz, with the injected light at 194.7963 THz and the lasing mode at 194.7887 THz. Four main peaks at 3.29, 6.57, 9.84, and 13.11 GHz are observed from the microwave spectrum in Figure 10.17b with magnitudes of 33, 29, 14, and 9 dB, respectively. In addition, minor peaks at 1.6, 4.9, and 8.2 GHz are observed. The P4 oscillation is a kind of quasi-period oscillation as a further development to subharmonic of P2 oscillation [46]. By decreasing the detuning frequency to 2.4 GHz, the microlaser transits into chaos again as shown in Figure 10.17.

The lasing spectra and the microwave spectra were also recorded for the optical injection at the negative detuning frequency $f_d = -0.6, -2.6, -5.1, -7.7, -10.5, -15.3$ GHz, under P1, injection locking, chaos, period-one, period-two, and four-wave-mixing states, respectively [43].

10.7 Integrated Twin-Microlaser with Mutually Optical Injection

In addition to external optical injection, we can also realize OIL for integrated microlasers with mutually optical injection. In this section, the twin-microdisk lasers with a radius of 10 μm connected by a 30-μm-long bridged waveguide are used to demonstrate the dynamical characteristics under mutually internal optical injection as in [42]. The scanning electron microscope (SEM) image of a twin-microdisk laser after the dry etching process and the microscopic picture of a twin-microdisk laser are shown in Figure 10.18a,b. The microdisk lasers were bonded p-side up on an AlN submount and mounted on a TEC. A SMF approaching individual microdisk resonator was utilized to measure the output power and the lasing spectra for the laser A and the laser B, respectively.

By applying currents to the two microdisk lasers at the same time, we can investigate the nonlinear dynamics resulting from the mutually internal optical injection

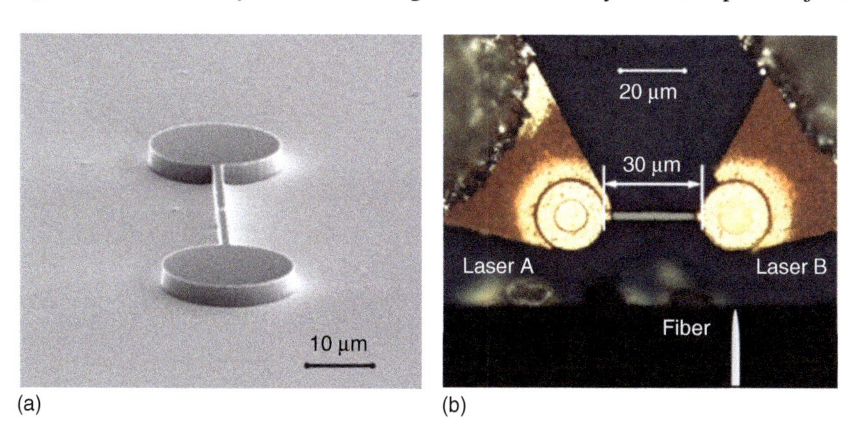

Figure 10.18 (a) The SEM image of a twin-microdisk resonator after the inductively-coupled-plasma (ICP) etching process, and (b) the microscopic picture of a twin-microdisk laser with a ring-pattern current injection window with a ring width of 5 μm. An optical fiber is drawn for collecting the light emission from the laser B. Source: Zou et al. [42]. © 2015, AIP Publishing LLC.

through the connected optical waveguide. The lasing spectra map for the dominant and the adjacent minor modes measured by coupling a SMF to the laser B are plotted in Figure 10.19 vs. the biased current I_a of the laser A at a biased current $I_b = 20$ mA. The lasing wavelength of the laser A increases with I_a caused by the heating effect, and that of the laser B is redshift at a smaller rate due to the thermal crosstalk. The lasing mode wavelength interval between the lasers A and B can be adjusted by varying I_a. At 24.0 mA $< I_a <$ 25.8 mA, one peak of four-wave-mixing state is obtained as the lasing wavelength difference is larger than 0.3 nm. More peaks at period-one state are observed around the dominant mode with smaller lasing wavelength difference as 25.8 mA $< I_a <$ 28.2 mA, and the injection locking state is realized as 28.2 mA $< I_a <$ 30.4 mA with a mode wavelength redshift due to the reduction of the threshold carriers. The chaos state is observed as 30.6 mA $< I_a <$ 31.0 mA. Furthermore, period-two oscillation is realized as 31.0 mA $< I_a <$ 31.3 mA with more peaks appeared at the middle points of the four-wave-mixing peaks. The detailed lasing spectra corresponding to the period-one state, injection locking state and period-two state are given in right side of Figure 10.19 for the laser B at $I_b = 20$ mA and $I_a = 27.6$, 29.0, and 31.2 mA, respectively. Different dynamic states are also observed for the adjacent minor mode around 1552 nm.

10.8 Discussion and Conclusion

We have summarized the dynamic characteristics of microdisk lasers subject to optical injection theoretically and experimentally. Rate equations with an injection term inversely proportional to passive mode Q factor are derived for microlasers under optical injection. Different dynamic states are studied based on the numerical simulation of the rate equations accounting the influence of the locking parameters, such as injection optical intensity, detuning frequency, and linewidth enhancement factor. The dynamic states including FWM, period-one, period-two, and period-four oscillations, and OIL are demonstrated experimentally for the microdisk laser subject to external optical injection and for integrated twin-microdisk laser with internally mutual optical injection. The greatly enhanced 3 dB bandwidth of the small signal modulation response is observed under the OIL states. In addition, the microwave signal is directly measured from a square microlaser subject to optical injection under different dynamic states, based on the active region carrier oscillation at the beating frequency.

Much high speed modulation is expected for integrated microdisk lasers under internal OIL with a high injection light intensity. Furthermore, in addition to realize internal optical injection as in the twin-microdisk laser, we can further integrate a PD with two microlasers and realize optical injection with optoelectronic feedback inside the integrated devices. Compact photonic microwave generation system with low noise is expected based on photonic integrated circuits with microlasers under the internal optical injection at the period one oscillation.

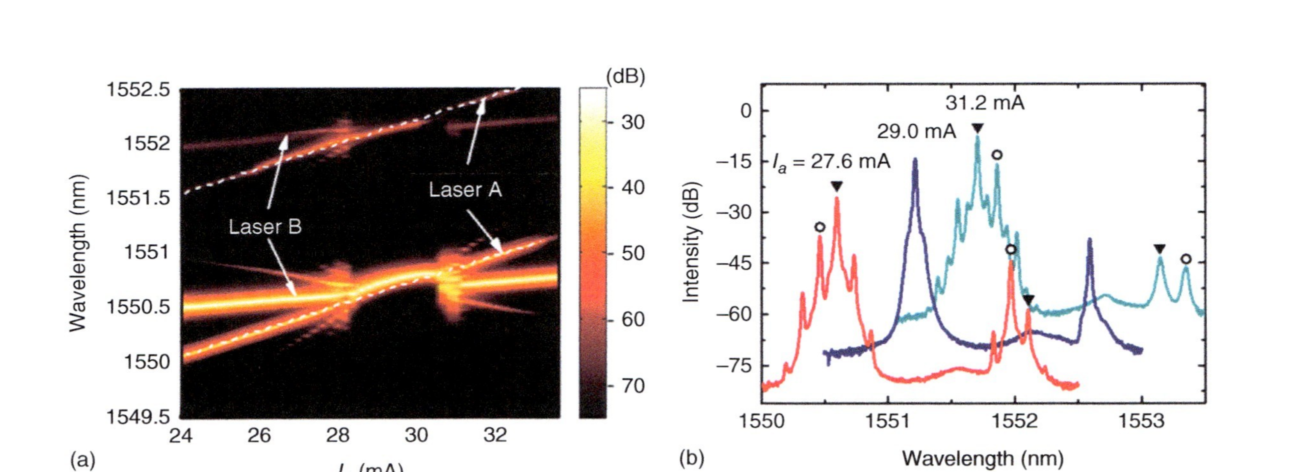

Figure 10.19 (a) Lasing spectra map for the laser B at at$I_b = 20$ mA vs. injection current I_a of the laser A. (b) Detailed lasing spectra are relatively shifted by 0.5 nm and 10 dB for clarity, and the original lasing modes of the lasers A and B are marked by the open circles and the solid triangles, respectively. Source: Zou et al. [42]. © 2015, AIP Publishing LLC.

References

1 Seeds, A.J. and Williams, K.J. (2006). Microwave photonics. *J. Lightwave Technol.* 24 (12): 4628–4641.

2 Capmany, J. and Novak, D. (2007). Microwave photonics combines two worlds. *Nat. Photonics* 1 (6): 319–330.

3 Yao, J.P. (2009). Microwave photonics. *J. Lightwave Technol.* 27 (1–4): 314–335.

4 Simpson, T., Liu, J., Huang, K., and Tai, K. (1997). Nonlinear dynamics induced by external optical injection in semiconductor lasers. *Quantum Semiclassical Opt.* 9 (5): 765.

5 Lau, E.K., Wong, L.J., and Wu, M.C. (2009). Enhanced modulation characteristics of optical injection-locked lasers: a tutorial. *IEEE J. Sel. Top. Quantum Electron.* 15 (3): 618–633.

6 Hurtado, A., Mee, J., Nami, M. et al. (2013). Tunable microwave signal generator with an optically-injected 1310 nm QD-DFB laser. *Opt. Express* 21 (9): 10772–10778.

7 Lin, F.Y. and Tsai, M.C. (2007). Chaotic communication in radio-over-fiber transmission based on optoelectronics feedback semiconductor lasers. *Opt. Express* 15 (2): 302–311.

8 Gross, N., Kinzel, W., Kanter, I. et al. (2006). Synchronization of mutually versus unidirectionally coupled chaotic semiconductor lasers. *Opt. Commun.* 267 (2): 464–468.

9 Uchida, A., Amano, K., Inoue, M. et al. (2008). Fast physical bit generation with chaotic semiconductor lasers. *Nat. Photonics* 2 (12): 728–732.

10 Sunada, S., Harayama, T., Arai, K. et al. (2011). Chaos laser chips with delayed optical feedback using a passive ring waveguide. *Opt. Express* 19 (7): 5713–5724.

11 Wang, A., Wang, B., Li, L. et al. (2015). Optical heterodyne generation of high-dimensional and broadband white chaos. *IEEE J. Sel. Top. Quantum Electron.* 21 (6): 1800710.

12 Li, N., Pan, W., Locquet, A. et al. (2015). Statistical properties of an external-cavity semiconductor laser: experiment and theory. *IEEE J. Sel. Top. Quantum Electron.* 21 (6): 1500908.

13 Goldberg, L., Taylor, H.F., Weller, J.F., and Bloom, D.M. (1983). Microwave signal generation with injection-locked laser-diodes. *Electron. Lett.* 19 (13): 491–493.

14 Gliese, U., Nielsen, T.N., Bruun, M. et al. (1992). A wide-band heterodyne optical phase-locked loop for generation of 3–18 GHz microwave carriers. *IEEE Photonics Technol. Lett.* 4 (8): 936–938.

15 Chen, X.F., Deng, Z.C., and Yao, J.P. (2006). Photonic generation of microwave signal using a dual-wavelength single-longitudinal-mode fiber ring laser. *IEEE Trans. Microwave Theory Tech.* 54 (2): 804–809.

16 Lo, Y.H., Wu, Y.C., Hsu, S.C. et al. (2014). Tunable microwave generation of a monolithic dual-wavelength distributed feedback laser. *Opt. Express* 22 (11): 13125–13137.

17 Hurtado, A., Henning, I.D., Adams, M.J., and Lester, L.F. (2013). Dual-mode lasing in a 1310-nm quantum dot distributed feedback laser induced by single-beam optical injection. *Appl. Phys. Lett.* 102 (20): 201117.

18 O'Reilly, J.J., Lane, P.M., Heidemann, R., and Hofstetter, R. (1992). Optical-generation of very narrow linewidth millimeter-wave signals. *Electron. Lett.* 28 (25): 2309–2311.

19 Neyer, A. and Voges, E. (1982). High-frequency electrooptic oscillator using an integrated interferometer. *Appl. Phys. Lett.* 40 (1): 6–8.

20 Zhuang, J.P. and Chan, S.C. (2013). Tunable photonic microwave generation using optically injected semiconductor laser dynamics with optical feedback stabilization. *Opt. Lett.* 38 (3): 344–346.

21 Lo, K.H., Hwang, S.K., and Donati, S. (2014). Optical feedback stabilization of photonic microwave generation using period-one nonlinear dynamics of semiconductor lasers. *Opt. Express* 22 (15): 18648–18661.

22 Simpson, T.B., Liu, J.-M., AlMulla, M. et al. (2014). Limit-cycle dynamics with reduced sensitivity to perturbations. *Phys. Rev. Lett.* 112 (2): 023901.

23 Hung, Y.-H. and Hwang, S.-K. (2015). Photonic microwave stabilization for period-one nonlinear dynamics of semiconductor lasers using optical modulation sideband injection locking. *Opt. Express* 23 (5): 6520–6532.

24 Juan, Y.S. and Lin, Y.F. (2011). Photonic generation of broadly tunable microwave signals utilizing a dual-beam optically injected semiconductor laser. *IEEE Photonics J.* 3 (4): 644–650.

25 Quirce, A. and Valle, A. (2012). High-frequency microwave signal generation using multi-transverse mode VCSELs subject to two-frequency optical injection. *Opt. Express* 20 (12): 13390–13401.

26 Hurtado, A., Henning, I.D., Adams, M.J., and Lester, L.F. (2013). Generation of tunable millimeter-wave and THz signals with an optically injected quantum dot distributed feedback laser. *IEEE Photonics J.* 5 (4): 5900107.

27 Zhu, N.H., Zhang, H.G., Man, J.W. et al. (2009). Microwave generation in an electro-absorption modulator integrated with a DFB laser subject to optical injection. *Opt. Express* 17 (24): 22114–22123.

28 Zanola, M., Strain, M.J., Giuliani, G., and Sorel, M. (2013). Monolithically integrated DFB lasers for tunable and narrow linewidth millimeter-wave generation. *IEEE J. Sel. Top. Quantum Electron.* 19 (4): 1500406.

29 Liu, D., Sun, C.Z., Xiong, B., and Luo, Y. (2013). Suppression of chaos in integrated twin DFB lasers for millimeter-wave generation. *Opt. Express* 21 (2): 2444–2451.

30 Schneider, G.J., Murakowski, J.A., Schuetz, C.A. et al. (2013). Radiofrequency signal-generation system with over seven octaves of continuous tuning. *Nat. Photonics* 7: 118–122.

31 Lin, C.Y., Chi, Y.C., Tsai, C.T. et al. (2015). 39-GHz millimeter-wave carrier generation in dual-mode colorless laser diode for OFDM-MMWoF transmission. *IEEE J. Sel. Top. Quantum Electron.* 21 (6): 1801810.

32 Radziunas, M., Glitzky, A., Bandelow, U. et al. (2007). Improving the modulation bandwidth in semiconductor lasers by passive feedback. *IEEE J. Sel. Top. Quantum Electron.* 13 (1): 136–142.

33 Lau, E.K., Zhao, X., Sung, H.K. et al. (2008). Strong optical injection-locked semi lasers demonstrating >100-GHz resonance frequencies and 80-GHz intrinsic bandwidths. *Opt. Express* 16 (9): 6609–6618.

34 Zhu, N.H., Li, W., Wen, J.M. et al. (2008). Enhanced modulation bandwidth of a Fabry–Perot semiconductor laser subject to light injection from another Fabry–Perot laser. *IEEE J. Quantum Electron.* 44 (5–6): 528–535.

35 Chen, C.H., Takeda, K., Shinya, A. et al. (2011). 40-Gb/s directly-modulated photonic crystal lasers under optical injection-locking. *Opt. Express* 19 (18): 17669–17676.

36 Sun, C.Z., Liu, D., Xiong, B. et al. (2015). Modulation characteristics enhancement of monolithically integrated laser diodes under mutual injection locking. *IEEE J. Sel. Top. Quantum Electron.* 21 (6): 1802008.

37 Alharthi, S.S., Hurtado, A., Korpijarvi, V.M. et al. (2015). Circular polarization switching and bistability in an optically injected 1300 nm spin-vertical cavity surface emitting laser. *Appl. Phys. Lett.* 106 (2): 021117.

38 Almulla, M. and Liu, J.M. (2015). Stable periodic dynamics of reduced sensitivity to perturbations in optically injected semiconductor lasers. *IEEE J. Sel. Top. Quantum Electron.* 21 (6): 1801708.

39 Zou, L.X., Huang, Y.Z., Lv, X.M. et al. (2014). Modulation characteristics and microwave generation for AlGaInAs/InP microring lasers under four-wave mixing. *Photonics Res.* 2 (6): 177–181.

40 Zou, L.X., Huang, Y.Z., Liu, B.W. et al. (2015). Nonlinear dynamics for semiconductor microdisk laser subject to optical injection. *IEEE J. Sel. Top. Quantum Electron.* 21 (6): 1800408.

41 Huang, Y.Z., Zou, L.X., Liu, B.W. et al. (2015). Dynamic and mode characteristics for AlGaInAs/InP microdisk lasers subject to optical injection. *Opt. Eng.* 54 (7): 076109.

42 Zou, L.X., Liu, B.W., Lv, X.M. et al. (2015). Integrated semiconductor twin-microdisk laser under mutually optical injection. *Appl. Phys. Lett.* 106 (19): 191107.

43 Liu, B.W., Huang, Y.Z., Long, H. et al. (2015). Microwave generation directly from microsquare laser subject to optical injection. *IEEE Photonics Technol. Lett.* 27 (17): 1853–1856.

44 Lang, R. (1982). Injection locking properties of a semiconductor laser. *IEEE J. Quantum Electron.* 18 (6): 976–983.

45 Erneux, T., Kovanis, V., Gavrielides, A., and Alsing, P.M. (1996). Mechanism for period-doubling bifurcation in a semiconductor laser subject to optical injection. *Phys. Rev. A* 53 (6): 4372–4380.

46 Wieczorek, S., Krauskopf, B., Simpson, T.B., and Lenstra, D. (2005). The dynamical complexity of optically injected semiconductor lasers. *Phys. Rep.* 416: 1–128.

47 Ohtsubo, J. (2012). *Semiconductor Lasers: Stability, Instability and Chaos*. Springer.

48 Kovanis, V., Gavrielides, A., Simpson, T.B., and Liu, J.M. (1995). Instabilities and chaos in optically injected semiconductor lasers. *Appl. Phys. Lett.* 67 (19): 2780–2782.

49 Simpson, T.B. (2003). Mapping the nonlinear dynamics of a distributed feedback semiconductor laser subject to external optical injection. *Opt. Commun.* 215 (1–3): 135–151.

50 Simpson, T.B., Liu, J.M., and Gavrielides, A. (1996). Small-signal analysis of modulation characteristics in a semiconductor laser subject to strong optical injection. *IEEE J. Quantum Electron.* 32 (8): 1456–1468.

51 Yabre, G. (1996). Effect of relatively strong light injection on the chirp-to-power ratio and the 3 dB bandwidth of directly modulated semiconductor lasers. *J. Lightwave Technol.* 14 (10): 2367–2373.

52 Simpson, T.B. and Liu, J.M. (1997). Enhanced modulation bandwidth in injection-locked semiconductor lasers. *IEEE Photonics Technol. Lett.* 9 (10): 1322–1324.

53 Wieczorek, S., Krauskopf, B., and Lenstra, D. (1999). A unifying view of bifurcations in a semiconductor laser subject to optical injection. *Opt. Commun.* 172 (1–5): 279–295.

54 Hwang, S.K. and Liang, D.H. (2006). Effects of linewidth enhancement factor on period-one oscillations of optically injected semiconductor lasers. *Appl. Phys. Lett.* 89 (6): 061120.

55 AlMulla, M. and Liu, J.M. (2014). Effects of the gain saturation factor on the nonlinear dynamics of optically injected semiconductor lasers. *IEEE J. Quantum Electron.* 50 (3): 158–165.

56 Li, L. (1994). A unified description of semiconductor lasers with external light injection and its application to optical bistability. *IEEE J. Quantum Electron.* 30 (8): 1723–1731.

57 Chan, S.C. (2010). Analysis of an optically injected semiconductor laser for microwave generation. *IEEE J. Quantum Electron.* 46 (3): 421–428.

58 Liu, D., Sun, C., Xiong, B., and Luo, Y. (2014). Nonlinear dynamics in integrated coupled DFB lasers with ultra-short delay. *Opt. Express* 22 (5): 5614–5622.

59 Henry, C.H. (1982). Theory of the linewidth of semiconductor lasers. *IEEE J. Quantum Electron.* 18 (2): 259–264.

60 Mogensen, F., Olesen, H., and Jacobsen, G. (1985). Locking conditions and stability properties for a semiconductor laser with external light injection. *IEEE J. Quantum Electron.* 21 (7): 784–793.

61 Coldren, L.A., Corzine, S.W., and Mashanovitch, M.L. (1995). *Diode Lasers and Photonic Integrated Circuits*. Wiley.

62 Wang, J., Haldar, M.K., Li, L., and Mendis, F.V.C. (1996). Enhancement of modulation bandwidth of laser diodes by injection locking. *IEEE Photonics Technol. Lett.* 8 (1): 34–36.

63 Schunk, N. and Petermann, K. (1986). Noise-analysis of injection-locked semiconductor injection-lasers. *IEEE J. Quantum Electron.* 22 (5): 642–650.

64 Ma, X.W., Huang, Y.Z., Long, H. et al. (2016). Experimental and theoretical analysis of dynamic regimes for optically injected microdisk lasers. *J. Lightwave Technol.* 34 (22): 5263–5269.

65 Iwashita, K. and Nakagawa, K. (1982). Suppression of mode partition noise by laser diode light injection. *IEEE J. Quantum Electron.* 18 (10): 1669–1672.

66 Mohrdiek, S., Burkhard, H., and Walter, H. (1994). Chirp reduction of directly modulated semiconductor-lasers at 10 GB/s by strong CW light injection. *J. Lightwave Technol.* 12 (3): 418–424.

67 Sattar, Z.A., Kamel, N.A., and Shore, K.A. (2016). Optical injection effects in nanolasers. *IEEE J. Quantum Electron.* 52 (2): 1200108.

68 Huang, Y.Z., Ma, X.W., Yang, Y.D., and Xiao, J.L. (2016). Review of the dynamic characteristics of AlGaInAs/InP microlasers subject to optical injection. *Semicond. Sci. Technol.* 31: 113002.

69 Liu, G., Jin, X., and Chuang, S.L. (2001). Measurement of linewidth enhancement factor of semiconductor lasers using an injection-locking technique. *IEEE Photonics Technol. Lett.* 13 (5): 430–432.

70 Bardella, P. and Montrosset, I. (2013). A new design procedure for DBR lasers exploiting the photon–photon resonance to achieve extended modulation bandwidth. *IEEE J. Sel. Top. Quantum Electron.* 19: 1502408.

71 Xiao, Z.X., Huang, Y.Z., Yang, Y.D. et al. (2017). Modulation bandwidth enhancement for coupled twin-square microcavity lasers. *Opt. Lett.* 42: 3173–3176.

11

Hybrid-Cavity Lasers

11.1 Introduction

Low-threshold and single-mode whispering-gallery-mode (WGM) microlasers have been demonstrated in various geometry designs, e.g. microdisks and microrings, microtoroids, and polygonal microresonators. Due to a short cavity length, it is easy to realize single longitudinal mode lasing for WGM microlasers. However, traditional WGM microlasers usually have a low-output power and low-coupling efficiency for a single-mode optical fiber. In this chapter, a hybrid square-rectangular laser (HSRL) composed of a Fabry–Perot (FP) cavity and a square microcavity is demonstrated for mode selection by enhancing mode Q factor for hybrid mode between WGM and FP mode [1, 2]. The mode Q-factor enhancements are numerically and experimentally demonstrated for the hybrid modes with a mode wavelength interval of the longitudinal mode interval of the microcavity. Stable single-mode operation with high coupling efficiency to a single-mode fiber (SMF) is realized by applying currents to the square microcavity and the FP-cavity sections at the same time. Furthermore, controllable optical bistability is realized due to mode competition as the square microcavity is unbiased, and all-optical set and reset operations are demonstrated [3].

In Section 11.2, the reflectivity spectra for a square microcavity as one side reflector of the hybrid cavity are simulated, which shows high reflectivity around WGMs and reveals the possibility for realizing mode selection. Hybrid mode behaviors are discussed by calculating the mode wavelengths and mode Q factors in Section 11.3, and the mode-field distributions are simulated and discussed for a hybrid-cavity laser with a square microcavity in Section 11.4. The fabrication process is briefly summarized in Section 11.5, lasing characteristics are presented in Section 11.6 with mode Q factor enhancement, and robustness of single-mode operation is verified in Section 11.7. In Section 11.8, optical bistability is tested for a hybrid-cavity laser, and high-speed all-optical switching and logic operations are verified in Sections 11.9 and 11.10. In Section 11.11, a hybrid-cavity laser with a deformed microcavity for improving mode-field pattern in the FP cavity is proposed and demonstrated, and summary is given in Section 11.12.

Microcavity Semiconductor Lasers: Principles, Design, and Applications, First Edition.
Yong-zhen Huang and Yue-de Yang.

11.2 Reflectivity of a WGM Resonator

Different from traditional coupled-cavity lasers such as cleaved and etched cavities [4, 5], the HSRL does not have a distinct mirror between the two cavities, which may result in self-consistent mode-field pattern in the whole cavity and make HSRL more stable. The square microcavity as one-side reflector of the FP cavity is numerically simulated, which shows high reflectivity around the instinct WGMs and reveals the possibility for realizing mode selection. A two-dimensional finite-difference time-domain (FDTD) method is utilized to calculate the mode reflectivity spectra of different microcavities [6, 7], for the fundamental transverse mode of the FP cavity as it impinges the interface between the WGM microcavity and the FP cavity. The reflectivity spectra are approximately calculated as the reflected/incident intensity spectra ratio $|E_r|^2/|E_i|^2$, where E_i and E_r are input and reflected field amplitudes at the midpoint of the FP waveguide [6]. A wide band exciting source with a Gaussian time profile is taken as $P(t) = \exp[-(t-t_0)^2/t_w{}^2]\cos(2\pi f_0 t)$ with $t_w = 1.195$ fs, $t_0 = 3\,t_w$, and $f_0 = 193.5$ THz. The reflectivity spectra for the transverse-electric (TE) mode are calculated by the Padé approximation [8], using the input and reflected fields at a distance of 18 μm away from the center of the microcavity. The incident field is taken to be the fundamental transverse mode of the FP waveguide, and the incident intensity spectra are calculated after the incident pulse transmits over a distance inside the FP waveguide. After transmitting over a rather long waveguide, the wide band exciting pulse can be fitted into the guided mode pattern and avoid radiation loss. The microcavities with a refractive index of 3.2 are confined by a bisbenzo cyclobutene (BCB) layer with a refractive index of 1.54.

The reflectivity spectra for an FP-cavity end face, a microdisk, a square, and a pentagon microcavity are calculated and plotted in Figure 11.1a–f with an FP-cavity width $d = 0.4$ μm. The reflectivity is about 0.10 for the FP-cavity end face as shown in Figure 11.1a, which is 20% smaller than the reflectivity of 0.123 determined by the refractive index difference for a vertical propagating plane wave. For a microdisk cavity with a radius of 5 μm, we have a lot of narrow reflectivity peaks as shown in Figure 11.1b, corresponding to high Q-coupled WGMs in the microdisk connecting with a waveguide [10]. The reflectivity spectra show a series of wide peaks for a square microcavity with a side length of 10 μm as shown in Figure 11.1c,d, which shows much higher reflectivity peaks as the FP cavity is connected to a vertex of the square resonator. As shown in Figure 11.1e,f, less and lower reflectivity peaks than the square microcavity are observed for a pentagon microcavity with a side length of 7 μm. The results indicate that square microcavity can have much higher resonant reflectivity peaks than the other WGM microcavities.

To avoid the influence of the high-order transverse mode on the approximation method by using the field amplitudes at the midpoint of the FP waveguide, we choose a narrow FP cavity for calculating mode reflectivity in the above simulation. However, such narrow waveguide cannot be fabricated by using contact photolithography technique. In the following part, we simulate mode reflectivity for a hybrid cavity with a square microcavity and an FP waveguide width of $w = 1.5$ μm. The calculated reflectivity spectra for a square microcavity with a side length

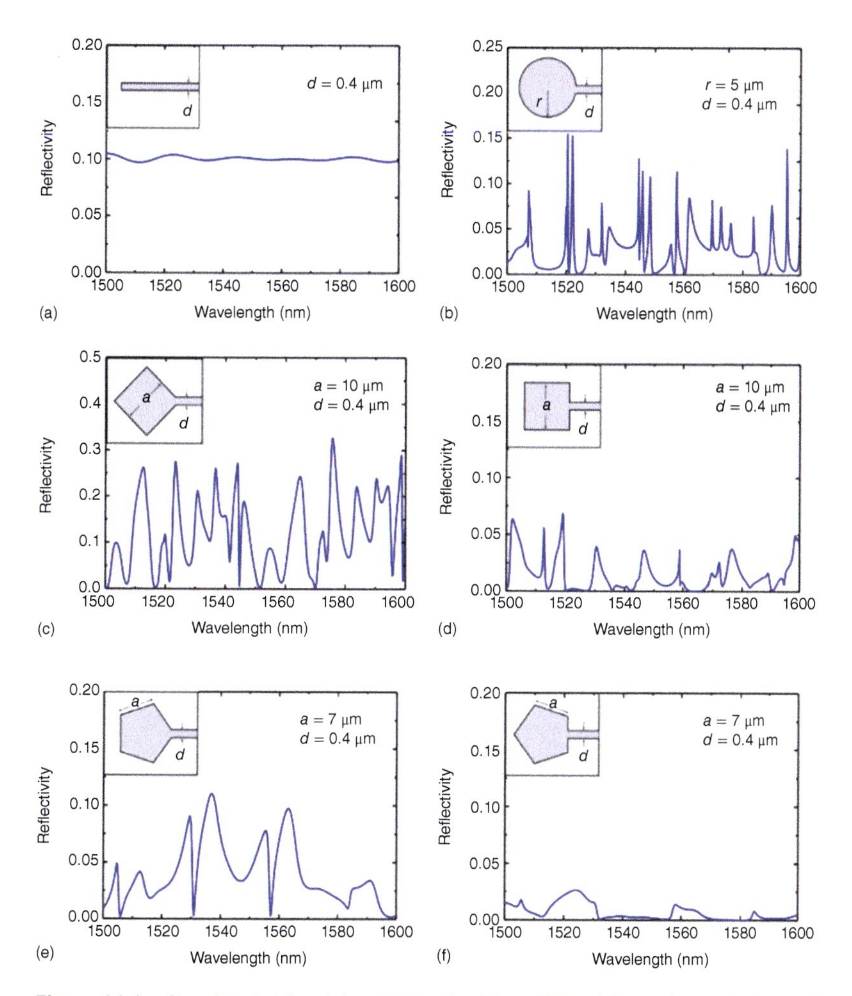

Figure 11.1 Simulated reflectivity spectra for (a) an FP end face with a width $d = 0.4\,\mu\text{m}$, (b) a microdisk cavity with a radius of $r = 5\,\mu\text{m}$, (c) a vertex and (d) the middle point of one side of a square microcavity with a side length $a = 10\,\mu\text{m}$, and (e) a vertex and (f) the middle point of one side of a pentagon microcavity with a side length $a = 7\,\mu\text{m}$. Source: Huang et al. [9]. © 2018, Springer Nature Switzerland AG.

$a = 10\,\mu\text{m}$ are shown in Figure 11.2 with the gain of the square cavity as 0 and $8\,\text{cm}^{-1}$, respectively. In addition, the intrinsic WGM intensity spectrum at zero gain condition is also simulated and plotted as dashed–dotted line in Figure 11.2, in which a symmetric wide band exciting source is placed within the square microcavity relative to the midline of the waveguide. Multiple modes are excited over a wide frequency range, and WGM intensity spectra are calculated from the FDTD output and the Padé approximation. The reflectivity spectra show maxima around the square WGM resonances with an interval of 54.37 nm. The maximum reflectivity increases sharply from 70% to 161% around the fundamental WG mode wavelength 1547.7 nm as the gain of the square cavity rises from 0 to $8\,\text{cm}^{-1}$, indicating that the square cavity can act as a resonant reflector, which modulates the end loss for efficient mode selection when varying the injection level in the square region. Comparing Figures 11.1 and 11.2, we can find that the reflectivity increases greatly as the FP waveguide width increases from 0.4 to 1.5 μm. However, the approximation of taking the reflected field amplitude at the midpoint of the FP waveguide as the field amplitude of the fundamental transverse mode can overestimate the reflectivity for a wide FP waveguide. A reflectivity larger than unity for the square microcavity without gain is calculated at $a = 15$ μm and $w = 2$ μm [2], because the reflected light from the square microcavity contains not only the fundamental waveguide mode but also some higher-order symmetric modes. If the reflected wave is composed of the fundamental and second-order transverse modes with the same amplitude, we have intensity distribution $(\cos ky + \cos 2ky)^2$ with the intensity of 4 at the center point $y = 0$, and the integrated intensity over the interface of the FP cavity is 1 for totally confined transverse modes in the FP cavity. Assuming the reflected wave only consisted of the fundamental transverse mode with the same intensity of 4 at the center, we can obtain the integrated intensity over the interface of 2 instead of 1. The reflectivity simply calculated using the intensity monitor at the center may be overestimated for an FP cavity with a wide waveguide.

11.3 Mode *Q*-Factor Enhancement for Hybrid Modes

In this section, the mode *Q*-factor enhancement for the HSRLs composed of a square microcavity with an FP cavity is numerically investigated using the two-dimensional (2D) finite element method (FEM). The hybrid cavity has a refractive index of 3.2, and the surrounding materials are BCB with refractive indices of 1.54 and the other side of the FP cavity is a cleaved mirror. By solving the eigenmodes with complex eigenfrequency, we can evaluate the mode *Q* factors, field distributions, and far-field behaviors. Symmetry boundary condition is taken along the center line of the FP cavity to investigate the mode coupling between the symmetric FP transverse modes and square WGMs. Here, the structure parameters are taken to be $a = 10$ μm, $w = 1.5$ μm, and the FP-cavity length $L = 300$ μm, which is measured from the center of the square microcavity as in [2], respectively.

The mode *Q* factors are shown in Figure 11.3a,b as a function of mode wavelength for the even parity TE modes in the HSRL. A complex refractive index $n = \mathrm{Re}(n) + i\,\mathrm{Im}(n)$ is utilized to describe the structure material. The FP cavity is set to be $\mathrm{Im}(n_{\mathrm{FP}}) = 0$, and the square cavity has $\mathrm{Im}(n_{\mathrm{SQ}})$ varied from 0.0002 to −0.0002, corresponding to the gain from −16 to 16 cm^{-1}, where n_{FP} and n_{SQ} are complex refractive indices of the FP and square cavities, respectively. The mode *Q* factors increase significantly with the gain of the square cavity around 1544.32 and 1541.43 nm, and the other modes show low sensitivity to the variation of $\mathrm{Im}(n_{\mathrm{SQ}})$ as the variation of mode reflectivity with the gain in Figure 11.2. Detailed mode *Q* factors vs. mode wavelengths are presented in Figure 11.3b by different symbols for the high *Q* hybrid modes. M1 and M2 are hybrid modes between the fundamental transverse mode of the WGMs and an FP mode, and M3 and M4 are those between the first-order transverse mode of the WGMs and an FP mode. Figure 11.4 shows the variation of mode *Q* factors with $\mathrm{Im}(n_{\mathrm{FP}})$ varied from 0.0002

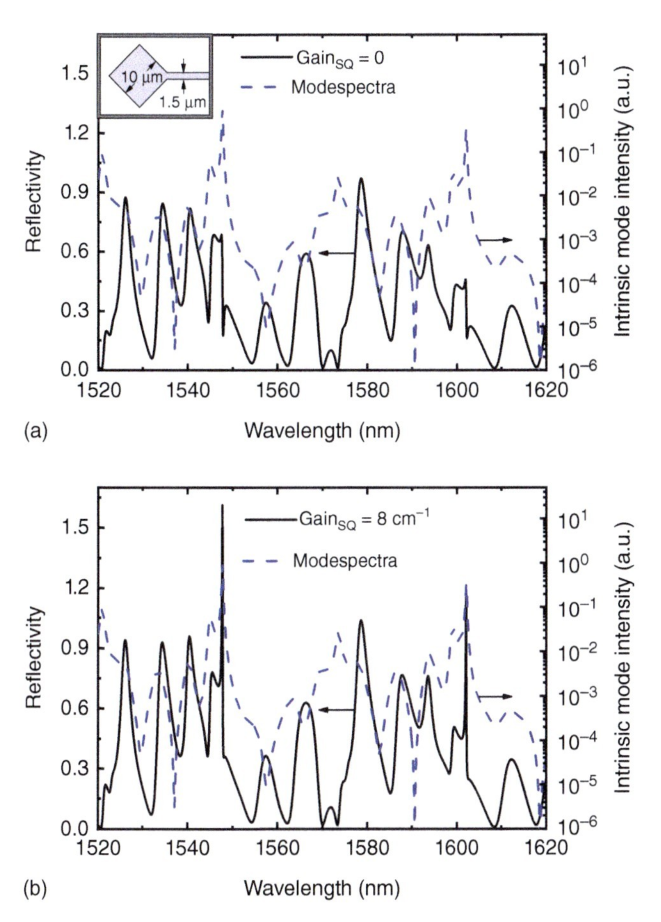

Figure 11.2 Simulated reflectivity spectra at different gain levels in the square cavity at the gain of square resonator of (a) 0 and (b) 8 cm^{-1}, and normalized mode-intensity spectrum obtained by the FDTD method, for a square microcavity connected to a vertex waveguide with a square side length $a = 10\,\mu$m and a waveguide width $w = 1.5\,\mu$m. Source: Based on Ma et al. [2]. © 2017, IEEE.

to −0.0002 at Im(n_{SQ}) = 0. We observe that all the mode Q factors increase with the FP-cavity gain.

The WG–FP mode coupling and the modulation of Q factors for the hybrid modes in an HSRL are further demonstrated numerically by calculating the mode wavelength and Q factor vs. the variations of refractive indices in the FP cavity Δn_{FP} and the square microcavity Δn_{SQ}, as plotted in Figure 11.5a,b. The red short-dashed lines are the mode wavelength of bare fundamental WGM in the square microcavity, and the black dashed lines are the mode wavelengths of the fundamental transverse mode in the bare FP cavity. Wavelength anticrossing and mode Q factor crossing occur due to the mode coupling between the bare WGM and FP modes. By tuning Δn_{FP}, the mode wavelengths of the high-Q hybrid modes are slightly tuned from 1544.2 to 1544.5 nm periodically as shown in Figure 11.5a, due to the mode coupling between the fundamental WGM and different-order FP longitudinal modes.

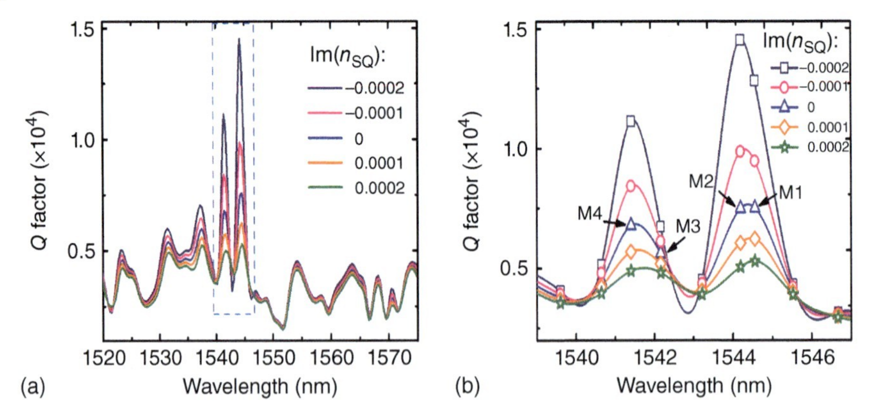

Figure 11.3 (a) Calculated mode Q factor vs. mode wavelength for the even TE modes in the HSRL at the different values of Im(n_{SQ}) as Im(n_{FP}) = 0, and (b) enlarged view of the high-Q modes. Source: Ma et al. [1]. © 2016, AIP Publishing LLC.

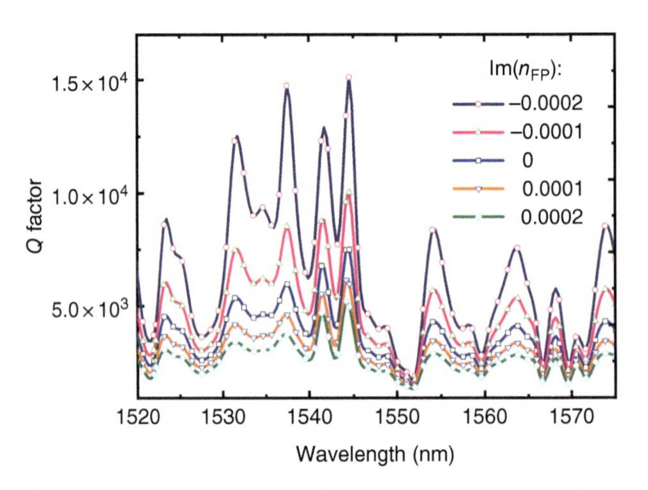

Figure 11.4 Calculated mode Q factor vs. mode wavelength for the even TE modes in the HSRL with different values of Im(n_{FP}) at Im(n_{SQ}) = 0. Source: Modified from Ma et al. [1].

Similar modulation of mode coupling can also be obtained by tuning Δn_{SQ} as shown in Figure 11.5b. The wavelength of the WGM increases linearly with the growing Δn_{SQ} and then anticrosses with different-order FP modes, which in consequence drives the wavelength redshift of high-Q hybrid mode and predicts the feasibility of lasing wavelength tuning for the HSRLs.

11.4 Hybrid Mode-Field Distributions

The mode-field distributions of the z-directional magnetic field H_z are calculated by 2D FEM for the TE hybrid modes. Furthermore, the full form of Fourier transform of the mode-field distributions in the square cavity is calculated using

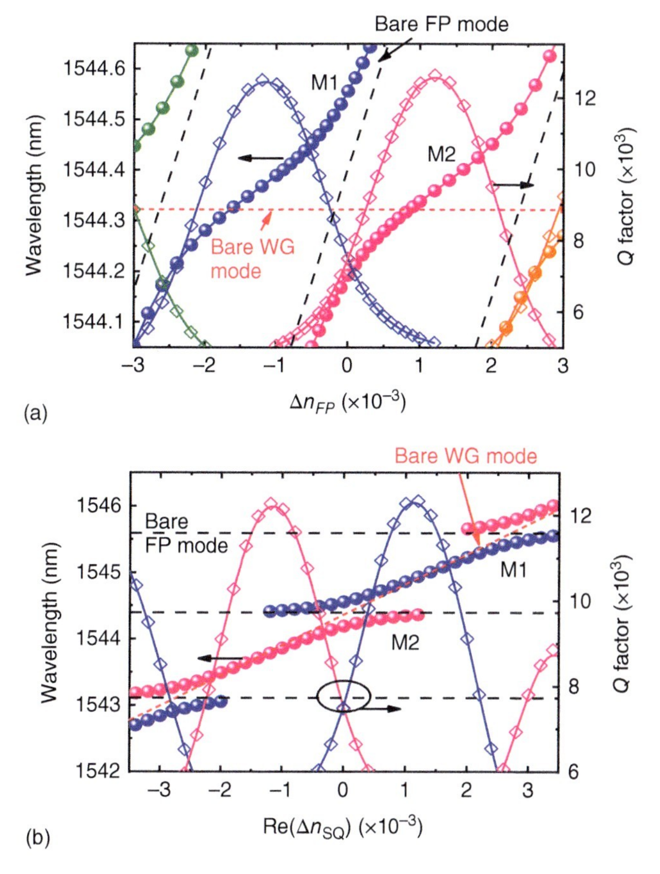

Figure 11.5 Mode wavelengths and Q factors vs. the variations of (a) Δn_{FP} and (b) Δn_{SQ}, for the even WG–FP hybrid mode doublets in an HSRL with $a = 10\,\mu m$, $w = 1.5\,\mu m$, and $L = 300\,\mu m$. Source: Ma et al. [2]. © 2017, IEEE.

the transformation Eq. (7.20) to distinguish the WGMs involved in the hybrid mode. The mode-intensity distributions of $|H_z|^2$ are shown in Figure 11.6a,b for the hybrid modes M1 and M2 at the wavelengths of 1544.56 and 1544.19 nm in the HSRL with $a = 10\,\mu m$, $w = 1.5\,\mu m$, and $L = 300\,\mu m$ at $\text{Im}(n_{FP}) = \text{Im}(n_{SQ}) = 0$, respectively, with the insets of the detailed profiles inside the square cavity and the FP cavity. The full Fourier transform patterns of Eq. (7.20) with the space as (x', y') and the integration domain of x', $y' \in [-8, 8]\mu m$ are expressed as $(k_{x'}a/\pi - 1$, $k_{y'}a/\pi - 1)$ in Figure 11.6b,e to indicate the mode numbers (p, q). However, they are smaller than (p, q) due to the slightly larger effective side length resulting from the evanescent field around the microcavity boundary. To clearly distinguish the mode indexes, the transform patterns of the last term in Eq. (7.20) with the wavefunction of $\cos(k_{x'}x')\sin(k_{y'}y')$ are plotted in Figure 11.6c,f. The results reveal that the high-Q mode doublets M1 and M2 behave as the fundamental WGMs in the square cavity formed by the superposition between $\text{TE}^{e,(28,29)}$ and $\text{TE}^{e,(30,27)}$, but with different composition ratios. $\text{TE}^{e,(p,q)}$ is used to describe TE WGMs in the square cavity, where the superscript "*e*" refers to modes with even parity relative to the diagonal line of the square microcavity. In addition, the mode hybridization also induces a nonpure fundamental transverse mode pattern in the FP-cavity region. In contrast, the intensity distributions and the corresponding Fourier transform patterns for hybrid modes M3 and M4 show that they are formed by mode hybridization between the first-order WG mode and the FP mode, and the first-order WG mode in square cavity origins from the superposition between $TE^{e,(30,27)}$ and $TE^{e,(26,31)}$ with different composition ratios [2].

Moreover, the far-field distributions are calculated from the complex near-field mode distributions using Eq. (6.25) with the integral along an arbitrary right perimeter of external region surrounding the coupled cavity. The ratio R of mode energy density integrals over the FP cavity to that over the whole HSRL is also calculated for the hybrid modes. Comparing the Q factor in Figure 11.5a and the ratio R in Figure 11.7a, we can find that the highest Q occurs simultaneously with the lowest FP-cavity energy ratio R. Mode-field patterns $|H_z|^2$ near the output facet are shown in Figure 11.7b,c, for the hybrid modes M1 and M2. As shown in Figure 11.7d at $L = 302\,\mu m$, the near-field patterns at the cleaved mirror vary greatly with the FP-cavity length because the mode-field patterns are not purified fundamental transverse mode, and a little narrow far-field pattern is observed.

11.5 Fabrication of Hybrid Lasers

The fabrication processes of the HSRLs are simply introduced in this section. An AlGaInAs/InP laser wafer grown by metal–organic chemical vapor deposition was employed to fabricate the HSRLs. The active region of the laser wafer consists of eight compressively strained 6-nm-thick quantum wells and nine 9-nm-thick barrier layers with a 1.52-μm photoluminescence peak. The upper layers are P-type cladding InP layer and a P^+-InGaAs contact layer with a total thickness of 1.6 μm. After growing a SiO_2 layer, standard contacting photolithography and inductively coupled plasma (ICP) etching techniques are used to transfer the coupled-cavity patterns with a deep etching depth of about 4 μm. The scanning electron microscope (SEM) image is shown in Figure 11.8a for a hybrid coupled cavity after ICP etching process. Then, the microcavity is confined by a 200-nm SiN_x layer to protect the active region from being oxidized and a BCB cladding layer is coated to create a planar surface followed by large area of reactive ion etching to expose the top of coupled-cavity resonators. To guarantee mutual electrical isolation, an isolation trench with a length of 20 μm is realized by another ICP etching technique to etch off the P-InGaAsP ohmic contact layer between the rectangular and square microcavity sections. After that, a patterned Ti/Pt/Au metal layer is evaporated as a top P-electrode using lifting-off technique and the substrate is mechanically lapped to a thickness of 130 μm with an Au–Ge–Ni metallization layer evaporated on the backside as the N-electrode. The microscope image of the fabricated HSRL is shown in Figure 11.8b after the deposition of P-type and N-type electrodes. Two patterned P-type electrodes are used for current injection into the square and the FP cavities separately.

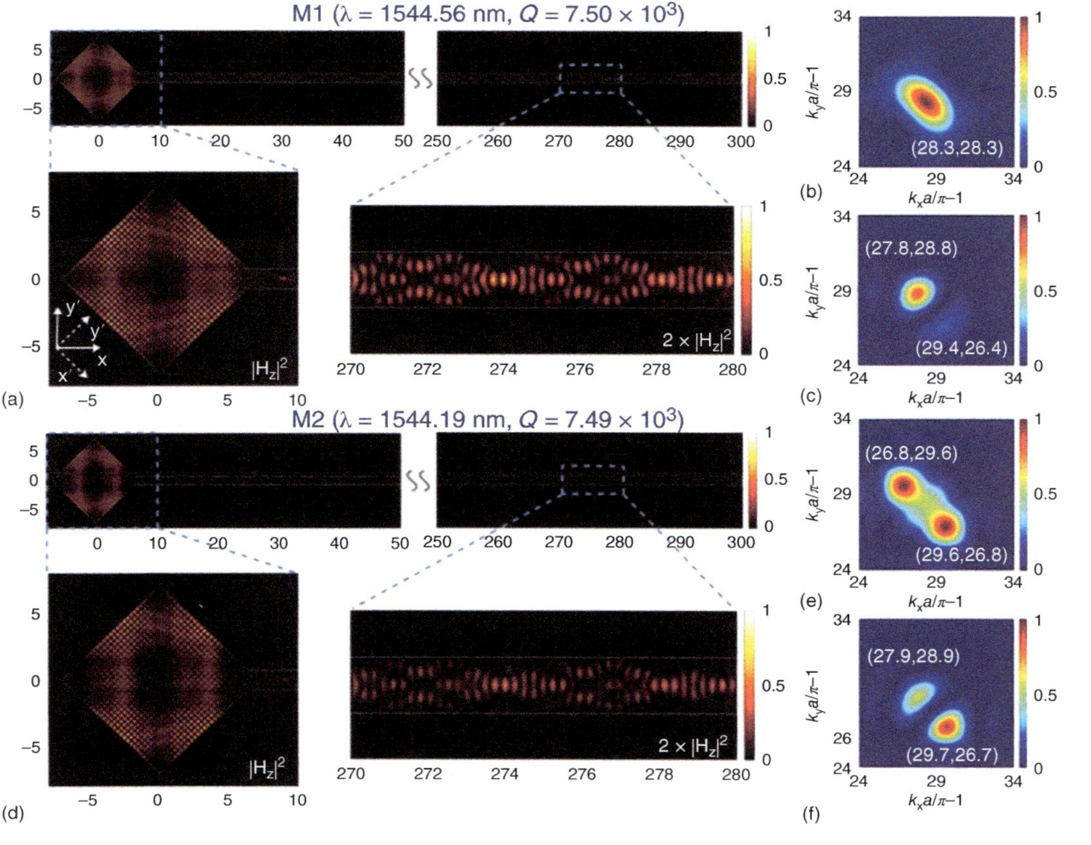

Figure 11.6 Mode-intensity profiles of $|H_z|^2$ for the high-Q modes (a) M1 and (d) M2, where the FP cavity from $L = 50$ to $250\,\mu m$ is omitted for a better view. Zoom in: enlarged views in the square cavity and the FP cavity from $L = 270$ to $280\,\mu m$. Fourier transform patterns in k-space for (b) M1 and (e) M2 in the square cavity, and one term of the Fourier transform patterns in (c) M1 and (f) M2. Source: Ma et al. [2]. © 2017, IEEE.

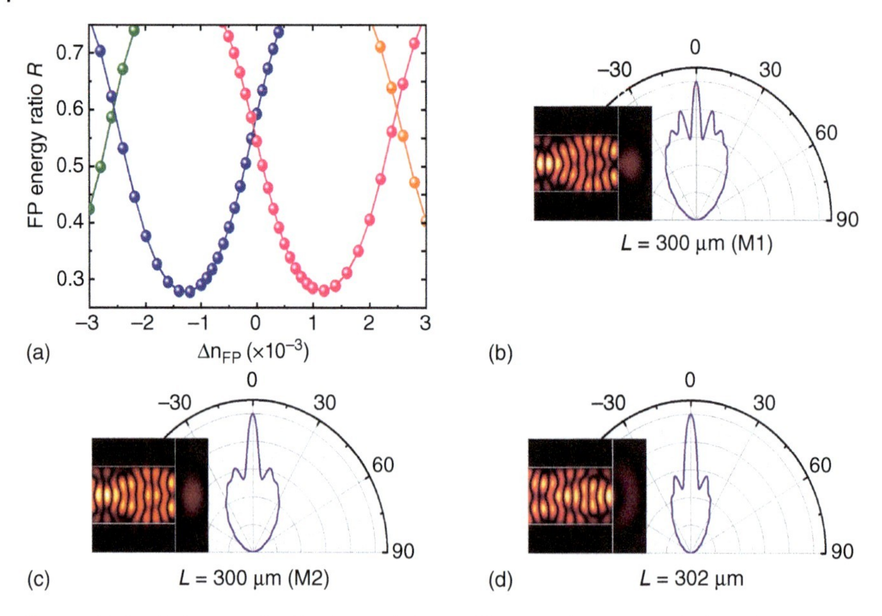

Figure 11.7 (a) Energy distribution ratio in the FP cavity R vs. the variation of Δn_{FP} for the even hybrid modes in the HSRL. Simulated far-field intensity profiles and mode-field patterns $|H_z|^2$ near the output facet for the high-Q hybrid modes (b) M1 and (c) M2 in the HSRL, and (d) at $\lambda = 1544.29$ nm in an HSRL with $L = 302\,\mu$m. Source: Ma et al. [2]. © 2017, IEEE.

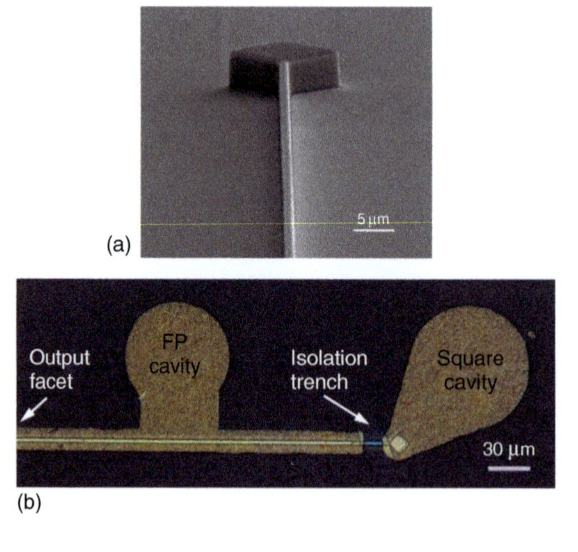

Figure 11.8 (a) SEM image of a square-rectangular coupled cavity after the ICP etching process. (b) Microscope image of the fabricated HSRL with BCB confinement after the deposition of the P-type and N-type electrodes. Source: Ma et al. [2]. © 2017, IEEE.

11.6 Q-Factor Enhancement and Lasing Characteristics

The fabricated devices were cleaved to provide a reflection facet for the FP cavity, with the FP-cavity length L of about 300 μm. Characterization measurements are performed after the laser chips are mounted on an AlN submount with

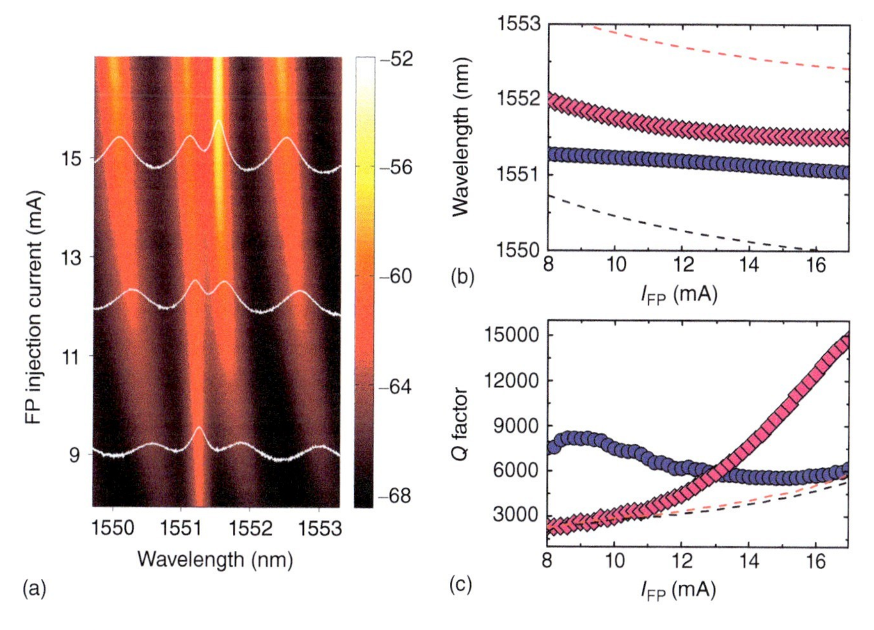

Figure 11.9 (a) Lasing spectra map exhibits the WG-FP mode hybridization with the variation of FP-cavity injection current, where the white curves are the lasing spectra at $I_{FP} = 9, 12, 15$ mA, respectively. (b) The corresponding mode wavelengths and (c) mode Q factors are plotted as functions of I_{FP}, for the HSRL with $a = 10\,\mu m$ and $w = 1.5\,\mu m$ at $I_{SQ} = 4$ mA. Source: Ma et al. [1]. © 2016, AIP Publishing LLC.

the temperature controlled by a thermoelectric cooler (TEC). Continuous-wave injection currents I_{FP} and I_{SQ} were applied to the square and the FP-cavity regions separately. The lasing spectra are measured at the optical spectrum analyzer resolution of 0.02 nm.

For an HSRL with $a = 10\,\mu m$ and $w = 1.5\,\mu m$, the lasing spectra map with the variation of the FP-cavity current I_{FP} at the square cavity current $I_{SQ} = 4$ mA are measured and plotted in Figure 11.9a at a TEC temperature of 287 K. Below the threshold, the FP modes shift to shorter wavelengths with the increase of I_{FP} due to the free carrier dispersion effect, i.e. the reduction of refractive index with the increase of carrier density. The corresponding mode wavelengths and Q factors for the hybrid mode doublets are plotted in Figure 11.9b,c, with the Q factors deduced from the full-width at the half-maximum (FWHM) of each resonance peak. The results indicate the anticrossing mode coupling between WGM and the FP mode as predicted in Figure 11.5a. The Q factors for the hybrid mode doublets are modulated significantly from 8.15×10^3 to 5.56×10^3 and 2.63×10^3 to 9.99×10^3, respectively, as I_{FP} is increased from 9 to 15 mA due to the mode coupling. The mode wavelengths and Q factors of two adjacent FP modes are marked by the dashed lines in Figure 11.9b,c, where the mode Q factors increase monotonously from 2.46×10^3 to 4.24×10^3 and 2.63×10^3 to 4.52×10^3 as the current increases from 9 to 15 mA, which correspond to the gain variation of the FP cavity. The mode Q-factor enhancement due to the mode coupling can result in single-mode operation for the HSRL.

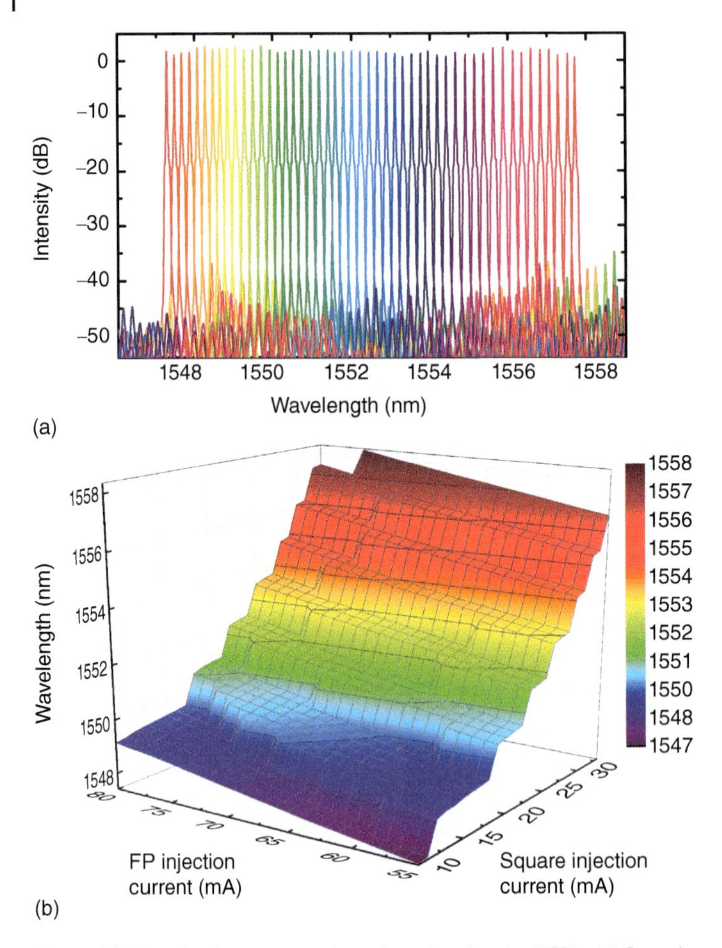

Figure 11.10 Continuous wavelength tuning for the HSRL. (a) Superimposed lasing spectra showing wavelength tuning range over 10 nm by varying the injection currents of two cavities simultaneously. (b) Lasing wavelength tuning map vs. injection currents. Source: Ma et al. [1]. © 2016, AIP Publishing LLC.

Single-mode operation with a continuous wavelength tuning range over 10 nm at side-mode suppression ratios (SMSRs) over 37 dB was demonstrated by varying the injection currents of the two cavities simultaneously. The lasing spectra and the lasing wavelength tuning map are shown in Figure 11.10a,b, respectively. The results further support the feasibility of excellent mode selection and wavelength tuning by integrating a WG mode cavity with an FP cavity. The output powers collected by an SMF and an integrated sphere at room temperature are plotted as functions of the FP-cavity current in Figure 11.11a at the square cavity current of 18 mA for the HSRL. Near-symmetric near-field patterns are observed below and above threshold as shown in Figure 11.11b,c, respectively, measured from the cleaved FP facet by an infrared-charged coupled device. The far-field profiles in Figure 11.11d demonstrate the near-symmetric output pattern, with the FWHM of 45° and 42° in the in-plane and out-of-plane directions, respectively. The intensity oscillation on the substrate (positive angle) side for the out-of-plane profile is caused by the

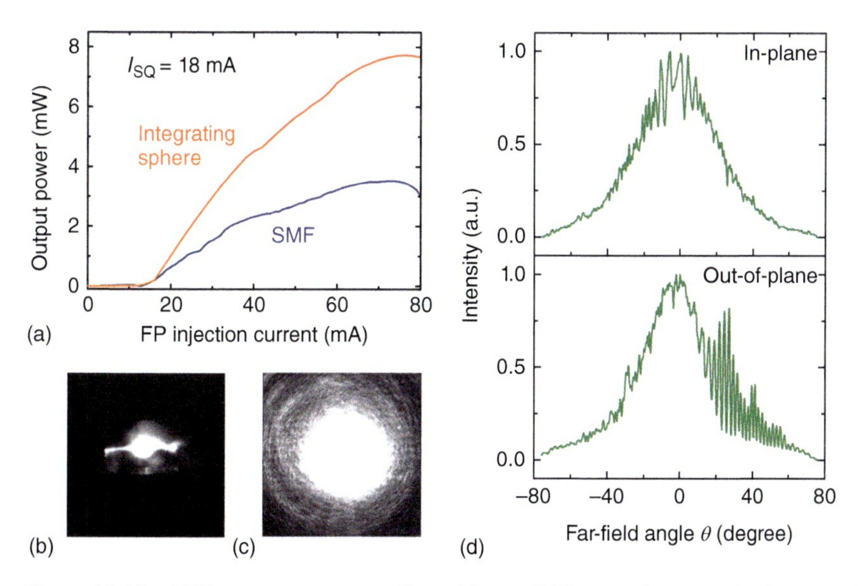

Figure 11.11 (a) The output powers collected by an SMF and an integrated sphere vs. the FP-cavity current, near-field patterns (b) below threshold and (c) above threshold, and (d) in-plane and out-of-plane far-field profiles of the HSRL.

scattering of the submount. The results indicate that good mode matching and hence a high-coupling efficiency can be obtained between the laser and an SMF; even the resonance mode in the FP cavity contains high-order transverse-mode components as shown in Figure 11.7.

11.7 Robust Single-Mode Operation

In this section, we demonstrate the stability of single-mode operation for the HSRLs, which is a big problem for traditional coupled-cavity lasers. The lasing spectra were measured for an HSRL with a = 15 μm, w = 2 μm, and L = 300 μm, under different injection currents. Single-mode operation at 1545.69 nm with an SMSR of 47 dB is realized as shown in Figure 11.12a at I_{FP} = 40 mA and I_{SQ} = 10 mA. The Q factor of the dominant mode is estimated to be 8.1×10^3 based on the full-width at the half-maximum just at the threshold. Single-mode lasing with an SMSR of 32 dB is observed as I_{FP} = 40 mA and I_{SQ} = 0 as shown in Figure 11.12b. In Figure 11.12c, the lasing of a bare WGM at 1541.57 nm is observed as I_{FP} = 1 mA and I_{SQ} = 10 mA, with an SMSR of 31 dB and a free spectral range of 15.3 nm. From this perspective, HSRLs can work as a laser amplifier as the FP cavity is set below threshold and serves as an optical amplifier. Lasing spectra were measured and plotted in Figure 11.12d around the lasing threshold at I_{SQ} = 10 mA and I_{FP} = 7 and 11 mA. Two mode doublets around 1540.4 and 1541.5 nm are observed around bare fundamental and first-order WGMs; however, the final lasing mode is about 1544.8 nm. In addition, mode wavelength can be blueshift or redshift with the increase of injection current, as the refractive index decreases with the increase of carrier density due to free carrier dispersion and increases with the current because of heating effect.

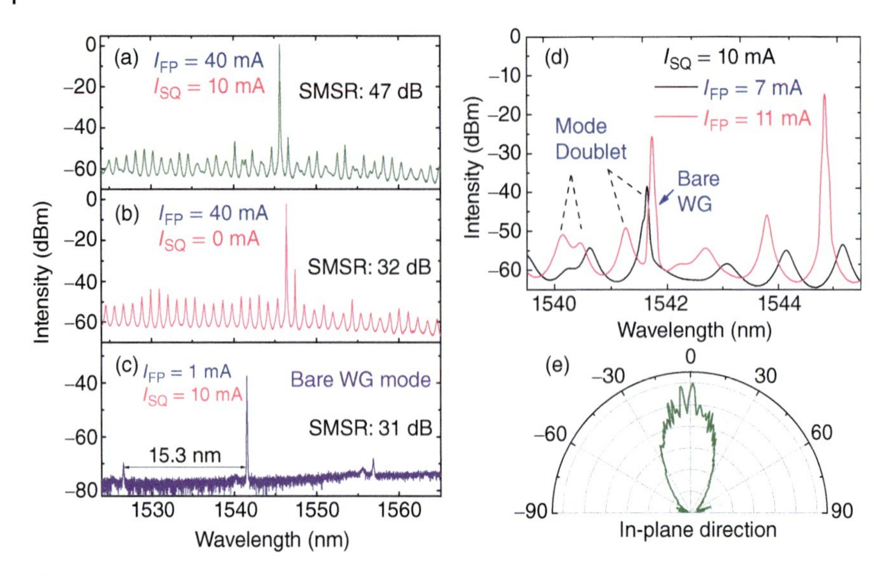

Figure 11.12 The lasing spectra of the device at (a) I_{FP} = 40 mA and I_{SQ} = 10 mA, (b) I_{FP} = 40 mA and I_{SQ} = 0, and (c) I_{FP} = 1 mA and I_{SQ} = 10 mA. (d) Detailed spectra measured around lasing threshold at I_{SQ} = 10 mA and I_{FP} = 7 and 11 mA, respectively. (e) Far-field pattern for the HSRL at injection currents of I_{FP} = 40 mA and I_{SQ} = 10 mA. Source: Ma et al. [2]. © 2017, IEEE.

Moreover, the far-field profile for the HSRL is presented in Figure 11.12e with an in-plane FWHM of about 45°.

The detailed lasing spectra maps are shown in Figure 11.13a,b vs. the injection currents of the FP cavity and the square microcavity, respectively. Bare WGMs are observed at low I_{FP} and hybrid modes appear as I_{FP} increases. Stable single-mode operations with SMSRs over 40 dB around 1546 nm as I_{FP} increases from 28 to 70 mA and 38 to 70 mA are demonstrated in Figure 11.13c at I_{SQ} = 10 and 13 mA, respectively. The SMSRs decrease at higher I_{FP} because long-wavelength modes increase with the redshift of gain spectrum. The lasing mode wavelength redshifts from 1544.77 to 1547.5 nm as I_{FP} increases from 12 to 80 mA at I_{SQ} = 10 mA. Temperature rise of 27 K is expected for the FP-cavity region by taking the mode wavelength vs. temperature at a rate of 0.1 nm K^{-1}. Similarly, lasing mode wavelengths and SMSRs vs. I_{SQ} are plotted in Figure 11.13d. Mode hopping to an adjacent shorter wavelength accompanied by the dips in SMSR is observed around I_{SQ} = 5 mA, while the subsequent redshift above threshold is mainly due to heating effect.

In Figure 11.14, we summarize threshold current densities of the FP cavity at I_{SQ} = 10 mA, for HSRLs with a square side length a = 10, 15, and 20 μm and an FP-cavity width w = 1.5 μm and 2 μm. With the increase of a from 10 to 20 μm, the lowest threshold current density decreases from 2.4 to 1.5 kA cm^{-2}, because of the increase of resonant mode reflectivity and mode Q factor. In addition, the wide disparity of the threshold current densities at w = 1.5 μm, a = 10 and 15 μm indicates a large variation of the scattering loss for the narrow FP cavity, and more uniform threshold current densities are realized at w = 2 μm, a = 15 and 20 μm. The lasing

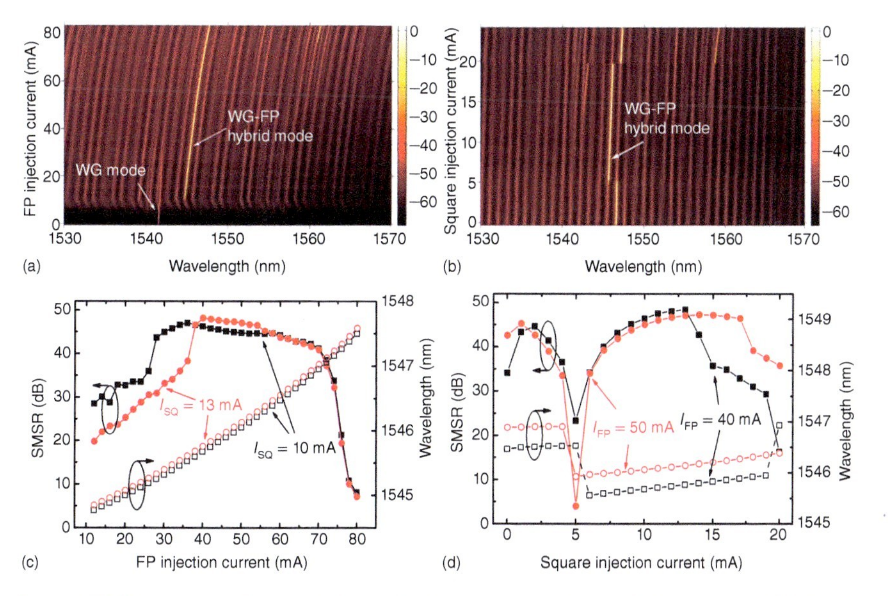

Figure 11.13 Lasing characteristics with the variations of I_{FP} and I_{SQ} for the HSRL. Lasing spectra (a) vs. I_{FP} at I_{SQ} = 10 mA, and (b) vs. I_{SQ} at I_{FP} = 40 mA. Dominant lasing mode wavelengths and corresponding SMSRs (c) vs. I_{FP} at I_{SQ} = 10 and 13 mA, and (d) vs. I_{SQ} at I_{FP} = 40 and 50 mA, respectively. Source: Ma et al. [2]. © 2017, IEEE.

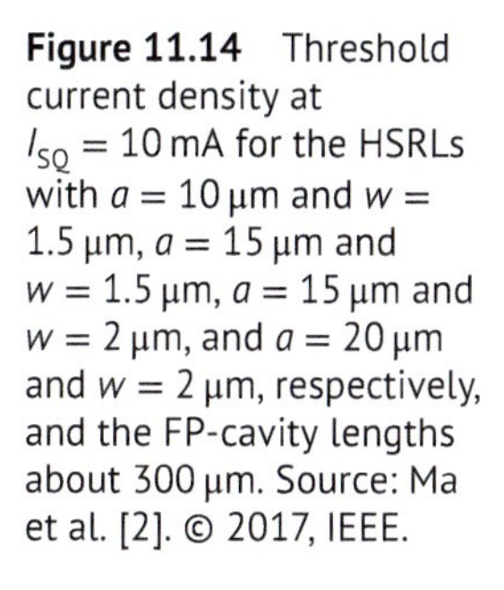
Figure 11.14 Threshold current density at I_{SQ} = 10 mA for the HSRLs with a = 10 μm and w = 1.5 μm, a = 15 μm and w = 1.5 μm, a = 15 μm and w = 2 μm, and a = 20 μm and w = 2 μm, respectively, and the FP-cavity lengths about 300 μm. Source: Ma et al. [2]. © 2017, IEEE.

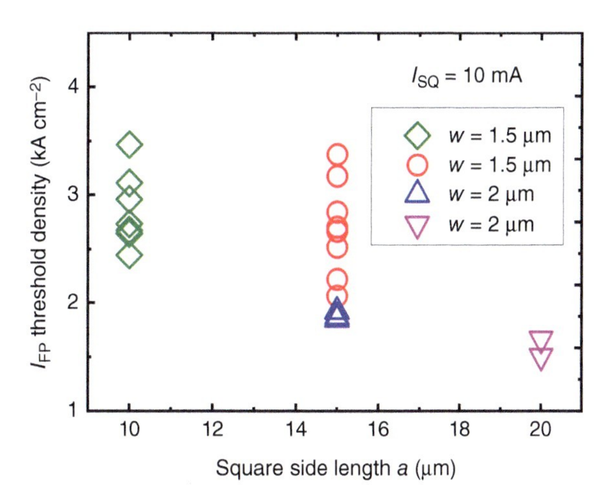

spectra for all these HSRLs show excellent single-mode operation with SMSRs above 42 dB. The results indicate that HSRL is a highly robust hybrid-cavity laser for stable single-mode operation.

11.8 Optical Bistability for HSRLS

Optical bistable semiconductor lasers, with the ability to switch between two output states under certain triggering pulses, have been extensively investigated based on

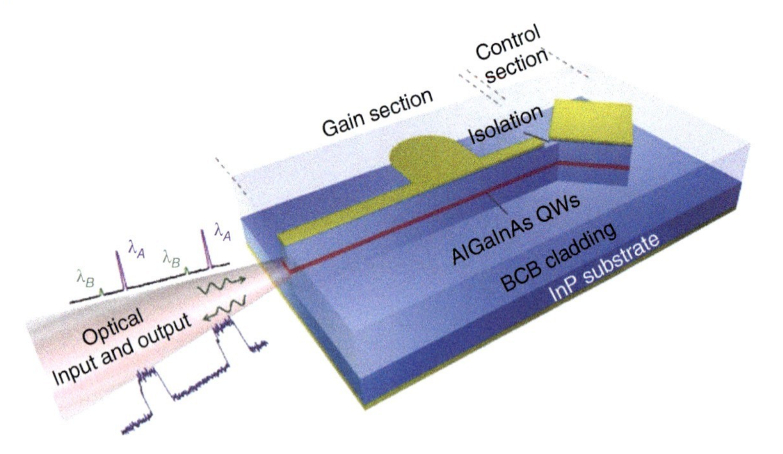

Figure 11.15 Schematic diagram of an HSRL with an isolation trench to ensure the electrical isolation between the FP cavity and square cavity as gain and absorptive section, respectively. A single-mode fiber is coupled to the FP-cavity cleavage face to launch light into the cavity as well as to collect the laser output light. Source: Ma et al. [3]. © 2017, Optical Society of America.

saturable absorptive bistability and two-mode intensity or polarization bistability by nonlinear gain saturation [11–20]. Various geometries of bistable laser diodes have been proposed and demonstrated for applications in optical switch and memory, with the advantages in optical gain and low optical switching energy. The saturable absorption effect of the square microcavity enables the repeatable controllability of on–off bistability and mode competition bistability [3, 21, 22].

The optical bistable HSRLs are realized due to mode competition and saturable absorption at the square microcavity section. As shown in Figure 11.15, the laser output of a bistable HSRL can be switched by input optical pulses. Mounted on a heat sink with temperature controlled by a thermoelectric cooler, the HSRLs were tested under continuous-wave injection current. The output power coupled into an SMF vs. the FP current I_{FP} was measured and is given in Figure 11.16a, for an HSRL laser with square side length $a = 15\,\mu m$, and FP-cavity width $w = 1.5\,\mu m$ and length $L = 300\,\mu m$, at the temperature of 287 K and the square section current $I_{SQ} = 0$ and 10 mA.

The threshold currents are 23 and 10 mA, and the maximum output power are 2.4 and 3.8 mW at $I_{SQ} = 0$ and 10 mA, respectively. Bistable hysteresis loops are observed at $I_{SQ} = 0$ around $I_{FP} = 23.5$ mA and between $I_{FP} = 43.5$ to 51 mA, as shown in the enlarged views in Figure 11.16b,c. The bistable loop around the threshold is related to the saturable absorption effect in the square cavity owing to the accumulation of photon-generated carriers, just like common-cavity two-section bistable lasers [11, 12]. The practical temperature of the FP region is estimated to be 305 and 325 K at 50 and 70 mA as $I_{SQ} = 0$, by taking the redshift of mode wavelength vs. temperature at a rate of 0.1 nm K^{-1}. The lasing spectra of the upper and lower states at $I_{FP} = 23.5$ mA and $I_{SQ} = 0$ are plotted in Figure 11.17 with a lasing mode wavelength of 1530 nm in the upper state, which shows bistability switching between spontaneous emission state and lasing state.

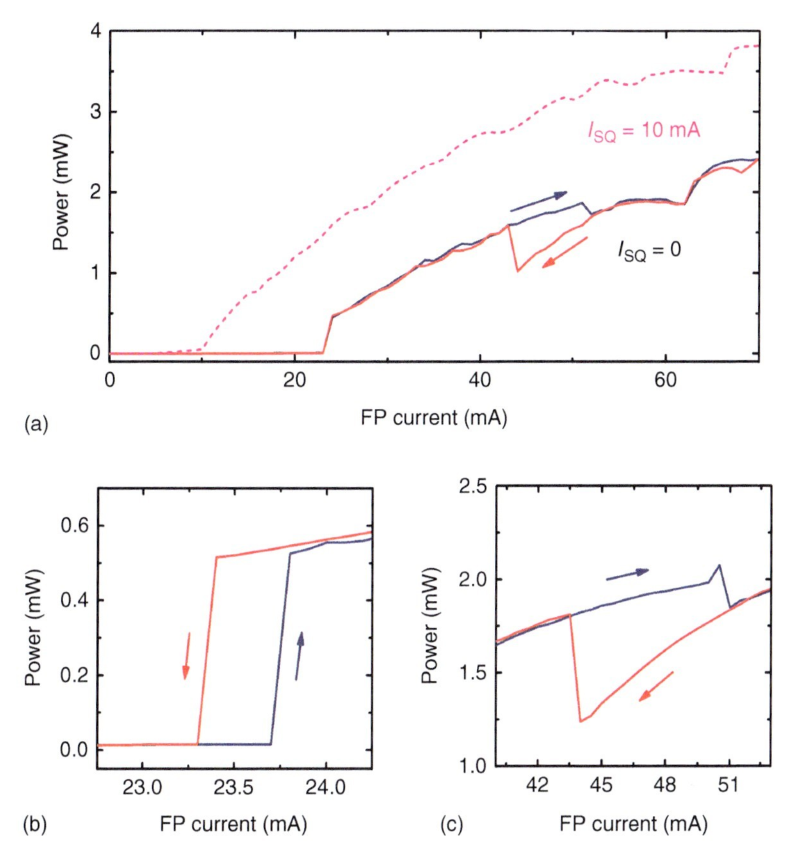

Figure 11.16 (a) Single-mode fiber-coupled output power vs. the FP-cavity injection current I_{FP} under $I_{SQ} = 0$ and 10 mA at TEC temperature of 287 K, for an HSRL with $a = 15$ μm, $w = 1.5$ μm, and $L = 300$ μm. Enlarged view of the bistable regions around (b) the lasing threshold and (c) the mode hopping between 43.5 and 51 mA under $I_{SQ} = 0$. Source: Ma et al. [3]. © 2017, Optical Society of America.

Furthermore, the lasing spectra for the bistable states at $I_{FP} = 48$ mA and $I_{SQ} = 0$ are presented in Figure 11.18a, which shows lasing mode jumping from 1530.21 nm of the upper state to 1560.44 nm of the lower state. The corresponding wavelength difference 30.2 nm is equal to twice the longitudinal mode interval of the square microcavity, i.e. the longitudinal mode interval of symmetric square WGMs due to the inherent even symmetry for hybrid WG-FP modes. The optical bistability around $I_{FP} = 48$ mA is mainly caused by mode competition. The SMSRs vs. I_{FP} are plotted in Figure 11.18b, which takes the highest value of 45 dB at $I_{FP} = 53$ mA and drops to 27 dB near the transition points of hybrid modes of different WGMs. In addition, the low SMSR values at 27 and 63 mA are caused by the transitions of mode couplings between WGMs with different FP modes with a wavelength interval of 0.9 nm, caused by a faster temperature rise in FP cavity than the square cavity due to the current applied to the FP cavity. Within the hysteresis loop, we further measure the mode net gains for the two competing modes at unlasing states as $g = -2\pi n_g/(Q\lambda)$

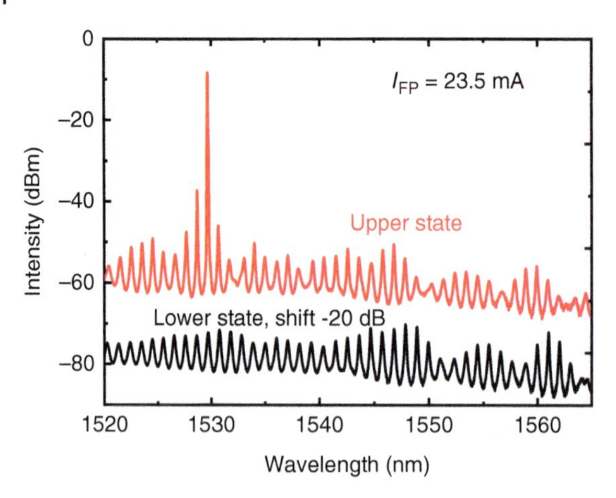

Figure 11.17 Lasing spectra of the bistability states at I_{FP} = 23.5 mA for the hybrid-cavity laser with square side length a= 15 μm, w = 1.5 μm, and L = 300 μm as I_{SQ} = 0 and 287 K. Source: Huang et al. [9]. © 2018, Springer Nature Switzerland AG.

and plot in Figure 11.18c, with the mode Q factors calculated by the FWHM of the mode resonance and the mode group index $n_g = 3.6$. With the increase of I_{FP}, the net gain of the unlasing mode λ_B rises from $-3.8\,\text{cm}^{-1}$($Q = 3.85 \times 10^4$) at 46 mA to $-1.7\,\text{cm}^{-1}$ ($Q = 8.64 \times 10^4$) at 50 mA. However, $Q = 8.64 \times 10^4$ is not quite precise because the fitted FWHM is beyond the resolution of the optical spectrum analyzer. When I_{FP} increases further, the lasing mode switches from λ_A to λ_B. However, as I_{FP} decreases, the net gain of the unlasing mode λ_A gradually rises from $-8.3\,\text{cm}^{-1}$ ($Q = 1.77 \times 10^4$) at 50 mA to $-5.5\,\text{cm}^{-1}$ ($Q = 2.67 \times 10^4$) at 46 mA, which shows a larger gain variation range than that of λ_B with increasing I_{FP}. The net gain of mode λ_B is observed to be higher than that of mode λ_A, due to the relatively higher level of absorption for the short-wavelength mode λ_A. Furthermore, we find that the optical bistability hysteresis varies greatly with temperature [3, 21], because optical bistabilities depend on the saturation absorption of the square cavity and the mode competitions between two modes with near-threshold gains. But gain spectrum peak wavelength and the mode wavelength shift with the temperature at different rates with a ratio about $3 \sim 5$. In addition, we experimentally find that the optical bistability is easy to realize as the square microcavity in an open circuit state than that in closed circuit state at $I_{SQ} = 0$ [21, 22]. The optical bistability characteristics were predicted based on rate equation simulations [21].

Comprehensive two-section dual-wavelength rate equations are established with a phenomenological gain spectrum considering the nonlinear gain and absorptive effect in a wide wavelength range [21]. Based on the steady solution of the constructed model, we affirmed that the bistability loops are caused by the two-mode competition and saturable absorptive effect in the two-section hybrid-coupled cavity. Bistable loops are observed from the laser output curves by the steady solution of the constructed model, which are in accordance with the experimental results. In addition, we found that compared with the proportion of the cavity volume, the proportion of photon number confined in each cavity is more efficient for the control of the width of bistable loops in a wide range, which makes the HSRL more particular than traditional two-section bistable lasers. The simulated output powers vs. the I_{FP} ranging from 47 to 59 mA at 295 K are plotted in Figure 11.19a for modes M1

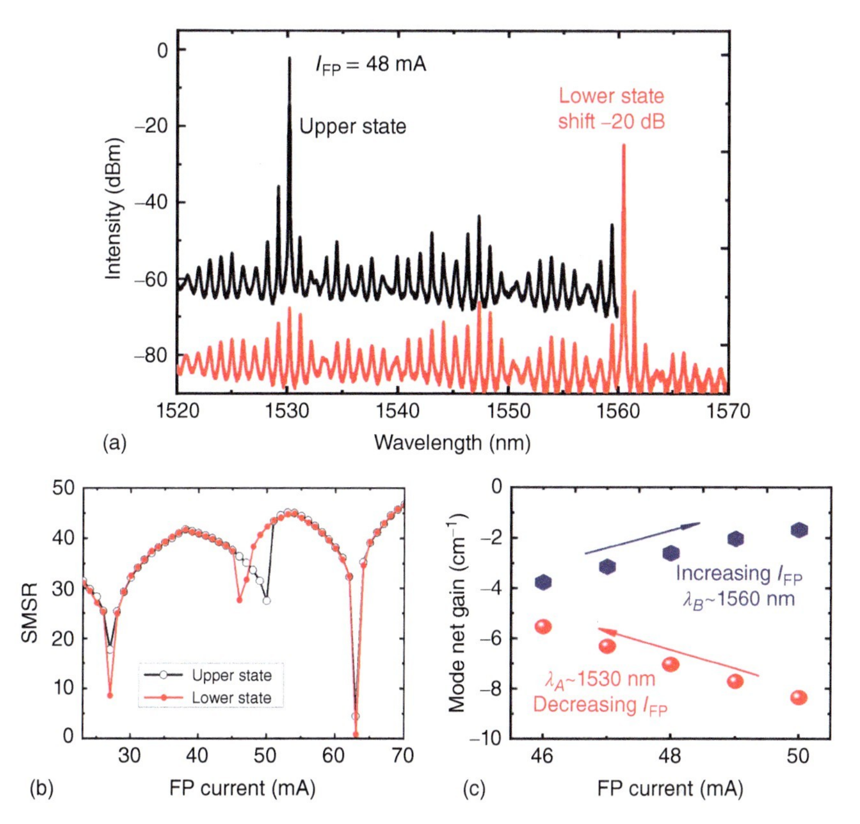

Figure 11.18 (a) Lasing spectra of the upper and lower states at I_{FP} = 48 mA, (b) the corresponding side-mode suppression ratio, and (c) measured mode net gain of unlasing modes vs. the variation of FP current within the hysteresis loop derived from the measured Q factors by fitting the FWHM of the mode peak, for the hybrid-cavity laser with square side length a = 15 μm, w = 1.5 μm, and L = 300 μm as I_{SQ} = 0 and 287 K. Source: Huang et al. [9]. © 2018, Springer Nature Switzerland AG.

and M2 at λ_1 = 1529 nm and λ_2 = 1560 nm, respectively. Mode hopping from M1 to M2 at 59 mA and inverse mode hopping from M2 to M1 at 47 mA are observed with the increase and decrease of I_{FP}, respectively. The corresponding carrier densities in the FP cavity and square microcavity sections are also plotted in Figure 11.19b to verify bistable operation mechanism. The lasing mode is first located at M1 with a higher carrier density near absorption saturation at λ_1. By further increasing I_{FP}, the lasing mode is switched to mode M2 at λ_2 because of the redshift of the FP gain spectrum caused by the heating effect. The carrier density in the square section decreases because of the lower absorption at long-wavelength mode M2. Inversely, with the decrease in the FP current, the mode gain around M1 increases and abrupt mode switching to M1 occurs as the absorption loss of M1 in the square section is compensated. The square section without electrical biasing provides a threshold gain difference required for optical bistability. In the case of no biasing, the carrier density in the square microcavity is smaller than the transparency carrier density at the pumping wavelength.

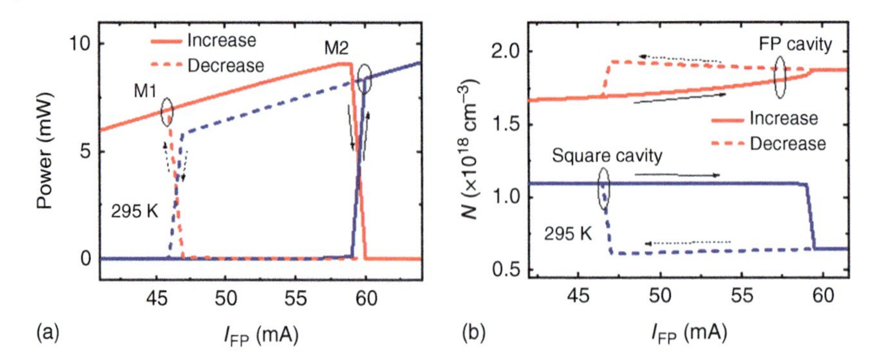

Figure 11.19 (a) Calculated output powers of modes M1and M2 and (b) carrier density in the FP and the square cavity varies with the increase (solid lines) and decrease (dash lines) of I_{FP} at temperature 295 K. Source: Wang et al. [21]. 2019, AIP Publishing LLC. Licensed under CC BY 4.0.

11.9 All-Optical Switching

The saturable absorption effect of the square section enables the repeatable controllability of bistability induced by mode competition. The dynamic switching operation of all optical flip-flop was demonstrated by setting the HSRL within the mode-competition bistable region at $I_{FP} = 48$ mA and $I_{SQ} = 0$ at 287 K. Two tunable lasers were used to generate the set/reset pulses. An optical circulator was used to inject the trigger pulses into the HSRL, and the injection polarization was carefully aligned with the laser mode by adjusting the polarization controller. In the low-speed switching measurement, the HSRL output was filtered by a tunable band-pass filter (BPF) with the bandwidth of 1.5 nm, and finally recorded by a high-sensitivity detector. In the high-speed switching measurement, the injected lights were modulated by two 10 Gits^{-1} lithium niobate modulators driven by a 12.5-Gits^{-1} pattern generator. The pulses were combined by a 50 : 50 coupler, with an optical delay line (ODL) inserted in one arm to adjust the arrival time of the set/reset pulses. Finally, the HSRL output was filtered and recorded by a high-speed sampling oscilloscope.

For the HSRL within the mode-competition bistable region of Figure 11.16 at $I_{FP} = 48$ mA and $I_{SQ} = 0$ at 287 K, the optical memory switching operation was measured and shown in Figure 11.20a,b, under injecting optical pulses at $\lambda_A = 1530.21$ nm and $\lambda_B = 1560.44$ nm with a pulse width of one second and peak power of 50 μW, respectively. The output powers traces in Figure 11.20a,b were measured at λ_A and λ_B wavelength channels, obtained by tuning the optical BPF. Initially, the HSRL was set at λ_B dominant state, so the power measured at the λ_A wavelength channel was low. As an optical set pulse at λ_A was injected into the HSRL, the dominant state of the HSRL was switched from λ_B to λ_A with a high-output power at the λ_A channel even after the pulse passes through. Similarly, a reset pulse at λ_B causes the dominant state of the HSRL switch from λ_A to λ_B. A high on/off contrast ratio of 36 dB was obtained. The switching interval time is 30 seconds as shown in Figure 11.20a,b, but memory states can last a long time. The sensitivity

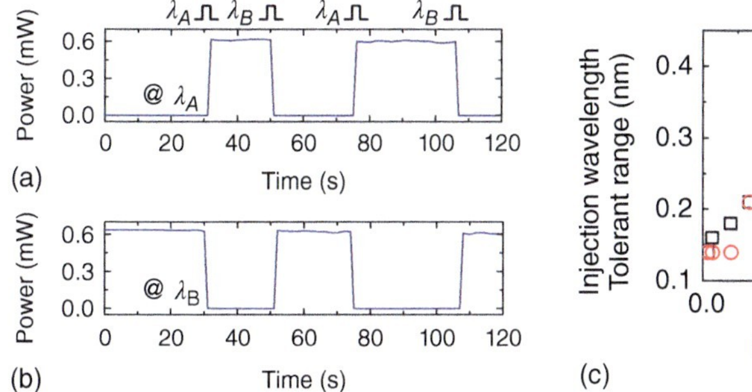

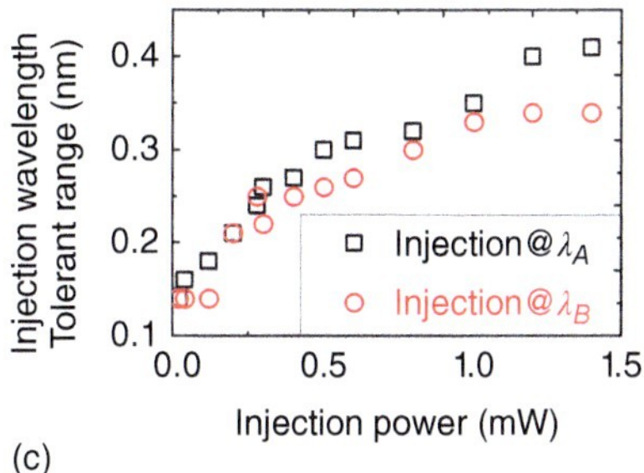

Figure 11.20 Optical memory measured at (a) λ_A and (b) λ_B wavelength channels, under trigger light pulses of 50 μW and a pulse width of 1 s. (c) Injection wavelength-tolerant range vs. injection power around λ_A and λ_B wavelengths. Source: Ma et al. [3]. © 2017, Optical Society of America.

of the flip-flop operation on the trigger signal wavelength was investigated by measuring the injection wavelength-tolerant range vs. injection power around λ_A and λ_B wavelengths, as shown in Figure 11.20c. Wavelength-tolerant range increases from ~0.14 to ~0.4 nm with the injection power increasing from 0.05 to 1.4 mW.

Furthermore, high-speed all-optical flip-flop operation was tested for the HSRL at I_{FP} = 48 mA and I_{SQ} = 0 at 287 K, using set/reset optical pulses with the width of 100 ps at the wavelengths of λ_A and λ_B as shown in Figure 11.21a. The oscilloscope traces of the output optical signals at λ_B and λ_A wavelength channels are shown in Figure 11.21b,c, respectively. Reliable switching between the bistable states was achieved with the lowest injected peak powers of 27 and 142 μW for λ_B and λ_A dominant states, after subtracting the injection coupling loss into the HSRL, with corresponding switching pulse energies of 2.7 and 14.2 fJ. Note that the required set pulse energy for λ_A is much higher than that for λ_B because of the higher loss of λ_A in the HSRL. Details of the state transitions are shown in Figure 11.21d–g for λ_B and λ_A channels. The rising and falling times are about 165 and 60 ps for λ_B dominant state, and 177 and 69 ps for λ_A dominant state, respectively. The rising edges in Figure 11.21d,f are partly attributed to the injected pulse, and the oscillation in the rising edge in Figure 11.21f is due to the relaxation oscillation induced by the large carrier density deviation from stable value under high injection level. The measured on/off contrast ratio is about 8 dB, which is limited by the modulated injection pulses with low signal-to-noise ratios of only 8 and 14 dB for λ_B and λ_A, respectively.

Optical bistability around threshold were also applied for realizing all-optical switching, by setting the square microcavity at an open circuit state as a saturable absorber. Output power coupled into an SMF vs. I_{FP} were measured and plotted in Figure 11.22a around threshold for an HSRL with the same parameters a = 15 μm, w = 1.5 μm, and L = 300 μm. The bistable ranges of 1.7, 3.6, and 6 mA were obtained for the HSRL at 285, 288, and 291 K, respectively. Biased within the bistable loop at I_{FP} = 25 mA and 288 K, the HSRL was injected by set/reset signals through a tapered SMF collimated at the FP facet. Taking set/reset signal power of 50 μW at 1529.9 and 1560.1 nm with a pulse width of one second, we measured output

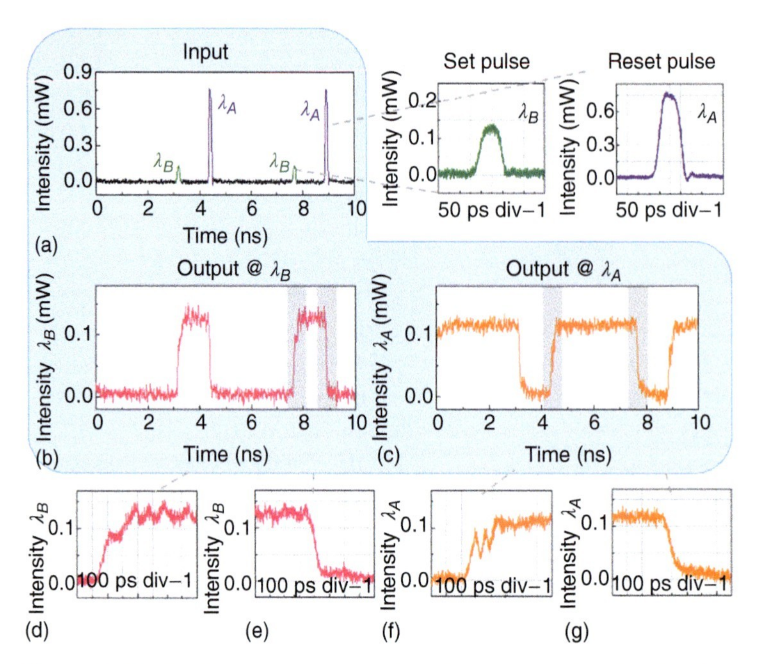

Figure 11.21 Oscilloscope traces showing the dynamic all-optical flip-flop operation. (a) Input trigger optical pulses for the set/reset signals at wavelengths of $\lambda_A = 1530.21$ nm and $\lambda_B = 1560.44$ nm. Inset: Details of the input set/reset pulses with the pulse width of 100 ps. Output optical signals at (b) λ_B and (c) λ_A wavelength channels. Details of the switch-on and switch-off transitions for the (d, e) λ_B wavelength and (f, g) λ_A wavelength channels. Source: Ma et al. [3]. © 2017, Optical Society of America.

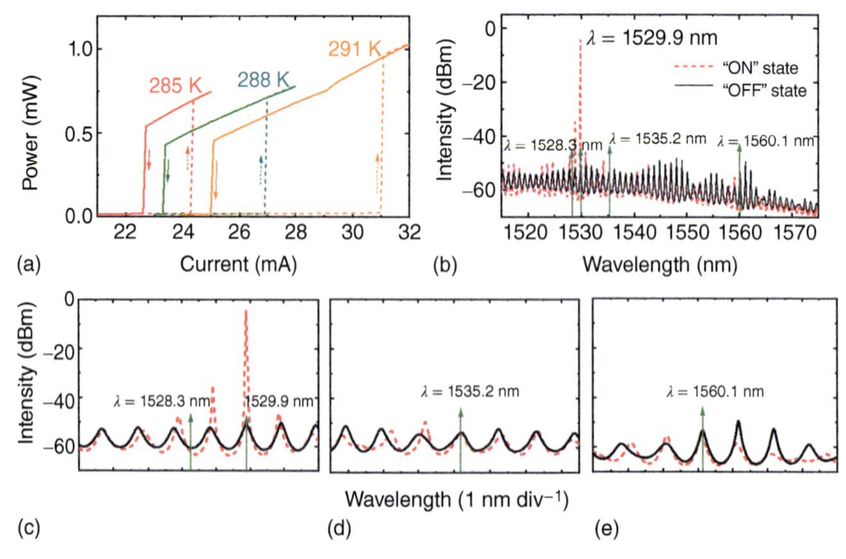

Figure 11.22 (a) Output powers coupled into an SMF vs. I_{FP} around threshold with increase and decrease of the I_{FP} at TEC temperature of 285, 288, and 291 K, (b) lasing spectra of "ON" and "OFF" states and enlarged views of lasing spectra around (c) 1528.3, (d) 1535.2, and (e) 1560.1 nm, at 288 K, for an HSRL at $I_{FP} = 25$ mA, as the square microcavity is at an open circuit. Source: Liu et al. [22]. © 2020, IEEE.

powers of the HSRL as 0.57 and 0.015 mW, corresponding to "ON" and "OFF" states, which can be maintained for a long time, after the trigger signal. The corresponding output spectra for the bistable "ON" and "OFF" states are plotted in Figure 11.22b at I_{FP} = 25 mA, for the HSRL at 288 K. Single-mode operation with an SMSR of 30 dB at 1529.9 nm was obtained for the "ON" state with a high on–off contrast ratio of 46 dB. Figure 11.22c–e is the enlarged lasing spectra around 1528.3, 1535.2, and 1560.1 nm. Trigger wavelength sensitivity was investigated by varying signal wavelengths under different signal powers around 1529.9 and 1560.1 nm for turn-on and turn-off operations. The turn-on trigger signal wavelength-tolerant range increases from 0.27 to 0.99 nm as the trigger signal power increases from 0.01 to 0.07 mW, and ranges from 1509 to 1552.86 nm at the trigger optical power of 0.2 mW. The turn-off trigger signal wavelength-tolerant range increases from 0.12 to 0.74 nm as the trigger signal power increases from 0.05 to 0.6 mW. Saturable absorptive effect is the main mechanism for the optical bistability around the threshold, and the corresponding all-optical switching operation is achieved only if carriers in the square microcavity are filled or consumed by trigger signals. So, the signal wavelength-tolerant ranges can be much larger than those for the all-optical flip-flop based on the mode competition in Figure 11.20, which is related to optical injection to one of the competed modes.

The dynamic operation of all-optical switching was demonstrated by setting the HSRL within the bistable region with I_{FP}= 25 mA at 288 K. Nonreturn-to-zero (NRZ)-bit signal with a pulse width of 500 ps was utilized as optical set/reset signal pulses at wavelengths of λ_{A1} and λ_{B1} as shown in Figure 11.23a. Lowest injected signal peak powers required for realizing reliable all-optical switching are 0.35 mW for set signal pulse at λ_{B1} = 1535.2 nm, and 1.28, 0.54, 0.31, 0.91 mW for reset signal pulses at λ_{A1} = 1560.0, 1560.1, 1560.2, 1560.3 nm, respectively. The corresponding

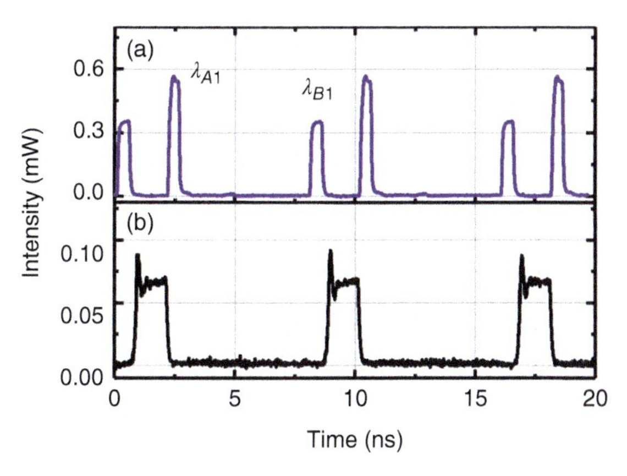

Figure 11.23 Oscilloscope traces showing all-optical flip-flop with injected trigger pulse width of 500 ps for the HSRL with I_{FP} = 25 mA and the square microcavity is an open circuit. (a) Input trigger optical pulses for the set/reset signals with wavelengths at λ_{A1} = 1560.1 nm and λ_{B1} = 1535.2 nm, and (b) output optical signal after filtered by BPF with center wavelength at 1529.9 nm and bandwidth of 1.5 nm. Source: Liu et al. [22]. © 2020, IEEE.

switching signal pulse energies are 0.18, 0.64, 0.27, 0.16, and 0.46 pJ without subtracting the optical injection coupling loss. Filtered by a BPF centered at the lasing wavelength 1529.9 nm, all-optical flip-flop output signal waveform is recorded by a high-speed sampling oscilloscope and plotted in Figure 11.23b. The bistable HSRL biased within bistable loop is switched between the "ON" and "OFF" states when triggered by the pulses at λ_{A1} and λ_{B1}. The input signal at short wavelength λ_{B1} is absorbed in the square microcavity and induces the enhancement of carrier density in the square microcavity. But the stimulated emission of input signal at long wavelength λ_{A1} results in the decrease of carrier density in the square microcavity. The decrease of the carrier density is not determined by the spontaneous emission process, so the switching speed is not limited by the spontaneous carrier lifetime but by the variation of carrier density inside the microcavity.

11.10 All-Optical Logic Gates

For an HSRL with $a = 15\,\mu\text{m}$, $w = 1.5\,\mu\text{m}$, and $L = 300\,\mu\text{m}$, single-mode operation of M0 with SMSR of 39 dB was realized as shown in Figure 11.24a under the injection currents of $I_{SQ} = 15\,\text{mA}$ and $I_{FP} = 40\,\text{mA}$ at 300 K. Mode competitions between the resonant modes M0, M1, and M2 at the wavelengths of 1541.58, 1540.305, and 1542.81 nm were used to realize all-optical multiple logic gates with NOT, NOR, and NAND functions under low-power optical pulses at M1 and M2. Filtered by an optical BPF with a central wavelength of M0 and bandwidth of 1.5 nm, the output power of the mode M0 represents the different output logic states. All-optical logic NOT gate was realized by injecting light beams at the side mode M1 or M2. Without optical injection, the HSRL is single-mode lasing at M0, which is a high-power state and represented as an output logic "1" for the dominant mode M0. With light injection of the minor mode M1 or M2 at a high-power state (logic "1"), the dominant lasing mode M0 is suppressed due to the minor mode that has a low threshold gain under optical injection mode, and transitioned from the logic "1" state to "0." Thus, the all-optical NOT operation is realized as shown in Figure 11.24b,c, with 0.43 and 0.27 mW optical injections at modes M1 and M2, respectively, where the dominant lasing mode M0 is suppressed by 34 dB relative to that in Figure 11.24a without the optical injection.

The injection wavelength-tolerant range for all-optical logic NOT gate to maintain normal operation (extinction ratio greater than 20 dB for the dominant mode M0) is shown in Figure 11.25a–c. When the suppression ratio is 20 dB for the dominant lasing mode, the injection wavelength tolerance range at M1 is 0.075, 0.15, and 0.2 nm under the injection optical power of 0.27, 0.54, and 0.81 mW, respectively, and the tolerance at M2 is 0.036, 0.12, and 0.21 nm at the injection optical power of 0.27, 0.54, and 0.81 mW. Figure 11.25c shows the injection wavelength-tolerant range vs. the input power at the modes M1 and M2. The wavelength-tolerant ranges increase with the injection optical power with the same trend. The output power of the mode M0 vs. the injection optical power at the modes M1 and M2 is shown in Figure 11.25d,e, respectively. The output power of the mode M0 decreases gradually

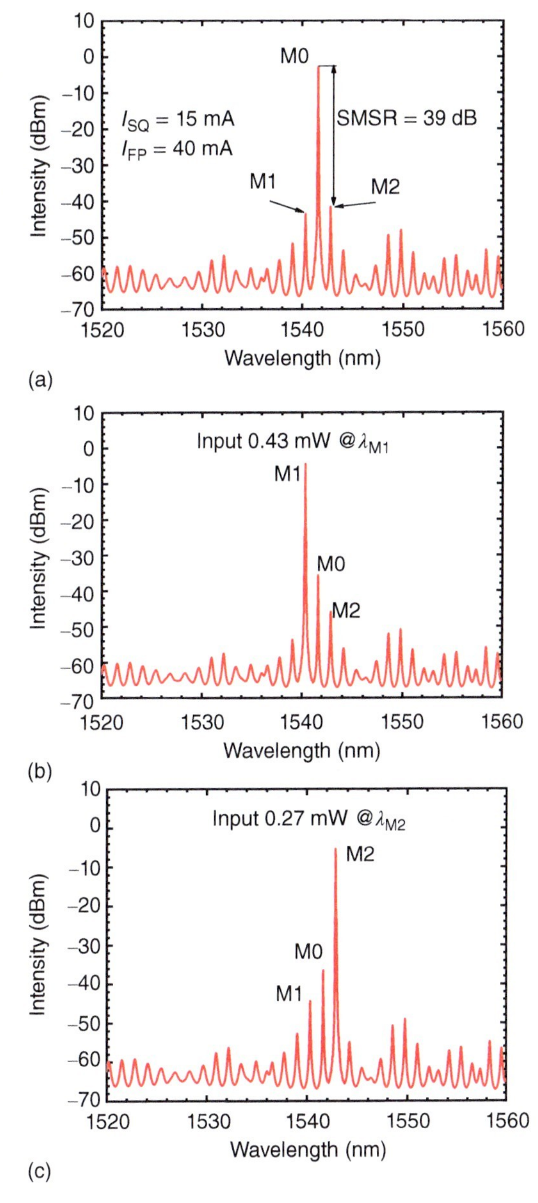

Figure 11.24 (a) Optical spectra for the HSRL at $I_{SQ} = 15$ mA and $I_{FP} = 40$ mA with the modes of M0, M1, and M2 marked. Optical spectra for the HSRL injected by optical with optical powers and wavelengths of (b) 0.43 mW at M1, and (c) 0.27 mW at M2, respectively.

with the increase of the injection optical power and approaches a constant level at high injection level. The minimum power to suppress the dominant mode completely is 0.28 and 0.3 mW at the modes M1 and M2, respectively. Figure 11.26 shows the all-optical logic NOT gate under trigger pulses from (a) the mode M1 at 10 Gb s^{-1} and (b) the mode M2 at 15 Gb s^{-1}. Suppose the rising and falling times are 10–90% of the output power, the rising and falling times triggered by M1 are 103 and 66.6 ps, and by M2 are 48.8 and 41.3 ps, respectively.

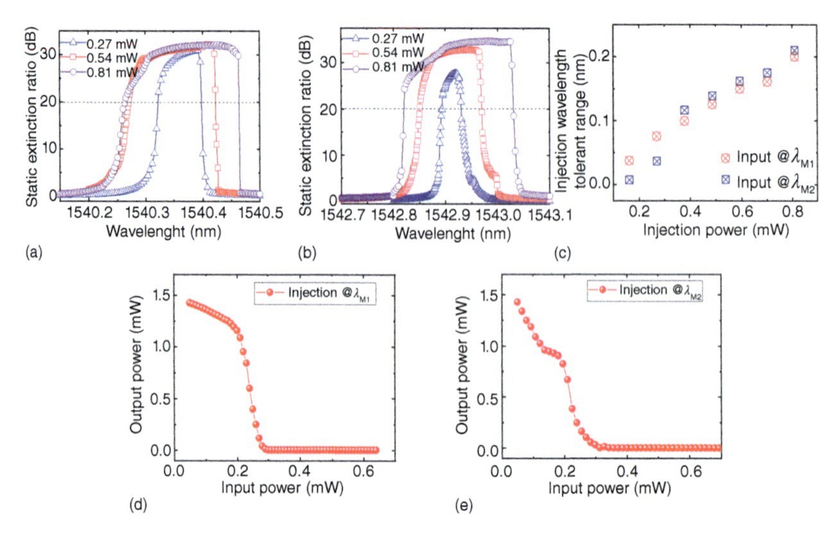

Figure 11.25 Extinction ratio vs. input wavelength (a) with input optical power of 0.27, 0.54, and 0.81 mW around the mode M1, and (b) with input optical power of 0.27, 0.54, and 0.81 mW around the mode M2. (c) Injection wavelength-tolerant range vs. input power at M1 and M2. Output power of the mode M0 after optical filter vs. injected optical power at the mode (d) M1 and (e) M2, respectively. Source: Liu et al. [22]. © 2020, IEEE.

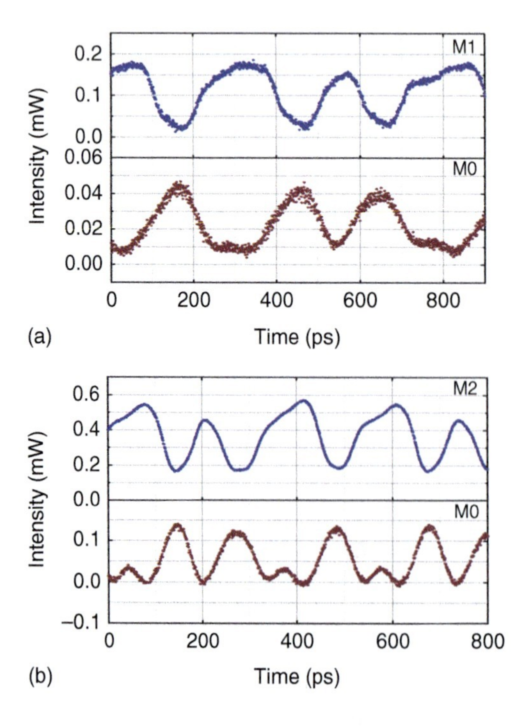

Figure 11.26 High-speed characteristic of all-optical logic NOT gate of mode M0 under (a) 10 Gb s^{-1} optical pulse signal at M1 and (b) 15 Gb s^{-1} optical pulse signal at M2. Source: Liu et al. [22]. © 2020, IEEE.

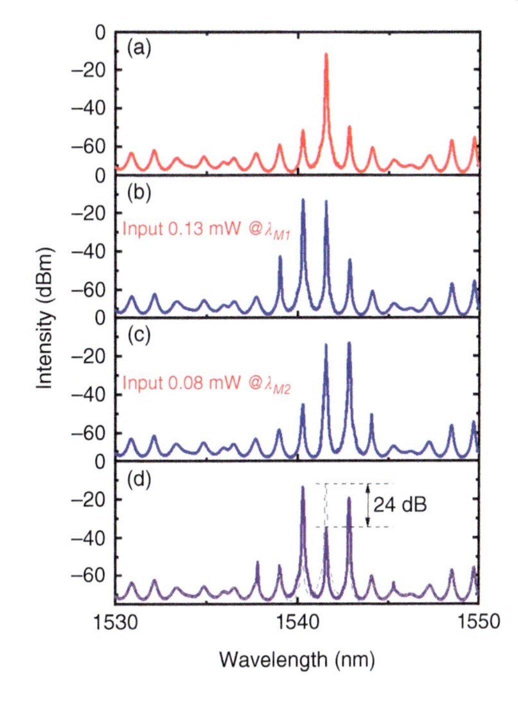

Figure 11.27 Output spectra for the HSRL at different optical injections at the modes M1 and M2 (a) {M1 : M2 : M0} = {0 : 0 : 1}, (b) {M1 : M2 : M0} = {1 : 0 : 1}, (c) {M1 : M2 : M0} = {0 : 1 : 1}; (d) {M1 : M2 : M0} = {1 : 1: 0}.

Furthermore, NAND gate with the static extinction ratios of 24 dB is demonstrated as the dominant lasing mode M0 is only suppressed when both input beams are at logic "1" value. The lasing spectra in four logic states for the all-optical logic NAND gate are shown in Figure 11.27 with the optical injection powers of 0.13 and 0.08 mW for the logic "1" of the modes M1 and M2 at 1540.28 and 1542.78 nm, respectively. The logic "0" corresponds to without optical injection for the modes M1 and M2. The output power of M0 is essentially at logic "1" state as shown in Figure 11.27a–c unless the optical injections of M1 and M2 are all at the logic "1." When the two side modes are injected at the same time, the dominant mode M0 is suppressed with a suppression ratio of 24 dB as shown in Figure 11.27d, which corresponds to the transition from logic "1" to logic "0." The high-speed responses of all-optical logic NAND gate with the injection optical pulse signal at 2 Gb s^{-1} are shown in Figure 11.28a. The two input columns are "11010110" and "10110101," which are 8-bit cyclic codes, and the oscilloscope trace with the output column of "01101011" is obtained. Figure 11.28b shows the enlarged views of the rising and falling edges of the output signal, and the rising and falling times are 352 and 190 ps, respectively.

11.11 Hybrid Square/Rhombus-Rectangular Lasers (HSRRLS)

Coupled mode-field pattern in the FP cavity of an HSRL is usually a mixture of fundamental transverse mode and symmetric high-order modes as shown in Figures 11.6 and 11.7, which may reduce the coupling efficiency into an SMF and

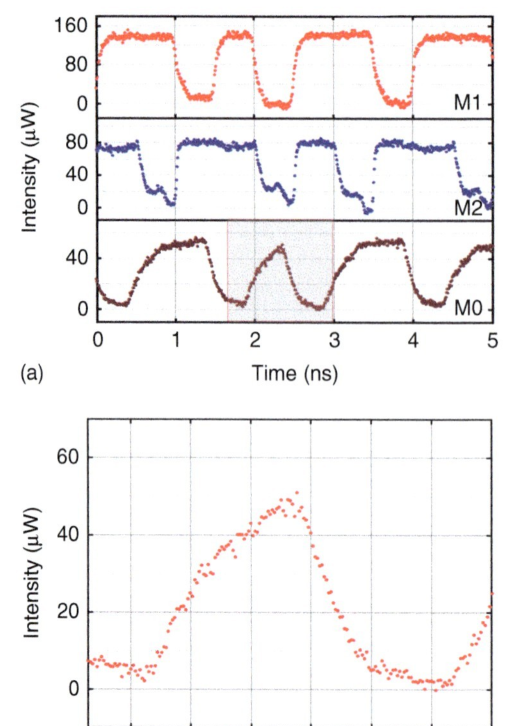

Figure 11.28 (a) high-speed characteristics of all-optical logic NAND gate when 2 Gb s^{-1} optical pulse signal is injected; (b) local magnification of the rising and falling edges of the output signal.

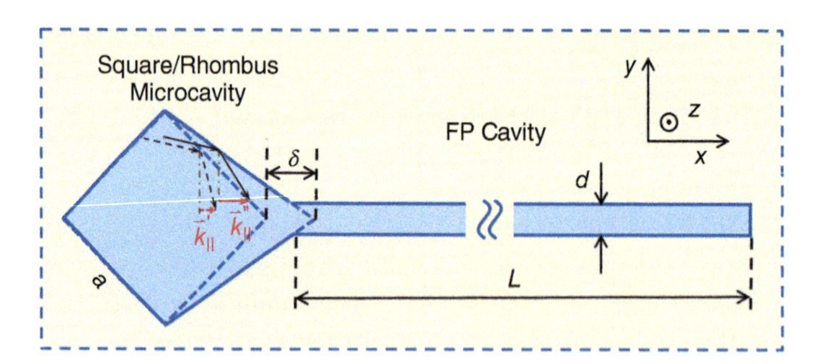

Figure 11.29 Schematic diagram of the coupled-cavity laser composed of an FP cavity and a square/rhombus microcavity as a deformed square microcavity with a vertex extended to a distance of δ, and the wavevectors for the mode light rays reflected from the sides of the SRM and the square microcavity. Source: Hao et al. [23]. © 2019, Chinese Laser Press.

increase the scattering loss in the side walls of the FP cavity. For improving the performance of the hybrid lasers, we designed and fabricated a hybrid-cavity laser with the square microcavity replaced by a square/rhombus microcavity (SRM) as shown in Figure 11.29 for improving the mode-field pattern in the FP cavity [23]. The mode-field distributions inside the hybrid cavity and the far-field patterns are numerically simulated with an optimized deformation of the microcavity. Squared

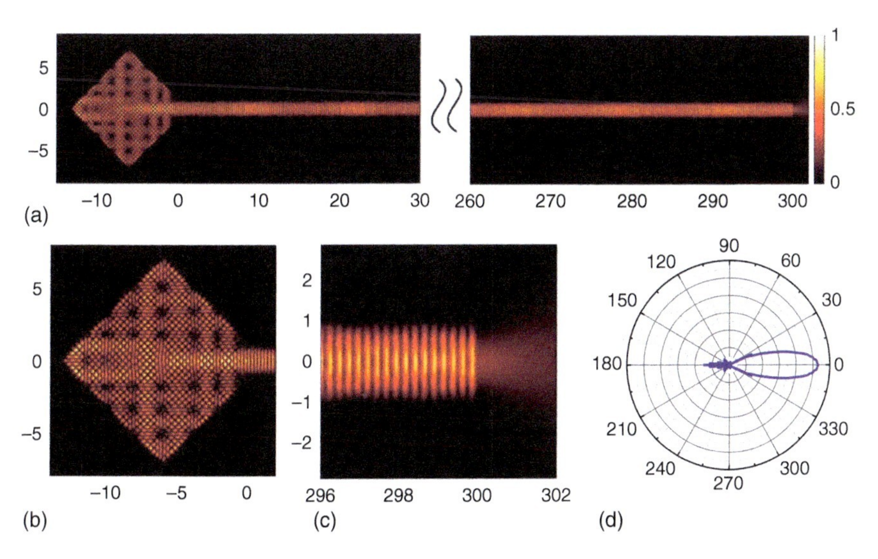

Figure 11.30 Squared *z*-direction magnetic field $|H_z|^2$ in the (a) HSRRL, (b) SRM and (c) FP cavity, and (d) simulated far-field intensity for a confined mode at 1541.5 nm. Source: Hao et al. [23]. © 2019, Chinese Laser Press.

z-direction magnetic field $|H_z|^2$ is simulated and plotted in Figure 11.30 for TE mode at 1541.5 nm in a hybrid square/rhombus-rectangular laser (HSRRL) with $a = 15\,\mu\text{m}$, $w = 2\,\mu\text{m}$, $L = 300\,\mu\text{m}$, and $\delta = 0.25\,\mu\text{m}$, and detailed distributions in the SRM and the FP cavity are illustrated in Figure 11.30b,c, respectively. Moreover, the far-field distributions are calculated from the complex near-field-mode distributions with a far-field angle of about 28° as shown in Figure 11.30d. The deformation induces an additional wavevector along the coupled waveguide *x*-axis for the mode light rays reflected from the two deviated sides of the SRM as shown in Figure 11.29, which forces the transfer of the mode-field pattern inside the SRM from the higher-order transverse mode to the fundamental mode in the FP cavity as shown in Figure 11.30b, because of the reduction of *y*-direction propagation constant with the increase of *x*-direction propagation constant. The results indicate a near fundamental-transverse-mode distribution in the FP cavity and improved far-field patterns compared to that with a square microcavity as shown in Figure 11.7.

The HSRRLs are fabricated by standard contact photolithography and ICP techniques as HSRLs. Output powers coupled into an SMF vs. I_{FP} at $I_{SRM} = 0$, 1, 5, and 30 mA were measured and plotted in Figure 11.31a for an HSRRL with $a = 10\,\mu\text{m}$, $\delta = 0.15\,\mu\text{m}$, and $w = 2\,\mu\text{m}$. The threshold current is about 15 mA under different I_{SRM}, due to the lower Q factor for the SRM with the mode reflectivity increasing slowly with the gain at $a = 10\,\mu\text{m}$. The maximum power is about 2.6 mW with nearly the same output power at $I_{SRM} = 0$, 1, and 5 mA. The output power is even lower at $I_{SRM} = 30$ mA owing to the thermal crosstalk between two cavities. For an HSRRL with $a = 15\,\mu\text{m}$, $\delta = 0.25\,\mu\text{m}$, and $w = 2\,\mu\text{m}$, the output powers collected by an SMF vs. I_{FP} are measured and plotted in Figure 11.31b at $I_{SRM} = 0$, 10, and 25 mA. The threshold current I_{th} is reduced from 17 to 13 mA when I_{SRM} is increased from 0 to

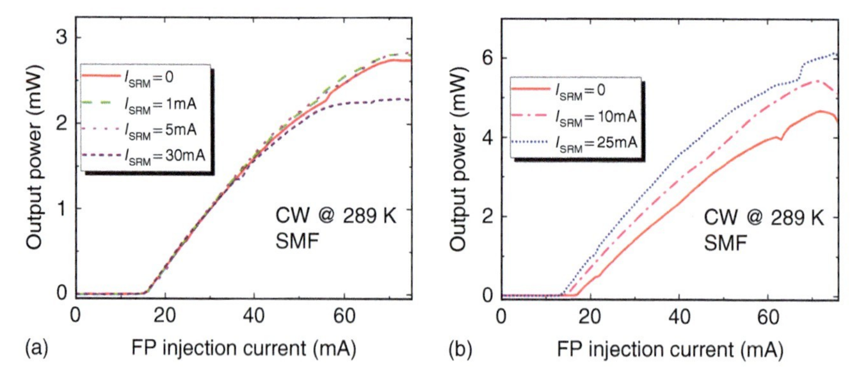

Figure 11.31 Output powers coupled into a single-mode fiber vs. I_{FP} as I_{SRM} is fixed at different currents for HSRRLs at $w = 2\,\mu m$, $L = 300\,\mu m$, (a) $a = 10\,\mu m$, $\delta = 0.15\,\mu m$, and (b) $a = 15\,\mu m$, $\delta = 0.25\,\mu m$. Source: Hao et al. [23]. © 2019, Chinese Laser Press.

25 mA, due to the higher Q factor for the SRM at $a = 15\,\mu m$. The maximum output power collected by the SMF is 6.14 mW. The slope efficiencies are estimated to be about 0.06 and 0.13 W A^{-1} before the saturation for the HSRRLs with $a = 10$ and 15 μm, respectively.

Lasing spectra are measured for the HSRRLs under different injection currents. As shown in the inset of Figure 11.32a, the laser operates in stable single mode at 1574.27 nm with an SMSR of 45.3 dB, as $I_{FP} = 57$ mA and $I_{SRM} = 40$ mA, for the HSRRL with $a = 15\,\mu m$, $\delta = 0.25\,\mu m$, $w = 2\,\mu m$, and $L = 300\,\mu m$, which verifies the excellent mode selection capacity for HSRRLs. Lasing spectra vs. I_{SRM} are shown in Figure 11.32a at $I_{FP} = 64$ mA. The lasing mode wavelength redshifts from 1567.7 to 1589.8 nm as I_{SRM} increases from 8 to 80 mA. The lasing wavelengths and corresponding SMSRs are plotted in Figure 11.32c and the interval between lasing modes is about 1.4 nm. Similarly, the lasing spectra, lasing mode wavelengths, and SMSRs vs. I_{FP} are plotted in Figure 11.32b,d at $I_{SRM} = 20$ mA. The wavelength can shift over 2.5 nm with high SMSR by changing I_{FP}. So, the lasing mode wavelength can be continuously tuned in a wide range by varying I_{SRM} and adjusting I_{FP} slightly. In addition, clear 25 and 35-Gb s^{-1} eye diagrams are observed for the HSRRL under an NRZ pseudo-random binary sequence (PRBS) of $2^{31}-1$ input into the FP cavity [23].

11.12 Summary

In summary, HSRLs composed of an FP cavity and a square microcavity are proposed and demonstrated as a stable single-mode semiconductor laser. The mode coupling between the WGM of the microcavity and FP modes enables the modulation of hybrid mode Q factors, which is capable of selecting and clamping the lasing mode around the intrinsic high-Q WGM of the square microcavity. High-performance single-mode laser is obtained with SMSR as high as 47 dB, and single-mode lasing with SMSR larger than 40 dB over wide variations of the injection currents indicates

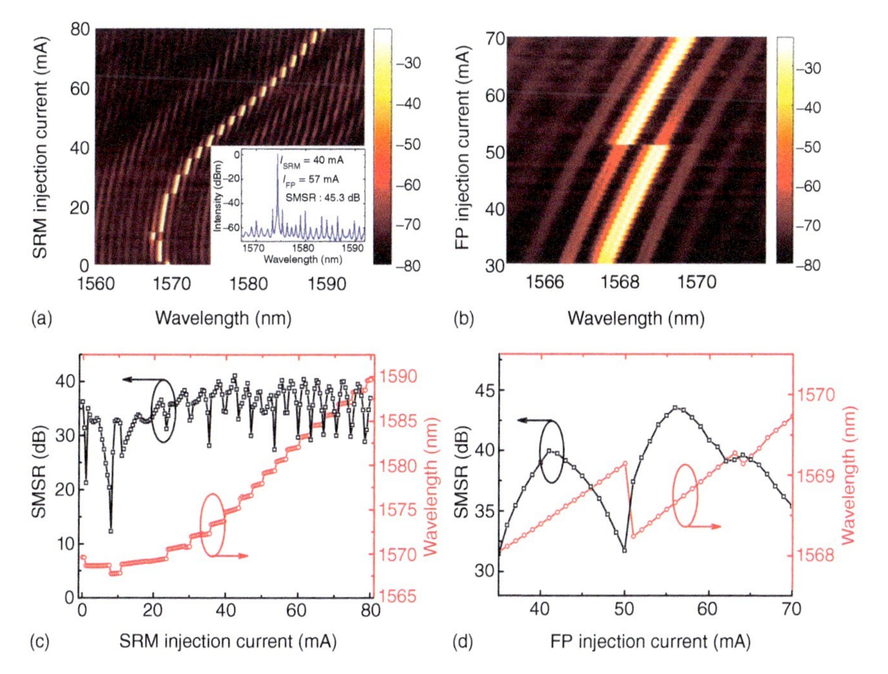

Figure 11.32 Lasing spectra map vs. (a) I_{SRM} at I_{FP}= 64 mA and (b) I_{FP} at I_{SRM} = 20 mA, where inset in (a) the lasing spectrum at I_{FP} = 57 mA and I_{SRM} = 40 mA with the SMSR = 45.3 dB, and lasing mode wavelengths and corresponding SMSRs vs. (c) I_{SRM} at I_{FP} = 64 mA and (d) I_{FP} at I_{SRM} = 20 mA, respectively, for the HSRRL with a = 15 μm, w = 2 μm, L = 300 μm, and δ = 0.25 μm. Source: Hao et al. [23]. © 2019, Chinese Laser Press.

the robustness of single-mode operation for the hybrid-cavity laser. Furthermore, two types of optical bistability are demonstrated for the hybrid-cavity lasers as the square microcavity is unbiased, especially in open circuit state, due to saturable absorption in the microcavity section and mode competitions, respectively. Ultra-fast all-optical flip-flop operations are realized using the on–off bistability around the threshold current and mode competition bistability above the threshold. In addition, all-optical multiple logic gates with NOT, NOR, and NAND functions are verified under low-power optical pulses. Furthermore, the hybrid lasers exhibit excellent noise property [24], which can contribute to high-mode Q factor.

Compared to traditional cleaved coupled-cavity lasers, the interface between the microcavity and the FP cavity is like a soft transition area for the hybrid-cavity lasers, and high-Q microcavity modes have a large wavelength interval, so the hybrid-cavity lasers are suitable for realizing stable single-mode operation. With simple fabrication process and comparatively low cost, the hybrid-cavity lasers with the characteristics of stable single-mode operation, and optical bistability are a potential compact and highly cost-efficient light source for applications like long-haul, high-speed optical communication system, on-chip photonic integrated circuits, and optical single processing.

References

1 Ma, X.W., Huang, Y.Z., Yang, Y.D. et al. (2016). Mode coupling in hybrid square-rectangular lasers for single mode operation. *Appl. Phys. Lett.* 109: Art. no. 071102.

2 Ma, X.W., Huang, Y.Z., Yang, Y.D. et al. (2017). Mode and lasing characteristics for hybrid square-rectangular lasers. *IEEE J. Sel. Top. Quantum Electron.* 23 (6): Art. no. 1500409.

3 Ma, X.W., Huang, Y.Z., Yang, Y.D. et al. (2017). All-optical flip-flop based on hybrid square-rectangular bistable lasers. *Opt. Lett.* 42 (12): 2291–2294.

4 Coldren, L.A., Miller, B.I., Iga, K., and Rentschler, J.A. (1981). Monolithic two-section GaInAsP/InP active-optical-resonator devices formed by reactive ion etching. *Appl. Phys. Lett.* 38: 315–317.

5 Tsang, W.T., Olsson, N.A., and Logan, R.A. (1983). High-speed direct single-frequency modulation with large tuning rate and frequency excursion in cleaved-coupled-cavity semiconductor lasers. *Appl. Phys. Lett.* 42: 650–652.

6 Hagness, S.C., Rafizadeh, D., Ho, S.T., and Taflove, A. (1997). FDTD microcavity simulation:design and experimental realization of waveguide-coupled single-mode ring and whispering-gallery-mode disk resonators. *J. Lightwave Technol.* 15 (11): 2154–2165.

7 Chen, Q., Yang, Y.D., and Huang, Y.Z. (2006). Distributed mode coupling in microring channel drop filters. *Appl. Phys. Lett.* 89 (6): Art. no. 061118.

8 Guo, W.H., Li, W.J., and Huang, Y.Z. (2001). Computation of resonant frequencies and quality factors of cavities by FDTD technique and Padé approximation. *IEEE Microwave Wirel. Compon. Lett.* 11: 223–225.

9 Huang, Y.Z., Ma, X.W., Yang, Y.D. et al. (2018). Hybrid-cavity semiconductor lasers with a whispering-gallery cavity for controlling Q factor. *Sci. China Inf. Sci.* 61: Art. no. 080401.

10 Yang, Y.D., Wang, S.J., and Huang, Y.Z. (2009). Investigation of mode coupling in a microdisk resonator for realizing directional emission. *Opt. Express* 17: 23010–23015.

11 Lasher, G.J. (1964). Analysis of a proposed bistable injection laser. *Solid State Electron.* 7: 707–716.

12 Harder, C., Lau, K.Y., and Yariv, A. (1981). Bistability and pulsations in cw semiconductor lasers with a controlled amount of saturable absorption. *Appl. Phys. Lett.* 39: 382–384.

13 Johnson, J.E., Tang, C.L., and Grande, W.J. (1993). Optical flip-flop based on two-mode intensity bistability in a cross-coupled bistable laser diode. *Appl. Phys. Lett.* 63: 3273–3275.

14 Kawaguchi, H. (1997). Bistable laser diodes and their applications: state of the art. *IEEE J. Sel. Top. Quantum Electron.* 3: 1254–1270.

15 Hill, M.T., Dorren, H.J.S., Vries, T.d. et al. (2004). A fast low-power optical memory based on coupled microring lasers. *Nature* 432: 203–206.

16 Takenaka, M., Raburn, M., and Nakano, Y. (2005). All-optical flip-flop multimode interference bistable laser diode. *IEEE Photonics Technol. Lett.* 17: 968–970.

17 Mezösi, G., Strain, M.J., Fürst, S. et al. (2009). Unidirectional bistability in AlGaInAs microring and microdisk semiconductor lasers. *IEEE Photonics Technol. Lett.* 21: 88–90.

18 Liu, L., Kumar, R., Huybrechts, K. et al. (2010). An ultra-small, low-power, all-optical flip-flop memory on a silicon chip. *Nat. Photonics* 4: 182–187.

19 Alharthi, S.S., Hurtado, A., Korpijarvi, V.M. et al. (2015). Circular polarization switching and bistability in an optically injected 1300 nm spin-vertical cavity surface emitting laser. *Appl. Phys. Lett.* 106: Art. no. 021117.

20 Wu, Y., Zhu, Y., Liao, X. et al. (2016). All-optical flip-flop operation based on bistability in V-cavity laser. *Opt. Express* 24: 12507–12514.

21 Wang, F.L., Huang, Y.Z., Yang, Y.D. et al. (2019). Study of optical bistability based on hybrid-cavity semiconductor lasers. *AIP Adv.* 9: Art. no. 045224.

22 Liu, J.C., Wang, F.L., Han, J.Y. et al. (2020). All-optical switching and multiple logic gates based on hybrid square-rectangular laser. *J. Lightwave Technol.* 38 (6): 1382–1390.

23 Hao, Y.Z., Wang, F.L., Tang, M. et al. (2019). Widely tunable single-mode lasers based on a hybrid square/rhombus-rectangular microcavity. *Photonics Res.* 7: 543–548.

24 Wang, F.L., Ma, X.W., Huang, Y.Z. et al. (2018). Relative intensity noise in high-speed hybrid square-rectangular lasers. *Photonics Res.* 6: 193–197.

Index

Microcavity Semiconductor Lasers: Principles, Design, and Applications, First Edition.
Yong-zhen Huang and Yue-de Yang.

n

o